The plantation of Ulster

Manchester University Press

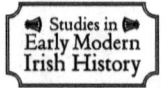

Series editors
DAVID EDWARDS & MICHEÁL Ó SIOCHRÚ

The plantation of Ulster

Ideology and practice

Edited by
ÉAMONN Ó CIARDHA
&
MICHEÁL Ó SIOCHRÚ

Manchester University Press
Manchester and New York
distributed in the United States exclusively by Palgrave Macmillan

Copyright © Manchester University Press 2012

While copyright in the volume as a whole is vested in Manchester University Press, copyright in individual chapters belongs to their respective authors, and no chapter may be reproduced wholly or in part without the express permission in writing of both author and publisher.

Published by Manchester University Press
Oxford Road, Manchester M13 9NR, UK
and Room 400, 175 Fifth Avenue, New York, NY 10010, USA
www.manchesteruniversitypress.co.uk

Distributed in the United States exclusively by
Palgrave Macmillan, 175 Fifth Avenue,
New York, NY 10010, USA

Distributed in Canada exclusively by
UBC Press, University of British Columbia, 2029 West Mall,
Vancouver, BC, Canada V6T 1Z2

British Library Cataloguing-in-Publication Data is available

Library of Congress Cataloging-in-Publication Data is available

ISBN 978 0 7190 9550 4 paperback

First published by Manchester University Press in hardback 2012

This paperback edition first published 2014

The publisher has no responsibility for the persistence or accuracy of URLs for any external or third-party internet websites referred to in this book, and does not guarantee that any content on such websites is, or will remain, accurate or appropriate.

Printed by Lightning Source

Contents

List of illustrations		*page* vii
List of contributors		viii
Series editors' preface		xi
Acknowledgements		xii
1	Introduction: the plantation of Ulster: ideas and ideologies *Éamonn Ó Ciardha & Micheál Ó Siochrú*	1
2	The 'British' crown, the earls and the plantation of Ulster *Jenny Wormald*	18
3	Civilising Gaelic Scotland: the Scottish Isles and the Stewart empire *Martin MacGregor*	33
4	Plantation and civil society *Phil Withington*	55
5	The city of London and the Ulster plantation *Ian W. Archer*	78
6	Success and failure in the Ulster plantation *Raymond Gillespie*	98
7	The Catholic Church in Ulster under the plantation, 1609–42 *Brian Mac Cuarta*	119
8	Randal MacDonnell and early seventeenth-century settlement in northeast Ulster, 1603–30 *Colin Breen*	143
9	Educating the colonial mind: Spenser and the plantation *Andrew Hadfield*	158

10	Responses to transformation: Gaelic poets and the plantation of Ulster *Marc Caball*	176
11	The plantation of Ulster: aspects of Gaelic letters *Diarmaid Ó Doibhlin*	198
12	Angling for Ulster: Ireland and plantation in Jacobean literature *Willy Maley*	218
13	'The Scottish inhabitants of that Province are actually revolted': John Milton on the failure of the Ulster plantation *Nicholas McDowell*	238

Index 255

Illustrations

FIGURES

1	The number of ESTC texts with 'society' or 'company' in the title, 1500–1700	*page* 66
2	The % of ESTC texts with 'society' or 'company' in the title, 1500–1700	67
3	The % of ESTC texts with 'society' and 'company' compared, 1500–1700	67
4	State, society/company and household/family, 1500–1700	68
5	Rates of incorporation in England, Wales, Scotland and Ulster, 1540–1640	70
6	Rates of urban incorporation in Ulster and English regions, 1540–1640	70

TABLE

1	Details of Randal MacDonnell's 1603 grant	148

Contributors

Ian Archer has been Fellow and Tutor in Modern History at Keble College, Oxford since 1991. He has published widely on the social and political history of early modern London, and recently produced with Douglas Price a major edited collection entitled *English Historical Documents, 1558–1603* (Routledge, 2011). He is a Literary Director of the Royal Historical Society.

Colin Breen is a Senior Lecturer in archaeology at the University of Ulster. His recent publications include *The Gaelic Lordship of the O'Sullivan Beare* (Dublin, 2005) and *An Archaeology of Southwest Ireland, 1570–1670* (Dublin, 2007). He is currently working on a programme of major excavation and landscape survey of later medieval sites across north Ulster and the western Isles.

Marc Caball is Director of UCD Humanities Institute and the UCD Graduate School of Arts and Celtic Studies. Among his recent publications is 'Cultures in conflict in late sixteenth-century Kerry: the parallel worlds of Tudor intellectual and Gaelic poets', *Irish Historical Studies*, 36:144 (2009), 483–501 and he is co-editor, with Andrew Carpenter, of *Oral and Print Cultures in Ireland 1600–1900* (Dublin, 2010).

Raymond Gillespie is Professor of History at NUI Maynooth. His research interests include economic, social and cultural change in early modern Ireland. His books include *Reading Ireland: Print, Reading and Social Change in Early Modern Ireland* (Manchester, 2005) and *Seventeenth-century Ireland* (Dublin, 2006). He is also editor with Andrew Hadfield of *The Oxford History of the Irish Book*, vol. 3: *The Irish Book in English, 1550–1800* (Oxford, 2006). He is currently working on hagiography in sixteenth-century Gaelic Ireland.

Andrew Hadfield is Professor of English at the University of Sussex and visiting Professor at the University of Granada. He is the author of a number of works on early modern literature, including recently *Shakespeare and Republicanism* (Cambridge, 2008). He has also edited, with Raymond Gillespie, *The Oxford History of the Irish Book*, vol. 3: *The Irish Book in English, 1550–1800* (Oxford, 2006).

Brian Mac Cuarta SJ has written on the colonial and religious history of early seventeenth-century Ireland. Author of *Catholic Revival in the North of Ireland 1603–41* (Dublin, 2007), he has edited *Ulster 1641: Aspects of the Rising* (Belfast, 1993), and *Reshaping Ireland 1550–1700: Colonization and Its Consequences: Essays Presented to Nicholas Canny* (Dublin, 2011). He is Director of the Archivum Romanum Societatis Iesu in Rome.

Martin MacGregor is a lecturer in Scottish history, with special emphasis on Gaelic Scotland, at the University of Glasgow. Publications include 'The view from Fortingall: the worlds of *The Book of the Dean of Lismore*', *Scottish Gaelic Studies*, 22 (2006), 35–85; 'The Statutes of Iona: text and context', *Innes Review*, 57 (2006), 111–81; and *Mìorun mòr nan Gall? Lowland Perceptions of the Highlands*, ed D. Broun and M. MacGregor (published on the web by the Centre for Scottish and Celtic Studies, University of Glasgow, 2009).

Willy Maley is Professor of Renaissance Studies at the University of Glasgow. He is author of numerous works including *Nation, State and Empire in English Renaissance Literature: Shakespeare to Milton* (Basingstoke, 2003) and editor, with Brendan Bradshaw and Andrew Hadfield of *Representing Ireland: Literature and the Origins of Conflict, 1534–1660* (Cambridge, 1993). Recent work includes the essay collection *Shakespeare and Wales: From the Marches to the Assembly* (Ashgate, 2010), edited with Philip Schwyzer.

Nicholas McDowell is Professor of English at the University of Exeter. He is the author of numerous works including *Poetry and Allegiance in the English Civil Wars: Marvell and the Cause of Wit* (Oxford, 2008). He is the co-editor, with Nigel Smith, of *The Oxford Handbook of Milton* (Oxford, 2009; paperback, 2011) and, with N. H. Keeble, of *The Oxford Complete Works of John Milton*, vol. 6: *Vernacular Regicide and Republican Tracts* (Oxford, 2012).

Éamonn Ó Ciardha is a Senior Lecturer in the School of English, History and Politics, University of Ulster. He has published and edited books, journals and articles on Jacobitism, the flight of the earls, the plantation of Ulster, law and order in early modern Ireland, the Irish outlaw and the use of Irish-language sources for seventeenth- and eighteenth-century Irish History. He recently completed a year as Gastprofessor Europaicum at the Universität des Saarlandes, Saarbrücken, Germany.

Diarmaid Ó Doibhlin was formerly a Senior Lecturer at the University of Ulster and Distinguished Visiting Professor at the Keough-Naughton Institute for Irish Studies, University of Notre Dame. He has published articles on numerous aspects of literary and local history as well as two well-received volumes of original poems *Briseadh na Cora* (1981) and *Dumaí Móra* (1997). At present he is engaged in a major research project on post-Tridentine devotional literature in the Irish language, with special emphasis on the sermon.

Micheál Ó Siochrú is Associate Professor of History at Trinity College Dublin. He is author of numerous books and articles on seventeenth-century Ireland, including *God's Executioner: Oliver Cromwell and the Conquest of Ireland* (London, 2009). He recently completed the online publication of the 1641 Depositions with colleagues at Aberdeen and Cambridge, and is currently working on a new edition of Oliver Cromwell's letters and papers for Oxford University Press.

Phil Withington is Professor of Early Modern History at the University of Sheffield. He is the author of *The Politics of Commonwealth: Citizens and Freemen in Early Modern England* (Cambridge, 2005) and *Society in Early Modern England: The Vernacular Origins of Some Powerful Ideas* (Cambridge, 2010). He also co-edited with Alexandra Shepard a volume entitled *Communities in Early Modern England* (Manchester, 2000) and is currently writing a book about early modern intoxicants.

Jenny Wormald is an Honorary Fellow of the University of Edinburgh and was formerly Fellow of St Hilda's College, Oxford. Her original work was in Scottish history, including the book *Lords and Men in Scotland: Bonds of Manrent 1442–1603* (Edinburgh, 1985). Work on James VI and I moved her into Anglo-Scottish history, but recently and essentially she has begun to explore Ireland under James VI and I. None of this has made the concept of 'British History' any easier for her; but she is at ease with the 'British Problem'.

Series editors' preface

The study of early modern Irish History has experienced something of a renaissance since the 1990s, with the publication of a number of major monographs, examining developments in Ireland during the sixteenth and seventeenth centuries from a variety of different perspectives. Nonetheless, these studies still tend to group around traditional topics in political or military history and significant gaps remain. The idea behind this new series is to identify key themes for exploration and thereby set the agenda for future research. Manchester University Press, a leading academic press with a strong record of publishing Irish-related material, is the ideal home for this venture.

This first volume in the series marks the 400th anniversary of the Ulster plantation, a key moment in the emergence of modern Ireland. It remains a hugely controversial topic today, though surprisingly one that has failed to attract substantial academic attention. The original and wide-ranging themes chosen for this volume, along with the high standard of the contributions from leading scholars in the fields of history and literature, should ensure that the collection becomes required reading for all those interested not only in the Ulster plantation, but also in the history of early modern Ireland more generally, as well as the foundations of modern English/British imperialism.

David Edwards
Micheál Ó Siochrú

Acknowledgements

This book emerged from the first two of three British Academy and IRCHSS-funded conferences hosted in 2009 by Goldsmiths, University of London, the University of Ulster and Trinity College, Dublin to mark the 400th anniversary of the plantation of Ulster. The outputs from the Trinity College conference will form the basis of another volume. Both editors would like to thank all the speakers, chairs and participants who contributed to three intellectually stimulating and thoroughly enjoyable events. In Dr Ariel Hessayon and Birgul Yavuz, the editors had first-rate collaborators for the conference in Goldsmiths and the success of these joint ventures and the resulting academic outputs provide a model for future cooperation between the respective institutions.

Nick Anstee, the Right Honourable Lord Mayor of London (2009) and Mr Paul Double, the City Remembrancer, kindly co-hosted Dr Jenny Wormald's keynote lecture and the subsequent reception in London's Guildhall. Councillor Paul Fleming, His Worshipful, the Mayor of Derry/Londonderry (2009) graciously welcomed conference delegates to the Plantation Citadel for Professor Jane Ohlmeyer's keynote address and to a social function in the Guildhall. This publication has been supported financially by Derry City Council through its Heritage and Museums Service. Craig McGuicken, Bernadette Walsh and Margaret Edwards (Museum Service, Derry City Council), Charles Fisher, Tom Hoffmann, Edward Montgomery, Sir David Lewis and Catherine McGuinness (the Honourable, the Irish Society) lent invaluable support to the London and Derry/Londonderry conferences. Dr Billy Kelly, Professor John Wilson and Professor Ian Thatcher (University of Ulster) also gave much-needed advice and assistance. Mr Andrew Robinson (University of Ulster) deserves enormous credit for his organisational skills, carried off with characteristic efficiency and good humour.

All the contributors to the volume not only submitted their essays in a timely fashion but dealt very patiently and effectively with all our subsequent requests. The job of editor can be difficult at times but not so in this instance and we are very grateful to the contributors for their cooperation. It was a singular pleasure to work with Manchester University Press, Emma Brennan in particular, and the editors would also like to thank Dr Ian Campbell and Lisa Scholey for their help in preparing the manuscript for publication.

Finally, the book is dedicated to the memory of the late Professor James Allen, Provost of Magee and Pro-Vice Chancellor for Student Affairs (University of Ulster). A proud son of Derry, a staunch defender of its rich history and heritage and a generous patron of these academic initiatives, Jim attended all three conferences on behalf of the university and hosted an enjoyable and memorable Conference Dinner in the *Aula Maxima* for speakers and delegates. *Ar shlí na fírinne ach i gcónaí inár gcuimhne.*

Micheál Ó Siochrú
Éamonn Ó Ciardha

1

Introduction
The plantation of Ulster: ideas and ideologies

ÉAMONN Ó CIARDHA & MICHEÁL Ó SIOCHRÚ

The pivotal importance of the Ulster plantation to the shared histories of Ireland and Britain would be difficult to overstate. It helped secure the English conquest of Ireland, and dramatically transformed Ireland's physical, demographic, socio-economic, political, military, religious and cultural landscapes. In effect, the plantation became the city of London's and England's first successful attempt at empire during the early modern period, providing a template for future colonial expansion in the Americas, the Caribbean and the Indian sub-continent. Moreover, the plantation's historical, political, cultural, environmental and visual legacies impacted heavily on developments in both Ireland and Britain for four hundred years and continue to do so today.

It is surprising, therefore, that the 400th anniversary of the plantation produced so few publications.[1] This crucial event remains much talked about but little studied or understood. Indeed, it did not attract much scholarly attention during the last century. Historians tended to concentrate instead on key military events such as the Nine Years War (1594–1603), the 1641 Rebellion and the Jacobite/Williamite Wars (1688–91). Studies of the plantation focused primarily on political history, from the perspective of the colonial administrations in both London and Dublin, as well as the English and Scottish planter communities. Theodore Moody's history of the Londonderry plantation, published in 1939, laid the foundation for later work by Michael Perceval-Maxwell on the Scots in Ulster, as well as a number of case studies by Robert Hunter on the fabric of plantation society across Ulster.[2] In 1976, Aidan Clarke, with Robin Dudley Edwards, synthesised and expanded on Moody's research in volume 3 of the *New History of Ireland*, which still remains the primary introduction on this topic for both scholars and students alike.[3]

Philip Robinson subsequently produced an overview of the entire plantation process in Ulster, the only modern scholarly monograph on the topic, but again exclusively from the perspective of the planters, while James Stevens Curl focused on the Londonderry plantation and wrote an institutional history of the Honourable the Irish Society, which oversaw this scheme on behalf of the London Companies.[4] Biographical studies by John McCavitt and John McGurk, on Sir Arthur Chichester and Sir Henry Docwra respectively, merely reinforced the emphasis on political and military histories from the top down.[5] The indigenous communities continued to be largely ignored, except by literary scholars working with Irish-language source material, such as Brían Ó Cuív, Tomás Ó Fiaich, Breandán Ó Doibhlin, Breandán Ó Buachalla and Marc Caball.[6] Few historians, however, have proved willing to engage with their literary colleagues on this topic, even after the publication of Ó Buachalla's masterpiece, *Aisling Ghéar*, in 1996. In an effort to redress the settler bias in plantation studies, Vincent Carey, David Edwards and Kenneth Nicholls have addressed the issue of the violence associated with the new regime, particularly the extensive use of martial law by the colonial authorities.[7]

Nicholas Canny surveyed the theory and practice of plantation during the Tudor and Stuart periods in his book, *Making Ireland British*. Through extensive use of historical and literary evidence in both English and Irish, he produced the most convincing explanation to date of the role of the plantation process in making Ireland English, or indeed Scottish (or perhaps even British) in certain parts of Ulster.[8] Canny, following a trail blazed by D. B. Quinn, has also been to the forefront of those seeking to place Ireland in a broader Atlantic context, comparing settlements in Munster and Ulster with those in North America. On the issue of ideology, Jane Ohlmeyer's article 'A laboratory for Empire? Early modern Ireland and English imperialism' provides an impressive appraisal of Ireland's role in Elizabethan state formation and the Stuart 'imperial' project. Moreover, her recently published research on the early modern Irish aristocracy and the plantation peerage in particular, points to their rapid Anglicisation (not Briticisation) during the course of the seventeenth century.[9]

Religious tensions played a key role in these political and constitutional developments, as demonstrated by the work of Brendan Bradshaw, Raymond Gillespie, Alan Ford, Henry Jefferies, and John McCafferty among others.[10] From its earliest inception, the Protestant Reformation in Ireland became inexorably linked to conquest, confiscation and the religious persecution of the majority Catholic population. Conversely, opponents of the English imperial project, from Silken Thomas in the 1530s to Hugh O'Neill in the 1590s, readily unfurled the papal banner to rally support and further their political ambitions. Protestant renewal through the Ulster plantation formed a key part of James's 'civilising' agenda in Ireland, which ensured that sectarianism continued to blight the Irish political and social landscape for the next four hundred years. In 1603 Catholics owned 90% of the land but by 1641, mainly as a result of the Ulster plantation, this had fallen to around 60%. In many ways, the plantation simply replaced one landed elite, the Gaelic nobility, with another, the new English and Scottish planters. This process, however, and the attendant socio-economic and cultural changes, did not happen without inflicting severe trauma on the native population. Although the 'deserving' Irish (those who had supported the crown against O'Neill or deserted his cause before the end) received land in the plantation settlement, few of them prospered under the new political, socio-economic and legal systems. Brian Mac Cuarta and Tadhg Ó hAnnracháin have focused on the difficulties encountered by the Irish Catholic clergy between the plantation and the 1641 rebellion, while the writings of Andrew Hadfield, Willy Maley, Richard McCabe, Clare Carroll and Patricia Palmer among others have explored the violent colonial rhetoric which infuses Tudor and Jacobean literature.[11] Much more research remains to be done, however, to uncover the experiences of the native Catholic population.

Historical geographers, such as J. H. Andrews, William J. Smyth, Patrick Duffy and Annaleigh Margey have produced some of the most innovative studies of the impact of the plantation on the demography and landscape of Ireland, exploiting the rich cartographical heritage of the early modern period.[12] Plantation emerged as a central plank of English policy in Ireland from the mid-sixteenth century, when Queen Mary's administration drove recalcitrant Gaelic families, such as the O'Mores and O'Connors, from the midlands, establishing military settlements in their place.[13] Her Protestant sister, Elizabeth I, sanctioned plantation projects in east and southeast Ulster by the Devereux earls of Essex and Sir Thomas Smith.[14] In the 1580s, in the aftermath of the Desmond rebellion, the large estates of the attainted earl formed the nucleus of the Munster plantation, which collapsed so dramatically in 1598, at the height of the Nine Years War, only to be resurrected in the aftermath of Hugh O'Neill's defeat.[15] The systematic dispossession of large numbers of Gaelic Irish landowners necessitated the creation of increasingly detailed maps to delineate the full extent of the confiscated estates. Significantly, in 1610 John Speed produced the first map to represent all three Stuart kingdoms on the same page in a graphic representation of the new political order.

Given the importance of the links between England, Scotland and Ireland during the early decades of the seventeenth century outlined above is it possible to write a 'British' history of the period? Proponents of the 'New British History' would certainly make this claim but John Morrill cautions that the historian's task is sometimes to acknowledge the incoherence of the past and not impose order upon it.[16] The main problem with the New British History is that it is not new, it is not British (certainly not in a seventeenth-century context), while even the use of the term history might be questioned given the overtly political agenda of its principal proponent, J. G. A. Pocock. It is some forty years since Pocock's call for a new approach to British history, primarily in response to the United Kingdom's accession to the European Economic Community. An army of 'neo-Britons' from the nineteenth-century colonial settlements were to lead the fight against what he called the foggy imperialism of 'European' ideology, and the threatened annexation of British history.[17] In an introduction to a major edited volume on the topic, Glenn Burgess acknowledges that the New British History has not developed in the manner originally envisaged by Pocock. The new historiography, Burgess argues, is more a reaction to the Whig view that English history was self-sufficient, and has played a role in shifting the anachronistic focus on the nation state, while at the same time respecting particular identities (meaning, in this instance, Irish, Scottish and Welsh).[18] According to this interpretation, the slow emergence of English historians from their splendid isolation does not constitute a form of academic imperialism. Pocock has also been at pains to stress that 'particular histories do not cease to exist when it is seen that they cannot be written in isolation'.[19]

Despite such assurances, Irish historians for the most part have responded negatively to the New British History, levelling charges against it of Anglo-centrism and anti-Europeanism. One of the most vocal critics, Nicholas Canny, believes the new historiography implies an integrity for 'these islands...probably in excess of any that ever existed', exaggerating unity at the expense of diversity and detracting from European and colonial comparatives.[20] In the preface to *Making Ireland British* Canny writes that his aim is to place Ireland in the history of British overseas expansion, an exercise in colonial or 'Atlantic' history. He concludes, however, with a caustic swipe in the direction of the New British History.

> My persistent efforts to connect developments in Ireland with simultaneous happenings in England and Scotland would seem to qualify the book for inclusion under the equally fashionable category of New British History. However, my concern to relate events in Ireland (and Britain) to happenings on the continent of Europe would suggest that I am indifferent to fashion when I treat the histories of these two islands as but parts of the history of Europe.[21]

Despite the difficulties associated with 'British' history, it is no more possible to write a detailed study of Ireland in the seventeenth century without making reference to developments in England and Scotland, than it is to write a meaningful history of England for the same period without examining Irish and Scottish affairs. This does not mean, however, that Irish, Scottish and English historians are writing British history. The 'Three Kingdoms' model is very much in vogue at the moment, but as Peter Lake has so succinctly stated, 'this is not so much a new subject (British history) as simply a more integrated reading of English, Scottish and Irish histories'.[22] The primary difficulty with an integrative approach involving Ireland is that much of the basic historical data required has yet to be produced. Unlike England, for example, there are few local or regional studies for Ireland in the seventeenth century, social groups such as the merchants and the gentry lack any in-depth analysis, while the first major study of the Irish aristocracy has just been published.[23] Without such specialist research, the integrative approach remains highly speculative and (it could be argued) in many instances meaningless.

The 'Three Kingdoms' model, although the best available, is far from perfect. Complaints that it excludes Wales need to be addressed and maybe 'Four Nations' might be more appropriate, as Hugh Kearney has proposed, while Jane Ohlmeyer makes a pitch for the 'Wars of the Five Kingdoms', to include France and Spain.[24] 'Five Kingdoms' probably overstates the case, given the peripheral involvement of the French and Spanish monarchies, but the suggestion highlights the importance of the continental dimension in an Irish context. Ireland retained strong links with continental Europe for centuries, mainly through the religious and mercantile communities. From the late sixteenth century, large numbers of mercenary troops also departed overseas, primarily to serve in the armies of France and Spain. The English Reformation and the subsequent Tudor conquest of Ireland reinforced these continental ties, with an increasing flood of religious, military and political refugees in the early decades of the seventeenth century.[25] The role of continental veterans in sustaining the revolt of 1641–42 has already been well documented, along with the influence of continentally trained clergy on the Catholic Church in Ireland and the confederate government in Kilkenny.[26] Much work remains to be done, however, in the continental archives on the Irish community abroad, and the diplomatic involvement of France, Spain, the papacy (and others) in Irish affairs. Fascinating comparative studies could also be undertaken between Ireland, Portugal, Catalonia, Bohemia and a host of other European regions. The broader European route, therefore, appears a more productive route than a narrowly based British one.

Is there any room, therefore, for the study of British history in seventeenth-century Ireland? John Morrill speculated that whereas Wales and Scotland were being institutionally and constitutionally Briticised at this time, Ireland on the other hand was being Anglicised.[27] It is true that loyalty to the crown did not inculcate a sense of Britishness, but rather as Jane Ohlmeyer suggests, qualified and refined people's sense of Irishness and Englishness.[28] The Protestant settlers in Ulster are the exception in this regard, and one would expect historians of Britishness to concentrate their research on Ireland here. The term British begins to appear in Irish records, albeit sporadically, from the time of the Ulster plantation in the early seventeenth century, usually employed, as Toby Barnard has illustrated, at moments of crisis such as the 1641 rebellion to promote Protestant solidarity among English and Scottish settlers.[29] Apart from Barnard, however, few others have addressed the question of Protestant identity in Ulster during the early decades of the seventeenth century.[30] This appears to be one particularly fruitful area of research for those interested in British history but one which nonetheless should heed Morrill's warning about attempting to impose order where none existed.

The sequence of events which culminated in the plantation of Ulster has its origins in the maelstrom of the final phase of the Tudor conquest in that province. The judicial murder of Hugh Roe McMahon in 1590 and subsequent partition of his Monaghan lordship, coupled with various attempts to garrison counties Armagh and Fermanagh, played a key role in driving the Ulster Irish into rebellion. Hugh O'Neill, second earl of Tyrone, emerged as the greatest single threat to English rule in Ireland during the Tudor period. He forged strategic political and marital alliances among the Gaels of Ulster to construct a powerful confederacy in the 1590s. After a number of stunning successes against Queen Elizabeth's forces, an untimely Spanish descent on Kinsale in 1601 forced O'Neill to march south to its aid from his hitherto impregnable heartland in Ulster. Decisively defeated in a pitched battle by Lord Deputy Mountjoy, he retraced his steps to Dungannon and waited in vain for additional Spanish support. Assailed on all sides by land and sea he accepted the Queen's generous terms and signed the Treaty of Mellifont three days after the death of the last Tudor monarch in March 1603.[31]

Pardoned and received at court in 1603 by the new king, James VI and I, O'Neill nevertheless felt besieged by crown officials and former soldiers who decried his lenient treatment, coveted his extensive lands and sought to further undermine his position within the province. Although James wished to integrate the rehabilitated O'Neill into his pan-Britannic imperial process, the relentless political and legal machinations of Lord Deputy Sir Arthur Chichester and Attorney-General Sir John Davies, alongside rumours of the earl's continuing contacts with Spain and an ominous royal summons to London, ultimately provoked his flight to the continent. He departed from Ulster in September 1607 along with Rory O'Donnell, first earl of Tyrconnell and Cúchonnacht Maguire, lord of Fermanagh, together with their wives, families and followers, in one of the most iconic and significant events in Irish history.[32] After an epic journey by sea and land, O'Neill reached Rome in 1608 and remained there as a frustrated and embittered exile until his death in 1616.

This defeat and 'Flight' effectively marked the collapse of an independent Gaelic Ulster and prepared the way for Ireland's full incorporation into the new tripartite Stuart monarchy. The servitors, soldiers and settlers who flooded into Ulster in the wake of the earls' departure, continually lambasted as usurpers and interlopers in the contemporary Gaelic literature, saw themselves as social engineers, harbingers of King James's self-styled policy to 'civilize these rude partes'.³³ They would make an indelible mark on the politics, economics and material culture of Ulster. The 'Flight' and plantation also precipitated the emergence of an Irish Catholic military, religious and intellectual diaspora on the European continent.³⁴ The government itself attempted to promote military migration among those who had served O'Neill, and Sir Arthur Chichester earmarked up to 6,000 demobilised Irish kern for the service of Gustavus Adolphus, King of Sweden. Thousands more joined the ranks of the Irish regiments in Spain and Spanish Flanders, while others migrated to France and the Habsburg empire.³⁵ Those who remained in Ulster often failed to adapt to changing circumstances, creating a legacy of bitterness which simmered below the surface before exploding in an orgy of violence in 1641.

The decades after the 'Flight' witnessed the decline in the fortunes of Gaelic poets, scribes, brehons, historians, genealogists and chroniclers. The wholesale destruction of manuscripts and the carelessness of future generations have deprived us of the evidence to tabulate the full extent of the influence of the *aos dána* (learned classes). Nevertheless, surviving material vastly outstrips similar literary sources in either Scotland or Wales and sheds invaluable light on contemporary Irish society. The traditional conservatism often attributed to this literary caste masks their appreciation of, and reaction to, contemporary events. The doom-laden reaction to the 'Flight', their heart-rending laments at the sorry plight of Ireland and despair at the scattering of her native aristocracy, co-existed with an emerging cult of the House of Stuart.³⁶

Despite the incessant wars and political turmoil of the Tudor conquest and Jacobean plantation, Gaelic Ireland witnessed a remarkable flowering of literary activity during the early decades of the seventeenth century. The Franciscan and Jesuit orders, utilising scions of the tradition learned families such as the Uí Chléirigh, Uí Mhaoilchonaire, Uí Eodhasa, Mhic Chathmhaoil and Uí Dhuibhgheannáin, strove both to stem the tide of Protestantism and preserve the nation's literary and cultural heritage. A stream of confessional and theological works, religious primers and catechisms emanated from Irish continental colleges, directed for the most part towards the clergy rather than the largely illiterate laity. These writings reflected the continental training of their authors and drew heavily on contemporary post-Tridentine, Counter-Reformation works in Spanish, French, Latin and Italian. 'Annála Rioghachta Éireann' ('The Annals of the Kingdom of Ireland'), and Seathrún Céitinn's (Geoffrey Keating's) 'Foras Feasa ar Éirinn' ('Foundation of Knowledge of Ireland') addressed the nation's cultural needs. Céitinn targeted Anglo-Norman and English writers such as Giraldus Cambrensis, Edmund Spenser, Richard Stanihurst, Edmund Campion, Meredith Hanmer and other polemicists who had cast aspersions on Ireland's literary and cultural heritage. Other Irish scholars such as Peter Lombard, Luke Wadding, David Rothe, Richard

Creagh, Cornelius O'Deveney, Richard O'Farrell and Robert O'Connell directed their Hiberno-Latin histories and theological tracts towards a continental audience, putting Ireland's case at the Catholic courts of Europe.

Closer to home, the union of the English, Irish and Scottish crowns in the person of James VI, self-styled king of Great Britain and Ireland, heralded a monumental shift in 'English' crown policy. Since the Scottish Wars of Independence of the late thirteenth/early fourteenth century, successive English kings and queens had endeavoured to keep the Scots, both settlers and mercenaries (*gall óglaigh*/galloglass), out of Ulster. Elizabeth's comprehensive defeat of O'Neill and his confederates, and their subsequent 'Flight', facilitated the crown's seizure of nearly 3.8 million acres of land for a comprehensive plantation project. It also enabled James to address a series of key political, religious, strategic, socio-economic and financial challenges in his three realms. The systematic influx of English and Scottish settlers was designed to hamper any attempts by hostile Catholic powers to use Ireland as a back door through which they could invest the king's Protestant realms. Similarly, the king was able to confront the socio-economic, political and religious problems that had plagued the Scottish–English borders through a wholesale transplantation of people to his Irish kingdom. Furthermore, the resulting capital investment in the plantation, including the foundation of new counties, towns and villages, provided a much-need financial boost to his depleted treasury.

Elizabeth I's Monaghan plantation in 1591 and the successful 'private' plantation scheme initiated by Hugh Montgomery and Sir James Hamilton on the lands of Conn O'Neill in south Clandeboye (Counties Antrim and Down) in 1606, provided suitable templates for the much larger and more ambitious plantation project across the northern province.[37] Much of the land in the remaining six Ulster counties of Armagh, Cavan, Donegal, Tyrone, Coleraine/Londonderry and Fermanagh passed to the crown, thus facilitating a plantation that far outstripped previous ventures in Ireland and compared in size and scope with contemporary English, Portuguese and Spanish initiatives in the Americas.

An Irish committee of the English Privy Council undertook extensive cartographical surveys and stocktaking exercises, before publishing detailed instructions for the ensuing scheme. Lands were divided among 'servitors' (government officials and soldiers who had served the crown during the Nine Years War), 'undertakers' (English and Scottish venture capitalists and men of property, who undertook to plant their newly acquired lands with English and Scottish settlers) and those 'deserving Irish' who had supported the crown in the 1590s. Undertakers received up to 30% of the allocated lands, parcelled out in units of 2,000, 1,500 and 1,000 acres, on condition that they removed the natives, encouraged English and Scottish settlers, founded small towns and villages and erected castles or 'bawns' (fortified dwellings). The 'servitors' received approximately 20% of the allocated lands, as did the 'deserving' native Irish, with the latter group often transplanted from their homes to new estates elsewhere in the province. Generous grants to the Church of Ireland, Trinity College, Dublin, and the newly founded 'free' or 'royal' schools at Cavan, Armagh, Dungannon, Newry and Enniskillen, furthered the king's plans to advance the Protestant Reformation and to 'civilise' his Irish realm.

Finally, the crown assigned 'O'Cahan's Country', re-named Coleraine and later Londonderry, to the livery companies (trade guilds) of the city of London, in return for the necessary capital to sustain the plantation. For a figure of £20,000, which would treble by the end of King James's reign, the London companies undertook to construct two new towns of 200 and 100 houses (Londonderry and Coleraine respectively) and plant their new possessions with London's surplus population. In return, The Honourable the Irish Society, the company set up to oversee the plantation, received over half a million acres. This was divided among the livery companies and their subsidiaries in lots of between 10,000 and 40,000 acres, based around what would become the major urban settlements in the newly escheated county, King James's 'lanterns of civility'.

The onset of the Ulster plantation also coincided with the establishment of a small English colony on the Jamestown River, in North America, a precarious toehold for a British North American empire to rival France and Spain. Indeed, The Honourable the Irish Society resembled the East India or Virginia joint-stock companies, which would oversee British colonial ventures in Asia and North America. London capital also bankrolled these imperial and commercial ventures. In fact, the decision to transfer funds from the Virginian project to Ulster nearly spelt disaster for the fledgling American colony. Many of those early American planters had been involved in the Ulster scheme or had learned harsh lessons from the earlier collapse of the Munster plantation. Thomas Holme, William Penn's surveyor in Philadelphia, was a Cromwellian officer and settler in Ireland and appears to have designed the city's central square directly from Londonderry's plan, while George Berkeley and Sir John Percival, who assisted in laying out the Georgia colony, corresponded regularly about the Londonderry plantation.[38] These seventeenth-century colonial ventures paved the way for the eighteenth-century exodus to the 'Land of Caanan' by countless thousands of Irish Catholics and Protestants, who would subsequently play a key role in both defending the British crown and founding the American Republic.

The essays in this volume address many of the historical and historiographical issues discussed above. Jenny Wormald shows how James VI, on ascending the English throne, brought a distinctively Scottish perspective to ruling Ireland. His abortive plantation of the Isle of Lewis in the Outer

Hebrides and his more successful 'civilising' policies directed against Scottish Gaeldom, paved the way for a widespread plantation and anglicising programme in Ulster. In imitation of his policy of ruling the Scottish Highlands through the earls of Argyll and Huntly, he sought to use Hugh O'Neill to effectively govern Ulster on behalf of the crown and chose to adopt a more relaxed attitude than his predecessor towards his Irish Roman Catholic subjects. These policies were frustrated by the 'Gunpowder Plot', the hostility of key crown officials in Ireland and the Flight of the Earls in 1607. The latter event presented a unique opportunity to both extend the remit of the plantation and solve a series of James' pressing socio-economic, strategic, political, legal and financial problems. Similarly, Martin MacGregor seeks to view the background, practice and rhetoric of the Ulster plantation in the context of concurrent developments in the Scottish Highlands and Islands, and through a common literature and rhetoric of colonisation. In doing so, he also brings a timely corrective to the one-Gaeldom thesis, showing that the native beneficiaries of James policies in Scotland, clans like the Campbells and MacKenzies, viewed themselves not as a Scottish-Gaelic 'other' but as pivotal actors in the British imperial project.

If James provided the imperial vision for the Ulster plantation, Philip Withington reminds us that 'English monarchical republicanism' or 'corporatism', the ideology which underpinned an expanding, participatory 'commonwealth' in Tudor and Stuart England, also played a crucial role. Without underestimating the importance of early modern state formation to the plantation process, Withington highlights a civic humanist ideology and prevalent corporate vocabulary, which characterised late sixteenth- and early seventeenth-century elites and percolated into plantation literature and discourse. A determined king and potent ideology still required a reluctant city of London to bankroll a plantation of this magnitude. Ian W. Archer argues that the city of London made a real effort to comply with its obligations. Nevertheless, as the largest undertaker in the Ulster plantation, it struggled to recruit the necessary quota of English settlers, thereby failing to remove the native Irish from their portions. Waning royal interest in the plantation, political factionalism and Charles I's straitened financial circumstances in the 1630s would culminate in the city's indictment at the Court of Star Chamber in 1635. These developments had enormous implications for the collapse of the tripartite Stuart monarchy in the 1640s.

The immediate impact of the plantation, however, was most keenly felt in Ulster itself. Raymond Gillespie's exploration of the Irish political contexts bewails the loss of key contemporary sources, not least the English Privy Council Registers, which hampers any attempt to write a definitive political history. Nonetheless, surviving evidence shows that in addition to the rearrangement of landholdings, the venture involved a more fundamental redistribution of power both regionally and nationally. This had major ramifications for the political fortunes of the various ethnic and confessional factions within the kingdom. Colin Breen's essay highlights Ulster's neglected archaeological heritage, which has the potential to fill the yawning gap in the written records. To date, excavations on Ulster's plantation settlements has been piecemeal – either focusing exclusively on isolated fortified houses, town and village settlements, or comprising rescue and restoration operations. It is hoped that Dr Breen's ongoing excavations at Dunluce in County Antrim, part of the 'unofficial' McDonnell plantation, might provide a template for major state-sponsored schemes elsewhere in the province.

Marc Caball and Diarmaid Ó Doibhlin's contributions assess the impact of the plantation on the Irish literati, a professional cadre of Gaelic praise-poets and of Hiberno-Latin writers based on the continent. Their literary works and theological writings provide a frank, cogent and despondent appraisal of unprecedented political, socio-economic, cultural and demographic upheaval. The defeat of O'Neill and his confederates heralded the collapse of a system which had sustained a classical Irish literature for half a millennium. Many of the alumni of the bardic schools subsequently migrated to Europe, trained in Irish continental colleges and became the foot-soldiers of a new Catholic revival in Ulster. Brian Mac Cuarta's article complements these literary studies by examining how the Irish Catholic Church coped in the immediate aftermath of the plantation. The post-Tridentine Church responded energetically to the threat of English Protestant expansionism by ensuring the loyalty of the clergy and laity, establishing a shadowy Catholic diocesan structure and launching Franciscan and Jesuit missions to the province. The poverty of the Ulster Irish Catholics, however, limited the extent of this revival and any progress was totally undone in the wake of the Cromwellian re-conquest.

Andrew Hadfield explores the context of Spenser's prose dialogue and argues that his work was the product of the Ramist reforms of rhetoric and logic that swept through Europe in the mid to late sixteenth century. Spenser's Ramist education equipped him to write his hugely influential tract on Ireland that lies behind so trenchant a defence of the Ulster plantation as Sir John Davies's *A discouerie of the true causes why Ireland was neuer entirely subdued* (1612). Willy Maley's essay examines Ireland's evolving role in Jacobean literature. While Elizabethan writers such as Edmund Spenser, Barnaby Rich and Fynes Moryson served as cheer-leaders for conquest, their Jacobean and Caroline successors adopted a more subtle approach, using the Irish experience to develop theories of commonwealth. Nonetheless, the recalcitrant Catholic Irish remained firmly outside the political fold. By the late 1640s, however, Nicholas McDowell shows that for John Milton at least the 'blockish Presbyters of Clandeboye', represented by the Scottish and the Presbyterian faction in the Westminster Assembly, had temporarily replaced the Catholic Irish as the major perceived threat to the English state.

On the quarto-centenary of the plantation of Ulster, and a quarter of a century since the appearance of Philip Robinson's book, an integrated, interdisciplinary history of this defining moment in Ulster, Ireland and Britain's history is still lacking. David Dickson's *Old World Colony: Cork and South Munster, 1630–1830*, a masterful survey of the socio-economic and political history of south Munster, provides the obvious template for any such undertaking.[39] Similarly, Jane Ohlmeyer's research on the plantation peers could be extended to look at the plantation gentry, Anglican clergy, Presbyterian ministry and the mercantile classes. This collection forms part of a series and the next volume to appear will deal directly with the impact of the Ulster plantation in Ireland, up to and including the outbreak of the rebellion in October 1641. Recent historiography and literary criticism has done much to address the insularity of previous writings on early modern Ireland, recognising the full extent of Irish engagement with Britain, Europe and the wider world. The importance of a comparative framework for the study of the Irish colonial experience is underlined in the forthcoming 1641 volume by contributions on the colonial experiment in the Americas by both England and Spain.[40] Furthermore, a number of articles will examine the native reaction to the arrival and establishment of the settler community in Ulster, and the systematic use of violence by the colonial authorities to maintain order. The outbreak of the 1641 rebellion, and the attendant massacres on both sides, are explored in a broader European context, with a focus on the French Wars of Religion, the Dutch Revolt and the Thirty Years War.

While not unique, the extent of political, cultural and economic upheaval in Ireland during the seventeenth century should never be underestimated. The legacy of the Ulster Plantation is still contested to this day. The divisions between the descendants of the native and settler communities continue to underpin Irish and British politics. As the Peace Process evolves and the violence of the previous forty years begins to recede into memory, vital space has been created for a timely reappraisal of the plantation process and its role in identity formation within Ulster, Ireland and beyond. This collection offers an important redress in terms of the previous coverage of the plantations, moving away from an exclusive colonial perspective, to include the native Catholic experience, and in so doing will hopefully stimulate further research into this crucial episode in Irish and British history.

NOTES

1 J. Lyttleton and C. Rynne (eds), *Plantation Ireland: Settlement and Material Culture, c.1550–1700* (Dublin, 2009) and a general survey by Johathan Bardon entitled *The Plantation of Ulster* (Dublin, 2011) are the only major academic works to have appeared so far.

2 T. W. Moody, *The Londonderry Plantation, 1609-1641: The City of London and the Plantation of Ulster* (Belfast, 1939); idem, *The Bishopric of Derry and the Irish Society of Londonderry, 1602-1705* (Dublin, 1968); M. Perceval-Maxwell, *Scottish Migration to Ulster in the Reign of James I* (Belfast, 1973); R. J. Hunter, 'The end of O'Donnell power', in William Nolan, Liam Ronayne and Máiread Dunleavy (eds), *Donegal: History and Society* (Dublin, 1995), pp. 229-67; J. J. Silke, 'Plantation in Donegal', in ibid., pp. 267-83; R. J. Hunter, 'The Fishmongers' Company of London and the plantation of Ulster, 1609-41', in Gerard O'Brien (ed.), *Derry and Londonderry: History and Society* (Dublin, 1999), pp. 205-59; idem, 'County Armagh: a map of plantation', in William Nolan and A. J. Hughes (eds), *Armagh: History and Society* (Dublin, 2001), pp. 265-95; idem, 'Sir William Cole, the town of Enniskillen and Plantation County Fermanagh', in E. M. Murphy and W. J. Roulston (eds), *Fermanagh: History and Society* (Dublin, 2004), pp. 105-47; idem (ed.), *The Strabane Barony during the Ulster Plantation, 1607-41* (Belfast, 2011); Hunter's collected essays, edited by John Morrill and entitled *Ulster Transformed: essays on Plantation and Print Culture 1590-1641*, will be published by the Ulster Histoprical Foundation in 2012; Philip Robinson, 'The Ulster plantation and its impact on settlement patterns in County Tyrone', in Charles Dillon and H. A. Jefferies (eds), *Tyrone: History and Society* (Dublin, 2000), pp. 233-67; William Roulston, 'The Ulster plantation in the manor of Dunnalong', in ibid., pp. 267-91.

3 Aidan Clarke, with R. D. Edwards, 'Pacification, plantation and the Catholic question, 1603-33', in T. W. Moody, F. X. Martin and F. J. Byrne (eds), *The New History of Ireland*, iii: *Early Modern Ireland, 1534-1691* (Oxford, 1976), pp. 187-231.

4 Philip Robinson, *The Plantation of Ulster: British Settlement in an Irish landscape, 1600-1670* (Dublin, 1984); J. S. Curl, *The Londonderry Plantation* (Chichester, 1986); idem, *The Honourable the Irish Society and the Ulster Plantation, 1608-2000: The City of London and the Colonisation of County Londonderry in the Province of Ulster in Ireland. A History and Critique* (Chichester, 2000).

5 John McCavitt, *Sir Arthur Chichester, Lord Deputy of Ireland, 1605-1616* (Belfast, 1998); John McGurk, *Sir Henry Docwra, 1564-1631: Derry's Second Founder* (Dublin, 2006).

6 Brían Ó Cuív, 'The Irish language in the early modern period', in Moody, Martin and Byrne (eds), *A New History of Ireland*, pp. 509-42; B. Ó Buachalla, *Aisling Ghéar: na Stíobhartaigh agus an t-Aos Léinn 1601-1788* (Dublin, 1996); Marc Caball, *Poetry and Politics: Reaction and Continuity in Irish Poetry, 1558-1625* (Cork, 1998). Tomás Ó Fiaich, *The O'Neills of the Fews* (Armagh, 2003); Breandán Ó Doibhlin, *Manual do Litríocht na Gaeilge*, fasc. I-V (Dublin, 2003-9).

7 Vincent Carey, 'John Derricke's *Image of Irelande*: Sir Henry Sidney and the massacre at Mullaghmast, 1578', *Irish Historical Studies*, 31 (1999), 305-27; idem, *Surviving the Tudors: Gerald the 'Wizard' Earl of Kildare and English Rule in Ireland, 1537-1586* (Dublin, 2002); David Edwards, Pádraig Lenihan and Clodagh Tait (eds), *Age of Atrocity: Violence and Political Conflict in Early Modern Ireland* (Dublin, 2007), specifically articles by Edwards, Carey, John McGurk and Kenneth Nicholls.

8 Nicholas Canny, *Making Ireland British, 1580–1650* (Oxford, 2001).
9 Jane Ohlmeyer, 'A laboratory for empire? Early modern Ireland and English imperialism', in Kevin Kenny (ed.), *Ireland and the British Empire* (Oxford, 2004), pp. 26–60. See also '"Civilizinge of those rude partes": the colonization of Ireland and Scotland, 1580s–1640s' in Nicholas Canny (ed.), *The Oxford History of the British Empire*, vol. 1 (Oxford, 1998), pp. 124–47. For her latest research on the aristocracy, see *Making Ireland English: The Irish aristocracy in the seventeenth century* (New Haven, 2012).
10 Brendan Bradshaw, 'Sword, word and strategy in the reformation in Ireland', *Historical Journal*, 21 (1978), 475–502; idem, 'Robe and sword in the conquest of Ireland', in C. Cross, D. Loades and J. J. Scarisbrick (eds), *Law and Government under the Tudors* (Cambridge, 1988), pp. 139–63; Alan Ford, *The Protestant Reformation in Ireland, 1590–1641* (2nd edn, Dublin, 1997); idem, 'James Ussher and the creation of an Irish Protestant identity', in Brendan Bradshaw and Peter Roberts (eds), *British Consciousness and Identity: The Making of Britain, 1533–1707* (Cambridge: 1998), pp. 185–212; Alan Ford and John McCafferty, (eds), *The Origins of Sectarianism in Early Modern Ireland* (Cambridge, 2005); Raymond Gillespie, *Devoted People: Belief and Religion in Early Modern Ireland* (Manchester, 1997); idem, *Reading Ireland: Print, Reading and Social Change in Early Modern Ireland* (Manchester, 2005); Raymond Gillespie and Andrew Hadfield (eds), *The Oxford History of the Irish Book*: iii: *The Irish Book in English, 1550–1800* (Oxford, 2006); R. J. Hunter, 'John Franckton (d.1620), printer, publisher and bookseller in Dublin', in Charles Benson and Síobhan Fitzpatrick (eds), *That Woman!: Studies in Irish Bibliography: A Festshrift for Mary 'Paul' Pollard* (Dublin, 2005), pp. 1–26; Henry Jefferies, *The Irish Church in the Tudor Reformations* (Dublin, 2010).
11 Brían Mac Cuarta, *Catholic Revival in the North of Ireland, 1603–41* (Dublin, 2007); T. Ó hAnnracháin, 'The survival of the Catholic church in the era of the flight of the earls and the plantation of Ulster', in David Finnegan, Éamonn Ó Ciardha and Marie-Claire Peters (eds), *Imeacht na nIarlaí: The Flight of the Earls* (Derry, 2010), pp. 221–6.
12 J. H. Andrews, *The Queen's Last Map-Maker: Richard Bartlett in Ireland, 1600–03* (Dublin, 2009); W. J. Smyth, *Map-Making, Landscapes and Memory: A Geography of Colonial and Early Modern Ireland c.1530–1750* (Cork, 2006); Patrick Duffy, *Exploring the History and Heritage of Irish Landscapes* (Dublin, 2007); Annaleigh Margey, *Mapping Ireland c.1550–1640: An Illustrated Catalogue of the Plantation Maps of Ireland* (Dublin, forthcoming).
13 Brendan Bradshaw, 'Native reaction to the westward enterprise: a case-study in Gaelic ideology', in K. R. Andrews, N. P. Canny and P. E. H. Hair (eds), *The Westward Enterprise: English Activities in Ireland, the Atlantic and America 1480–1650* (Liverpool, 1978), pp. 65–80; idem, 'The bardic response to conquest and colonisation', *Bullán*, 1 (1994), 119–22; Vincent Carey, 'The end of the Gaelic political order: the O' More lordship of Laois', in P. G. Lane and William Nolan (eds), *Laois: History and Society* (Dublin, 1999), pp. 213–57; Christopher McGinn, *'Civilizing' Gaelic Leinster: The Extension of Tudor Rule into the O'Byrne and O'Toole Lordships* (Dublin, 2004).

14 Hiram Morgan, *Tyrone's Rebellion: The Outbreak of the Nine Years War in Tudor Ireland* (Woodbridge, 1993); idem (ed.), *The Battle of Kinsale* (Bray, 2004); Patrick Duffy, 'Patterns of landownership in Gaelic Monaghan in the late sixteenth century', *Clogher Record*, 10 (1981), 304–22; idem, 'Farney in 1634: an examination of John Raven's survey of the Essex estate', *Clogher Record*, 11 (1983), 245–63; idem, 'A lease from the estate of the Earl of Essex, 1624', *Clogher Record*, 13 (1990), 100.

15 Michael McCarthy-Murrough, *The Munster Plantation: English Migration to Southern Ireland 1583–1641* (Oxford, 1986).

16 John Morrill, 'The wars of the Three Kingdoms', in Glenn Burgess (ed), *The New British History: Founding a Modern State, 1603–1715* (London, 1999), p. 86.

17 J. G. A. Pocock, 'British history: a plea for a new subject', *New Zealand Historical Journal*, 8 (1974), 3–12. For his later views on the topic see 'The New British History in Atlantic perspective: an Antipodean commentary', *American Historical Review*, 104 (1999), 490–500. See also, John Morrill, 'Thinking about the New British History' in David Armitage (ed.), *British Political Thought in History, Literature and Theory 1500–1800* (Cambridge, 2006), pp. 23–46.

18 Glenn Burgess, 'Introduction: the New British History', in Burgess (ed.), *New British History*, pp. 1–29.

19 Morrill, 'The wars of the Three Kingdoms', p. 82.

20 Ibid., p. 79.

21 Canny, *Making Ireland British*, p. vii.

22 Peter Lake is quoted in Ohlmeyer, 'Seventeenth-century Ireland and the New British and Atlantic histories', p. 448.

23 Ohlmeyer, *Making Ireland English*. For an excellent example of a regional study see David Edwards, *The Ormond Lordship in County Kilkenny, 1515–1642: The Rise and Fall of Butler Feudal Power* (Dublin, 2003).

24 Hugh Kearney, *The British Isles: A History of Four Nations* (Cambridge, 1989). In fairness, Ohlmeyer only applied the 'five kingdoms' model in relation to the career of the marquis of Antrim. See Jane Ohlmeyer, *Civil War and Restoration in the Three Stuart Kingdoms: The Career of Randal MacDonnell, Marquis of Antrim, 1609–83* (Cambridge, 1993), p. 17.

25 There is rapidly increasing body of literature on the Irish Diaspora in the early modern period since Gráinne Henry, *The Irish Military Community in Spanish Flanders, 1586–1621* (Dublin, 1992). Scholars such as Tadhg Ó hAnnracháin, Tom O'Connor, Éamonn Ó Ciosáin, Mary Ann Lyons, Igor Pérez Tostado, Óscar Recio Morales and others have undertaken systematic research through the enormous manuscript collections in Brussels, Paris, Salamanca and Rome, uncovering material vital to our overall understanding of events in Ireland and of the wider European context. See for example, Thomas O'Connor and Mary Ann Lyons (eds), *Irish Communities in Early-Modern Europe* (Dublin, 2006).

26 Jerrold Casway, *Owen Roe O'Neill and the Struggle for Catholic Ireland* (Philadelphia, 1984); Tadhg Ó hAnnracháin, *The Catholic Reformation in Ireland: The Mission of Rinuccini, 1645–49* (Oxford, 2001).

27 John Morrill, 'The British problem c.1534–1707', in Brendan Bradshaw and John Morrill (eds), *The British Problem c.1534–1707: State Formation and the Atlantic Archipelago* (London, 1996), p. 3.
28 Ohlmeyer, 'Seventeenth-century Ireland and the New British and Atlantic histories', p. 454. See also, Breandán Ó Buachalla, *The Crown of Ireland* (Galway, 2006).
29 Toby Barnard, 'British and Irish history', in Burgess (ed.), *New British History*, p. 204.
30 David Stevenson, *Scottish Covenanters and Irish Confederates: Scottish–Irish Relations in the Mid-Seventeenth Century* (Belfast, 1981); Toby Barnard, 'The Protestant interest, 1641–1660', in Jane Ohlmeyer (ed.), *Ireland from Independence to Occupation, 1641–1660* (Cambridge, 1995), pp. 218–40.
31 J. J. Silke, *Kinsale: The Spanish Intervention in Ireland at the End of the Elizabethan Wars* (repr., Dublin, 2000); Morgan, *Tyrone's Rebellion*, passim.
32 John McCavitt, *The Flight of the Earls* (Dublin, 2002); Finnegan, Ó Ciardha and Peters (eds), *Imeacht na nIarlaí: The Flight of the Earls*, passim.
33 Jane Ohlmeyer, 'Civilizinge of those rude partes', passim.
34 Thomas O'Connor (eds), *The Irish in Europe 1580–1815* (Dublin, 2001); idem, *Irish Migrants in Europe after Kinsale 1602–1820* (Dublin, 2003); idem, *Irish Communities in Early Modern Europe* (Dublin, 2006); idem, *The Ulster Earls in Baroque Europe: Refashioning Irish Identities* (Dublin, 2009); Igor Pérez Tostado, *Irish Influence at the Court of Spain in the Seventeenth Century* (Dublin, 2008); Óscar Recio Morales, *Ireland and the Spanish Empire, 1600–1825* (Dublin, 2010); E. G. Hernán, Miguel Ángel de Bunes, Óscar Recio Morales and Barnardo J. García García (eds), *Irlanda y la Monarquía Hispanica: Kinsale, 1601–2001. Guerra Política, Exilioy Religión* (Madrid, 2002).
35 J. J. Silke, 'The Irish abroad, 1534–1691', in Moody, Martin and Byrne (eds), *New History of Ireland*, pp. 587–632; R. A. Stradling, *The Spanish Monarchy and Irish Mercenaries: The Wild Geese in Spain, 1618–68* (Dublin, 1994); Gráinne Henry, *The Irish Military Community in Spanish Flanders, 1586–1621* (Cork, 1992); David Worthington, *British and Irish Emigrants and Exiles in Europe, 1603–88* (Leiden, 2010); idem, *At 'the forepost of Christianitie': The British and Irish in Central Europe, 1560–1688* (London, 2012); Ciaran O'Scea, 'The significance and legacy of Spanish intervention in west Munster during the battle of Kinsale', in Thomas O'Connor and Mary Ann Lyons (eds), *Irish Migrants in Europe after Kinsale*, pp. 32–63; idem, 'Irish emigration to Castile in the opening years of the seventeenth century', in Patrick Duffy (ed.), *To and from Ireland: Planned Migration Schemes c.1600–2000* (Dublin, 2004), pp. 17–37; Éamon Ó Ciosáin, 'Hidden by 1688 and after: Irish Catholic migration to France, 1590–1685', in Worthington (ed.), *British and Irish Emigrants and Exiles in Europe*, pp. 125–41.
36 The best single survey is Ó Buachalla, *Aisling Ghéar*. See also Ó Cuív, 'The Irish language in the early modern period', in Moody, Martin and Byrne (eds), *A New History of Ireland*, pp. 509–42; Marc Caball, *Poetry and Politics: Reaction and Continuity in Irish Poetry, 1558–1625* (Cork, 1998); Mícheál Mac Craith, 'Literature in Irish, c.1550–1690: from the Elizabethan settlement to the Battle of the Boyne', in Margaret Kelleher and Philip O'Leary (eds), *The Cambridge History of Irish*

Literature (2 vols, Cambridge, 2006), pp. 191–231; Éamonn Ó Ciardha, 'Irish and Latin sources for the history of early modern Ireland', in Alvin Jackson (ed.), *The Oxford Handbook of Irish History* (Oxford, forthcoming).
37 Michael Perceval-Maxwell, *Scottish Migration to Ulster*, passim.
38 Robert Home, *Of Planting and Planning: The Making of British Colonial Cities* (London, 1997), pp. 16–17. See also Oscar Handlin and John Burchard (eds), *The Historian and the City* (Harvard, 1963), pp. 190–4. Thanks to Andrew Kincaid for both these references.
39 David Dickson, *Old World Colony: Cork and South Munster, 1630–1830* (Cork, 2005).
40 See Jane Ohlmeyer and Micheál Ó Siochrú, *Ireland 1641: Contexts and Reactions* (Manchester, forthcoming).

2

The 'British' crown, the earls and the plantation of Ulster

JENNY WORMALD

When, on 24 March 1603, James VI acquired the crown of England, it came with a good deal of baggage: the reasonably quiescent principality of Wales and the anything but quiescent kingdom of Ireland. Indeed, while poets and politicians and, in 1604, the king's own proclamation on the new 'Great Britain' heralded the blessed age of peace now established between 'two mightie, famous and ancient Kingdomes' formerly at war,[1] the grim reality was that James actually inherited two kingdoms, England and Ireland, which were at war with one another. Indeed, the new king – Rex Pacificus – succeeded a monarch who claimed to be as resolutely opposed to war as he, and yet who died leaving her English kingdom embroiled in two of them, with Ireland and Spain. So to the massive problems of his accession to the English throne itself James was faced in addition with a ruinous and inconclusive foreign war and a protracted and damaging civil one. That this was the situation he knew that he would inherit gives good reason for doubting the old and Anglocentric idea that the Scottish king had lived his life dreaming of his English inheritance and had therefore shaped his policies to please the English queen.[2] The reality, to Elizabeth's continual fury, had been very different. James's acceptance of the English throne was based not on any belief that life would be easier than in Scotland (although it would hopefully be richer) but on his confidence in his own abilities and long experience of royal rule and his certainty of God's purpose. That experience would now be put to use in England; it is a too often neglected clue to James's kingship after 1603, but it is in fact crucial to the understanding of that kingship.

The immediate nightmare of two wars quickly dissipated with the end of the Nine Years War in 1603 and the conflict with Spain in 1604. Peace with Spain helped the English exchequer and revolutionised English foreign policy, until the last years of James's reign. Scottish kings, well aware that they could not afford war, relied for the most part (with very rare exceptions) on diplomacy. James followed that practice, and with his enhanced position as king of three kingdoms, raised 'Britain' to become a major player in European affairs, strenuously maintaining peace through an alliance with Spain until the disastrous outbreak of the Thirty Years' War. Paradoxically, what looked like the greater of the two problems which awaited him in 1603, the foreign war with one of Europe's great powers, proved much easier to resolve.

Ireland presented an infinitely greater challenge, not least because it was not a foreign power. Cessation of outright hostilities did not solve the long-standing and very grievous problem of English mismanagement of Irish affairs. Ireland had not been well governed by the Tudors; their shifting policies were never sufficiently pressed, especially during Elizabeth's reign.[3] Henry VIII, in making himself king of Ireland, created an entirely new constitutional position for the native Irish which had offered a possible way forward. The Tudors, however, naively believed that the root of the problem lay in the fact that the original conquest of Ireland had not gone far enough – which was certainly right – and that the solution lay in the relatively easy matter of completing it – which was certainly wrong. The Nine Years War had left Ireland in a dire situation, a problem that did not go away, despite the focus shifting temporarily in 1603 to the coming together of England and Scotland.[4] Scotland became the new factor in a novel 'British' mix, and it seemed a sufficiently fulltime job for the new king and the political elites both north and south of the border to try to find a *modus vivendi*, which was to an extent achieved, despite the innate hostility between the peoples of the two kingdoms, by not trying to define anything too precisely.[5] This temporarily involved a change of priorities, as English politicians, hastily joining in James's desire to end the wars, devoted themselves to the urgent business of fixing their interest with their new king, while at the same time trying to persuade him not to give so much of his attention to the Scots. Scotland may have been seen by the English an unwelcome addition, but Ireland constituted an integral part of the English composite monarchy and could not be ignored. It was the duty of the king of England to control Ireland according to English principles, however vaguely and inconsistently these might be applied. Indeed, the difficulty in ruling the peoples of England and Ireland, whose hatred of one another made Anglo-Scottish hostility look positively mild, was compounded by the fact that James VI and I was not simply an English king.

It is something of an irony that while claims for and objections to James's desired style of king of Great Britain were couched in the context of his Anglo-Scottish kingship, 'Britishness' actually becomes meaningful only when all three kingdoms are brought into play. The hugely ambitious plantation of Ulster demonstrates this fact very clearly. Inevitably the plantation had to be British, because a Scottish king was not going to regard the process as purely English business. Those who planned and executed plantations under Mary and Elizabeth did not include the Scots. Indeed, a helpful offer by the Protestant earl of Argyll in the 1560s to support Thomas, earl of Sussex and Lord Deputy of Ireland, received short shrift, unwisely so, given Argyll's geographic proximity to Ireland and his enthusiasm for the Protestant cause.[6] This policy of exclusion, however, could not survive the Jacobean succession.

Ireland as a British problem after the union of the crowns became even more confused. Successive Tudor governments dealt with Gaelic lords, Old English and New English from the starting point of an intense dislike of the native Irish and 'degenerate' English. Now a fourth group, the Scots, were added to an already inflammable mix. The new British king further complicated matters, as from an exclusively English perspective, Ireland posed a problem for the English crown. Equally, and again from the English point of view, James's infuriating refusal to identify himself exclusively with the English crown caused more difficulties. To the dismay and astonishment of his leading English subjects, the new king refused to be straitjacketed into the role of being only monarch of England and, indeed, insisted on drawing on his Scottish experience. It was even worse because James, as a consequence of dealing with his own highly problematic Gaelic population, part of which had long political, military, marital and cultural contacts with the Gaelic Irish, thought that he understood Ireland better than the English government. Yet he may well have been right in this regard.

At first sight, the whole concept of plantation lends some credence to James's belief; indeed the very idea of planting 'civilised' outsiders among 'barbaric' insiders, and bumping up royal revenue in the process, not only drew on Tudor precedents in Ireland but also, albeit on an infinitely smaller scale, on policies pursued by James in Gaelic Scotland a decade before the plantation of Ulster. This Scottish monarch, however, with his particular style of kingship, did not confine himself to one approach. A consistent policy of civilising, confessionalising or commercial plantation might have found favour with his English government, providing a degree of continuity with their experience of Munster in the 1580s. Plantation, however, was neither his central nor his initial policy. His insistence on the advantages of Scottish kingship in England might have been contentious, but at least James and his opponents at Westminster fought out their differences over kingship, law and parliament in a language and a rhetoric both could understand.[7] In contrast, with regard to Ireland, he insisted on a style of kingship which the English deemed hopelessly backward and which they did not understand.

It could of course be argued that trying to incorporate the kingdom of Ireland into a multiple monarchy comprising Scotland, England and Wales was a hopeless prospect, given the large number of the Irish who did not wish to embrace an alien political, religious and secular culture. Yet this highly intelligent king of Scotland and England might have achieved a greater level of success in Ireland than his Tudor predecessors partly because he treated his Irish subjects with less contempt, and partly because of a scattergun approach which responded to problems as they arose rather than attempting to impose an Anglicising solution. Indeed, he adopted an equally malleable approach to the Scottish Highlands; in 1609, at the outset of the Ulster plantation, he abandoned the idea of plantation in Scotland and turned to a new, more successful policy. Flexibility was undoubtedly one of King James's great strengths, in each of his three kingdoms.

Apparently, the idea of plantation in Ireland came easily to James as a result of his Scottish experience, although his three attempts to 'civilise' at least a small part of the Scottish Highlands, the island of Lewis, by planting lowlanders from Fife, the 'gentlemen adventurers', failed utterly.[8] In May 1598 the Scottish Privy Council ordained that burghs should be established in Lewis, Lochaber and Kintyre, while in June the king directed the Fife Adventurers 'to plant policy and civilisation in the hitherto most barbarous Isle of Lewis', at their own expense (naturally), and 'to develop the extraordinarily rich resources of the same for the public good and the King's profit'.[9] Initially at least, the project gained ground, encouraged by that self-opinionated elder statesman Sir James Melville of Halhill, who trumpeted the untapped wealth of the Highlands.[10] In July 1602, however, the Scottish Privy Council recorded the dismal fact that the some of the unfortunate gentlemen had been murdered, and the few who survived had gone native. A subsequent proclamation summoned musters for a stillborn expedition against Lewis, while the king stated, as he did on a number of occasions, his determination to go in person to the Highlands.[11] His great-grandfather and grandfather, James IV and V, had done so, to notable effect. James VI had also travelled there in 1596 to resolve – successfully, as it turned out – the problem of Angus Macdonald of Dunivaig.[12] Unfortunately, he never returned to the Highlands, any more than he ever went to Ireland. The years 1605 to 1607 witnessed another abortive attempt, bedevilled by spies professing friendship to the lowlanders but actually working for the Highlanders. In July 1606 the 'gentlemen adventurers' complained to the council about the 'tirrany and oppression' of the inhabitants of Lewis, the Macleods and others, and the council weakly told the 'schereffes in that pairt' to outlaw and escheat them. The reality of the situation became clearer in 1607. In August, the council heard that Rory Macleod of Dunvegan had seized the castle of Stornoway and other fortalices in Lewis which belonged to the planters. The council pathetically and uselessly charged him to hand them over within six hours on pain of rebellion; this hopeless demand, along with that to the 'schereffes in that pairt' begs the question whether the council actually cared at this stage. It certainly cannot have helped when in March James VI, all too typically, shifted the blame and berated the council itself for its failure in Lewis.[13] Following another failed attempt in 1609, the adventurers simply gave up – the wonder is that they stuck it for so long.

The comparisons with the meticulous preparations by Sir Thomas Smith for a plantation in the Ards in the 1570s and the enormously detailed planning of the Munster plantation in the 1580s could not be more stark. It is not clear whether anyone even bothered to map Lewis. The king thought that he and his council could simply send people out with the instruction to civilise and to gain riches, and expect miraculous results. That James indulged this fantasy might give credence to the idea that he saw plantation as a route to improvement. But James the formulator of policy, often on a grand scale, always assumed that the responsibility to make the trains run on time lay with other lesser mortals. In the case of Lewis, the lesser mortals did not do so. Fortunately, the detailed planning for the plantation of Ulster was not left to a Scottish king and his Scottish advisers.

James's traditional and conservative-minded approach, however, both in Scotland and Ireland, involved a great deal more than plantation. Unlike Tudor monarchs, anxiously seeking to curtail the power of the great landed aristocracy, James viewed his aristocrats in a positive light and described them in *Basilikon Doron* as the 'armes and executers of your lawes' in the localities.[14] Their power, if sometimes abused by individuals, was nevertheless crucial to the successful running of the kingdom. Pre-1603 Jacobean government had a markedly more aristocratic bent than its Elizabethan counterpart. As in England, lairds and cadet branches of aristocratic families enthusiastically pushed their way into central government, a process which accelerated after the king crossed the Tweed. But James's own attitude to his territorial nobility and his *noblesse de robe* is clearly seen after the death of his secretary and chancellor John Maitland, laird of Thirlestane in 1595, that archetypal political laird pushing his way into the highest echelons of central government. While Holyrood hummed with enjoyable and energetic speculation about his successor, James retired to Linlithgow, bored with the din of the 'faccioneres'. Indeed he expressed greater concern at the death of the earl of Atholl, his chancellor; for who, asked the king, would now keep order in Perthshire?[15] More dramatically, James coped remarkably well with dissident earls, balancing their troublemaking with their usefulness to him. 'Good Lord! me thinke I doe but dreame: no king a weeke would beare this!' an impassioned Elizabeth burst out in 1589, when James refused to take strong action against Huntly and Erroll the great Catholic earls of the northeast.[16] Between 1589 and 1595, the northern earls, sometimes in contact with Spain, appeared to threaten the king's rule. Unlike the paranoid Elizabeth, however, James saw more to Huntly than a potential rebel. The Gordons of Huntly in the eastern Highlands and the Campbells of Argyll in the west had long been the crown's lieutenants. He often used George, the sixth earl of Huntly, along with Archibald seventh earl of Argyll, to impose control on the highlands, even after 1603. Argyll, for example, rooted out the notorious Macgregors in the southwest, culminating in the suppression of their very name in 1617, an act which duly gave them an honoured and largely fictional place in the romantic history of Scotland.[17]

Moreover, James and his landed aristocracy shared attitudes to law and order very much at odds with the aspirations of the lairds, the social class which increasingly provided recruits for an emerging lay legal profession. Much as he might complain about their 'deidlie feidis', he did not lead the attacks on this antiquated system of communal justice. He left that to the ministers of the reformed kirk and the lawyers, who championed the partially successful 'anent feidis' act of 1598.[18] The king by contrast underwrote the 'assythment' (compensation) crucial to the settlement of feuds, a traditional approach with important political and legal ramifications. For a century and a half, Scottish practice involved lords and their men writing personal bonds of maintenance and manrent – protection, allegiance and service. The lords also made bonds of friendship in which they promised to act together in a common cause. Already, from the mid-sixteenth century, such bonds had been taken over into the religious sphere, before being pressed into a wider service to deal with the problem of the Highlands and the Borders, in the form of the General Band of 1587. This made clan chiefs responsible, under financial penalties, for the good behaviour of their followers.[19]

This style of kingship was profoundly different from its Tudor counterpart and had a particular relevance for Ireland after 1603. It is impossible to imagine Elizabeth, at the end of the Nine Years War, treating O'Neill and O'Donnell as James did, before their sudden and dramatic flight in 1607 facilitated the plantation of Ulster. Their flight, analysed brilliantly by Nicholas Canny and others, still retains an element of mystery or more specifically, the flight of Tyrone does. Indeed it is arguably even more puzzling than has hitherto been thought, simply because of the king's policy before the flight.[20] In the first four years of his English rule, James adopted a typically Scottish solution in dealing with Elizabeth's Irish rebels, while at the same time rushing into a Scottish-sponsored plantation in Antrim and Down. Sir Henry Docwra, the governor of Derry, in an effort to prevent further rebellion in Ulster, urged the king to 'send some Scots to inhabit the country', while Lord Deputy Mountjoy would have preferred Dutchmen.[21] In any event, the king did neither. He confirmed Sir Randal MacDonnell in possession of the Route and the Glens of Antrim in May 1603, with subsequent instructions for an embryonic plantation. For the rest of Ulster, the king's plans were strikingly different.

The key to his policy was Tyrone; and the point about Tyrone was that, for James, he was the Irish equivalent of the earl of Huntly. Elizabeth's great rebel, the man who for nine years had threatened her rule of Ireland, tilted the balance against her ill-treated favourite Essex, and struck fear into English hearts now received ample reward. The king restored his earldom and lands, to the fury of Sir Arthur Chichester who coveted them for himself. From an English perspective, this Scoto-Irish policy of rewarding disloyalty appeared remarkably perverse. For centuries, failed rebellion in England meant execution, which begs the question why so many chose to rebel. Scotland had fewer full-scale rebellions, and nothing to match the Peasants' Revolt, the Glyndwr rising, Jack Cade's rebellion, the five great risings of the mid-sixteenth century, or the Nine Years War; and her rebellions were attended with far fewer executions. Politically and socially, Scotland proved more low-key, less fraught and its kings worried less about their magnates.

Consequently, despite the appalled reaction of men like Sir Robert Cecil, Sir Arthur Chichester and Sir John Davies, it is not surprising that James should import to Ulster his tried and tested belief in the efficacy of the great local lord. In addition to the restoration of O'Neill's titles and lands, Rory O'Donnell received an earldom, although he had to wait until his rival, Niall Garbh, fell out of favour. Lord Deputy Mountjoy, who took such pleasure in destroying the inaugural seat of the O'Neills at Tullaghoge in 1602, and accepted the earl's submission at Mellifont the following year, now emerged as Tyrone's great champion in Dublin and London, presumably in response to the new king's extraordinary attitude. According to Canny, the English government possibly sought to prevent O'Neill from capitalising on Elizabeth's death by claiming that his rebellion had been the result of her provocation.[22] The fact that James VI allowed Scottish aid to the Ulster rebellion may also have raised concerns about Tyrone's future relations with the new king of England. The English government soon received an unwelcome reminder that Elizabeth's successor took a distinctly unacceptable line towards not one but two of her rebels, Essex and Tyrone. Cecil discovered very quickly that James's accession meant the end of Cecilian dominance, as the king showed favour to Essex's supporters. Similarly, English opponents of Tyrone found themselves in an equally uncomfortable position. Members of James's English government now had the choice of trimming their sails or standing against the changing political winds. Mountjoy chose the first option.

After O'Neill's submission at Mellifont, the earl and Mountjoy both travelled to London where it turned out the king 'wished to humour him [Tyrone] from the start'.[23] Tyrone, therefore, got Mountjoy's backing when urging James not to appoint a lord-president of Ulster. Initially at least, the policy had the desired result and Sir John Davies reported to Cecil in 1604 that 'for the north there is no part so quiet and void of thieves as my lord of Tyrone's country', a situation which contrasted exceedingly favourably with the disorder in Tyrconnell.[24] Two years later, after the death of Mountjoy, Tyrone's 'very good lord', the earl wrote again to the king in June on the objectionable question of the lord presidency of Ulster. In September, a hostile Chichester had the humiliating task of not only protecting Tyrone but also assuring him that the king had no intention of establishing the lord-presidency.[25] Tyrone's direct appeals to the king in 1607 about his grievances and concerns brought another favourable response. James requested that he or an agent should come to London where the king and council would arbitrate; a typically Scottish approach.[26]

So while the situation in Ulster cannot simply be equated with Scottish Gaeldom, it does appear that James regarded Tyrone, and, to a lesser extent Tyrconnell, as the best solution for the province, just as he favoured the Huntly–Argyll nexus in the Scottish Highlands. And surely Tyrone knew this; in 1607, just months before the Flight, he arranged a marriage between his son and Argyll's daughter.[27] After 1603, James lavishly scattered peerages and titles on those Scottish lairds who, like Maitland, had moved into the vacuum of central government during the absentee monarchy and would become its backbone. One of them, Alexander Seton of Fyvie, ennobled in 1605 as the earl of Dunfermline, remarked smugly to Robert Cecil that the king's power was such that 'it will make the courses of all our great ydalgos the more temperate'.[28] That the crown should reduce its reliance on the great aristocracy may have suited ambitious lairds like Maitland and Seton as they emerged from their localities to positions of central power and influence. It certainly appealed to later generations of historians for whom the less aristocratically dominated southern kingdom was by definition more civilised and better governed. James, however, saw the advantages of 'new men' and great aristocrats operating in different spheres. No great aristocrat, after all, had ever done the donkey work of central government, where lesser laymen now replaced clerics. That did not obviate the need for control of the localities and in reality, the Scottish king's approach may have been much more acceptable than the Tudor policy of putting lesser men in to control Ireland which, like Scotland, expected to be dominated by its magnates.

The problem of the Flight of the Earls, therefore, is not just a matter of seeking to explain why Tyrone, instead of pushing his appeal to the king in London either in person or through an agent, as James encouraged him to do, appeared to panic and joined Tyrconnell in fleeing to the continent. It raises the question of why Tyrone abruptly threw away the opportunity offered by the new king to exercise power in Ulster, a power which he deemed rightfully his, but which Elizabeth and her officials had sought to deny him. Unlike Argyll and Huntly, however, Tyrone had little experience of the very different approach of the Scottish crown and lacked their extensive personal knowledge of James, which would have been of great consequence in the summer of 1607. In June, the king assured Tyrone of his 'favourable consideration', but the following month, he expressed his summons to Tyrone to come to London at the beginning of Michaelmas term in much more hostile terms.[29] While

Argyll and Huntly might have been able to shrug off the change of tone from a master given to sudden rages and a tendency to lash out even at his supporters, Tyrone, with a very different perception of monarchy, could not. Moreover, in his list of 20 grievances submitted after the Flight, the tenth recounted that 'although it pleased his Majesty to allow the Earl to be lieutenant of his country, yet he had no more command there than his boy'.[30] Elizabeth's late rebel and James's new lieutenant in Ulster had little opportunity to establish the role enjoyed by Campbells and Gordons in the Highlands of Scotland for a century. Indeed, it might never have been possible for him to do so. The Scottish earls of Huntly and Argyll, although by no means always in favour, dealt exclusively with the Scottish crown, whereas the Irish earl Tyrone had to work with the English crown, and the English officials in Ireland who also assailed the king's ears. Yet the Scot who wore the English crown saw in Tyrone the answer to Ulster's security problem, and possibly thought that, given time, his Scottish solution might have prevailed. His harsher approach in July 1607 might well have been temporary, but it clashed disastrously with Tyrconnell's intention to flee.

It arguably proved a huge personal mistake for Tyrone, as his subsequent career on the continent can hardly be deemed a success. It was no less so for King James, abruptly cutting off his Scottish solution to an Irish problem. With nobody to build up as the successor to Tyrone, the massively ambitious plantation of Ulster now provided the only way forward, and this presented entirely new problems. King James might have understood how to use his greatest nobility but his ability to embark on plantation was an altogether different question. The king was perfectly happy to see his crown in imperial terms and he commissioned a medal in 1603 showing himself as emperor of Britain. This was not about creating an empire but rather underwrote his claim to be king of Great Britain. His English subjects foiled the scheme, pointing out that England already had a 'British' empire, a blocking device conveniently thought up in the late sixteenth century. James as king of England ruled a people with colonising ambitions not only in Ireland but in the New World as well. As king of Scotland, James lacked any such colonial ambitions, despite granting a licence in 1622 to the Scottish court poet and in England courtier and secretary of state, William Alexander, earl of Stirling, to colonise Nova Scotia. Alexander does seem to have thought about colonisation. In 1625 he wrote *An encouragement to colonies*, just as in the same year a fellow Scot, Gordon of Lochinvar, produced his *Encouragement for all such as have intention to be undertakers in the new plantation of Cape Breton*. Two Scots at least, therefore, saw plantation as the laboratory of empire.[31]

These two exceptions prove the rule that the Scots themselves did not have an imperialist agenda. This is perhaps surprising in view of their later involvement in the second British Empire. Indeed, in the words of a medieval French proverb, 'Rats, lice and Scotsmen: you find them the world over'.[32] They tended to travel as fighting men, for which they were famed – hence the prestigious Scots Guard founded by Charles VII in 1446 – or as merchants and scholars who flooded into the markets and universities of Europe. Thus for all the opportunities opened up by the British crown in Ireland, Scots could be found in greater numbers in Poland and Scandinavian the first half of the seventeenth century: 30–40,000 compared to the 20–30,000 who went to Ireland between 1609 and 1641.[33] Fighting in the Thirty Years War seems to have appealed to the Scots as much as settlement in Ireland. Empire, for the Scots, lay in the future, when it offered the best opportunities. In the early seventeenth century it comprised only one of several options, and not necessarily the most attractive one. Now that the British crown sat on the head of a Scottish king, however, Ireland did offer alternative opportunities to the European theatre.

Yet the Scots in Ulster gained an enduring reputation in their religious impact on the plantation, or, more specifically, the Presbyterian Scots. This gave rise to the emotive and historically sloppy idea that the troubles of modern-day Northern Ireland can be traced back to King James's plantation of Ulster. It is important to reiterate, therefore, as Raymond Gillespie and Robert Armstrong have made clear, that the impact of the Scottish Presbyterians goes back to the mid and not the early seventeenth century.[34] Presbyterianism did not feature prominently among the influx of Scots in the Jacobean period. The king who proposed using a major Catholic earl as a possible solution to controlling Ulster, in the teeth of violently anti-Catholic English officials, was hardly likely to have favoured transferring what he regarded as extreme Presbyterians to Ireland as an intelligent way forward, except perhaps as a convenient way of getting them out of Scotland, which might have been understandable. In fact, James's contribution to the religious tensions in Ireland proved entirely at odds with what would happen after the mid-seventeenth century.

The use of Tyrone and the plantation of Ulster were two very different policies. The king's ecumenism, as a theologian who infinitely preferred debate to confrontation, provided the link between them.[35] He was that most unusual of early modern rulers, one who was not afraid of differing beliefs; rather, he was repelled by those who could not countenance differing beliefs. Bringing this approach to bear in a country where the determined and flourishing Catholicism of the inhabitants outraged his English Protestant subjects could only create further confusion and despair. The Protestant Elizabeth could be criticised for not giving enough support to those who, like Sir Henry Sidney, sought to build up the Church of Ireland.[36] The Protestant James gave out not insufficiently zealous signals, but rather thoroughly conflicting ones. Not only did James initially show favour to the Catholic Irish Hugh O'Neill and the Catholic Scot Randall MacDonnell, the Plantation of Ulster itself was by no means a solidly Protestant plantation. From southwest Scotland came Catholics planted in Strabane by the earl of Abercorn and Sir George Hamilton, while some Scots from Ayrshire converted to Catholicism.[37] Jacobean Ulster was not Protestant Ulster and was probably never intended to be.

In this regard, the 1613 Irish parliament, with its wonderful tussle between the Protestant Sir John Davies and the Catholic Sir John Everard on the Speaker's chair, cannot be seen as a straightforward put-down of Catholics.[38] On the other hand, the excessive and illegal claims on ecclesiastical lands in Ulster of Bishop George Montgomery – a Scot, though beneficed in the English church – seriously alienated Tyrone, O'Donnell and O'Doherty. Two more Scots, Andrew Knox and John Leslie, succeeded Montgomery as bishop of Raphoe. The British crown was not simply going to put in the English or the Protestant Irish. But the choice of Knox is of particularly significance, because of his role, in partnership with James, in jettisoning the policy of plantation in Scotland. In 1609, the Statutes of Iona marked the first step in the gradually successful policy of bringing clan chiefs into the culture and language of the lowlands.[39] At this point, the Scottish and Irish experience diverged widely as plantation emerged as the key policy for Ireland from 1609. Conversely, Scottish chiefs signed up to the Statutes in order to avoid plantation or extirpation, perhaps rather oddly in view of the abject failure of plantation in the Western Isles, but clearly not wanting to be forced to go the way of Ireland.[40] In both contexts, James clearly trusted Knox as a man who could work effectively through a mixture of gentle persuasion and blackmail with his Gaelic Irish and Scottish subjects. Perhaps the best clue to king's views came in his response, at the end of his life, to another bishop of the Isles, Thomas Knox, who wrote demanding that something be done about the Jesuits infiltrating the western seaboard. The king turned him down; if anyone could civilise the Highlanders, he responded, let them do it, even if they were papists.[41] Plantation did not constitute the only route to civilisation.

The reign of the first 'British' king, therefore, stands out as an unusual period in Irish history, one of relative calm and acceptance of royal rule from London, largely because of a king who brought a range of skills, experience and flexibility to bear, in both secular and religious terms. After 1609, plantation came to the fore and in terms of its Jacobean impact, Nicholas Canny is surely right to divide his discussion of plantation in 1622.[42] The aspirations and achievements of James's reign would not be repeated under Charles I. The dominant note would increasingly be sounded by Charles's inconsistent religious policy, whose determination to impose his own Arminian faith on both England and Scotland, while holding out hopes of increased toleration for his Catholic Irish subjects, took the issue away from plantation in Ireland to a much more dangerous, and ultimately fatal, 'British problem'. After 1649 'plantation' would operate in a different and more dreadful world. It would take much tougher and more horrible measures in the mid-seventeenth century for the Cromwellian government to get a grip of Irish affairs.

This outcome apparently leads to the conclusion that whereas James's use of his Scottish experience when he became king of England had much to recommend it, in the Irish context his success proved at best only temporary and superficial. This, however, would be far too negative and draw too clear a distinction. What James extended to the three kingdoms was his rule of the differing parts of Scotland, the lowland and the highland. Thus the English might not like being told that their parliaments were less efficient and more time-consuming, than Scottish ones, but the king certainly had a point. They might resent James's determined Scottishness, but at least for the Scots that staved off the chill of neglect and marginalisation which would later afflict them, and bought time for union to become an established fact. The problem in Ireland, however, became more complicated, more difficult to resolve, because his actions cut so dramatically across English *mores* and English expectations of how a monarch should behave and govern. Yet that is to look at the problem in terms of the British crown. It is arguable that there was no solution to Ireland, not because of the British crown but because of the English crown, which annexed Wales, failed to annex Scotland, and did not try to complete its annexation of Ireland until religious division and an intolerable level of contempt by the annexing power had made that annexation impossible. At least James regarded Ireland in the manner that the English crown had made it in theory, a kingdom, and not what the English crown made it in practice, a colony. Whatever muddle he created, the approach adopted by James was surely preferable to the proposals found in the spine-chilling writings of Sir Edmund Spenser and Sir John Davies.

The ultimate fascination of James's reign, however, is the unusual light it throws on the 'British problem'. That 'problem' is usually seen in terms of how the other two kingdoms, the satellites, interacted with the central kingdom of England. But there is another way of looking at it, a way possible only in the reign of James VI and I. Scotland and Ireland did have more in common than England and Ireland, and even, if in a limited sense, Scotland and England. That the first 'British' king was a Scot did not mean that he positively liked his Irish Gaelic subjects, any more than he particularly liked his Scottish Gaelic ones; in both cases, efforts had to be made to control them which went beyond the need to control his non-Gaelic subjects in any of his kingdoms. But simply because of his Scottish Gaels, he never saw Irish efforts in the English terms of suppression and ruthless Anglicisation. Moreover, he was not afraid of Catholics. Indulging for a moment in counter-factual history, which knocks out England and leaves James as king of Scotland and Ireland, it is just possible that in these circumstances the problem of Ireland, that ongoing English obsession, might have taken on a very different and gentler perspective. But it is not a matter of counterfactual history to suggest that what actually happened in terms of James's rule of Ireland before 1607 was that Scotland and Ireland became the centre, England the satellite, the one which kept getting in the way. That rule was therefore a very curious form of the British mix, one of considerable potential. Sadly, it was also unrealisable potential; for in 1607 the earls fled, and after 1625 'normality' was bound increasingly to assert itself, as England established its position as the dominant force in the composite monarchy.

On the shores of Rathmullan, from which the earls fled, stand the dramatic and deeply evocative statues by John Behan created for the fourth centenary of the Flight in 2007. There are the departing earls. There is the desolate group representing the defeat of Gaelic culture. One figure is missing: the king whose first solution to the problems of governing his new Irish kingdom was to invest in that culture.

NOTES

1 'A Proclamation concerning the Kings Majesties Stile, of King of Great Britaine', 20 October 1604, in J. F. Larkin and P. L. Hughes (eds), *Stuart Royal Proclamations* (Oxford, 1973), i, p. 95.
2 This goes back to James's day; writing to James before his accession the earl of Northumberland said that he had no doubt that the king would 'think that your honor in being reputed a king of England will be greater than to be king of scottes': John Bruce (ed.), *Correspondence of King James VI of Scotland with Sir Robert Cecil and Others in England* (London, 1861), p. 56. It was a mistaken assumption, but one with a very long life.
3 See, for example, the excellent discussion of sheer mess by Ciaran Brady and James Murray, 'Sir Henry Sidney and the Reformation in Ireland', in Elizabethann Boran and Crawford Gribben (eds), *Enforcing Reformation in Ireland and Scotland, 1550–1700* (Aldershot, 2006), pp. 14–39.
4 These complexities are described in detail by Ciaran Brady, 'East Ulster, the MacDonalds and the provincial strategies of Hugh O'Neill, earl of Tyrone, 1585–1603', in W. P. Kelly and J. R. Young (eds), *Scotland and the Ulster Plantations: Explorations in the British Settlements of Stuart Ireland* (Dublin, 2009), pp. 41–61.
5 Jenny Wormald, 'James VI, James I and the identity of Britain', in B. Bradshaw and J. Morrill (eds), *The British Problem, c.1534–1707: State Formation in the Atlantic Archipelago* (Basingstoke, 1996), pp. 148–71.
6 J. E. A. Dawson, *The Politics of Religion in the Age of Mary Queen of Scots: The Earl of Argyll and the Struggle for Britain and Ireland* (Cambridge, 2002), pp. 126–37.
7 Jenny Wormald, 'James VI and I, *Basilikon Doron* and *The Trew Law of Free Monarchies*: the Scottish context and the English translation' and J. P. Sommerville, 'James I and the divine right of kings: English politics and continental theory',

in L. L. Peck (ed.), *The Mental World of the Jacobean Court* (Cambridge, 1991), pp. 36–70.
8 Alison Cathcart, 'Scots and Ulster: the late medieval context', in Kelly and Young (eds), *Scotland and the Ulster Plantations*, pp. 62–83, provides an excellent discussion of both the context and the plantations themselves.
9 J. H. Burton et al. (eds), *The Register of the Privy Council of Scotland* (Edinburgh, 1877) (hereafter *RPC*), v, 455, 462–3.
10 Sir James Melville of Halhill, *Memoirs of His Own Life* (Bannatyne Club, Edinburgh, 1827), pp. 392–3.
11 *RPC*, vi, 420–1.
12 Cathcart, 'Medieval context', pp. 74–5, convincingly links this to the Lewis plantation policy.
13 *RPC*, vii, pp. 204, 229–30, 430, 511–12.
14 J. Craigie (ed.), *The Basilikon Doron of King James VI* (2 vols, Edinburgh, 1944–50), i, 87.
15 *Calendar of State Papers Scottish* xii, pp. 39–42.
16 D. Calderwood, *The History of the Kirk of Scotland* (Edinburgh, 1842–49), v, 8.
17 *Acts of the parliaments of Scotland* (Edinburgh, 1814–74) (hereafter *APS*), iv, 550–1; J. Wormald, 'Bloodfeud kindred and government in early modern Scotland', *Past and Present*, 87 (1980), 85–6. For a different view of James's approach, see K. M. Brown, *Bloodfeud in Scotland 1573–1625* (Edinburgh, 1986), pp. 241–3.
18 *APS*, iv, 158–9.
19 *APS*, iii, 461–7. Jenny Wormald, *Lords and Men in Scotland: Bonds of Manrent, 1442–1603* (Edinburgh, 1985).
20 Nicholas Canny, 'The flight of the earls', *Irish Historical Studies*, 17 (1971), 380–99; N. Canny, 'Taking sides in early modern Ireland: the case of Hugh O'Neill, earl of Tyrone', in V. P. Carey and U. Lotz-Heumann (eds), *Taking Sides? Colonial and Confessional Mentalités in Early Modern Ireland: Essays in Honour of Karl S. Bottigheimer* (Dublin, 2003), pp. 94–115; John McCavitt, *The Flight of the Earls* (Dublin, 2005).
21 Michael Perceval-Maxwell, *The Scottish Migration to Ulster in the Reign of James I* (Belfast, 1990), p. 46.
22 *Calendar of the State Papers Relating to Ireland* (24 vols, London, 1860–1911) (*CSPI*), 1603–6, pp. 13–14, 20–1; Nicholas Canny, 'The Treaty of Mellifont and the reorganization of Ulster, 1603', *Irish Sword*, 9 (1969), 249–62.
23 McCavitt, *Flight*, p. 55.
24 *CSPI*, 1603–6, p. 215.
25 Ibid., pp. 548–9.
26 *CSPI*, 1606–8, p. 194.
27 Ibid., pp. 272–3.
28 Historical Manuscripts Commission, *Calendar of the Manuscripts of the Marquess of* Salisbury (23 vols, London, 1883–1973), xvii, 149.
29 McCavitt, *Flight*, pp. 86–7.
30 *CSPI*, 1606–8, p. 377.
31 It was particularly dismal, therefore, that their king chose to send Alexander off to what was the French colony of Port Royal, founded in 1605. In 1632 Charles I

gave back to France what the Scots had taken, not exactly an inspiring imperial tale.

32 Quoted by David Armitage, 'The Scottish diaspora', in Jenny Wormald (ed.), *Scotland: A History* (Oxford, 2005), p. 272.
33 Armitage, 'Diaspora', p. 278.
34 Raymond Gillespie, 'Scotland and Ulster: a Presbyterian perspective, 1603–1700', and Robert Armstrong, 'Viscount Ards and the presbytery: politics and religion among the Scots of Ulster in the 1640s' in Kelly and Young (eds), *Scotland and the Ulster Plantations*, pp. 18–40 and 84–107.
35 W. B. Patterson, *King James VI and I and the Reunion of Christendom* (Cambridge, 1997).
36 Brady and Murray, 'Sir Henry Sidney and the Reformation in Ireland', pp. 14–39.
37 J. Michael Hill, 'The origins of the Scottish plantation in Ulster to 1625: a reinterpretation', *Journal of British Studies*, 32 (1993), 24–43.
38 Aidan Clarke with Robin Dudley Edwards, 'Pacification, plantation and the Catholic question, 1603–23', in T. W. Moody, F. X. Martin and F. J. Byrne (eds), *A New History of Ireland*, iii: *Early Modern Ireland 1534–1691* (Oxford, 1976), p. 214.
39 *RPC*, ix, 24–33. J. Goodare, 'The Statutes of Iona in Context', *Scottish Historical Review*, 77 (1998), 31–57; but see the critique by Martin MacGregor, 'The Statutes of Iona: text and context', *Innes Review*, 57 (2006), 111–81; also the wide-ranging analysis by Alison Cathcart, 'The Statutes of Iona: the archipelagic context', *Journal of British Studies*, 49 (2010), 4–27.
40 Cathcart, 'Statutes of Iona', pp. 25–7.
41 Cathaldus Giblin, *Irish Franciscan Mission to Scotland, 1619–1646* (Dublin, 1964), pp. 45–7.
42 Nicholas Canny, *Making Ireland British, 1580–1650* (Oxford, 2001), chs 4 and 5.

3

Civilising Gaelic Scotland: the Scottish Isles and the Stewart empire

MARTIN MACGREGOR

A monograph on Gaelic Scotland and James VI and I remains a desideratum, for all the evident richness of the theme. Donald Gregory's old narrative is still remarkably serviceable, but raises immediately the problem of the west Highlands and Islands standing proxy for the whole.[1] Gregory's focus on the west means that, to cite two obvious examples, he barely discusses the major Acts of Parliament of 1587 and 1594, which sought to ensure the answerability of clan society before the law, or the repression of the MacGregors. The strength of the scholarly assumption that the story of the west is the whole story is at its most acute in discussions of the Statutes of Iona, often described as Highland-wide in application when in fact they only dealt with the Isles and associated west coast territories, excluding Lewis. This was not the king's attitude. The traditional point of departure for any exploration of his vision of Gaelic Scotland is 1598 and *Basilikon Doron*, his treatise on kingship addressed to his son Prince Henry:[2]

> Here now speaking of oppressours and of iustice, the purpose leadeth me to speake of Hie-land and Border oppressions. As for the Hie-landes, I shortly comprehend them all in two sortes of people: the one, that dwelleth in our maine land, that are barbarous for the most part, and yet mixed with some shewe of civilitie: the other, that dwelleth in the Iles, and are alluterly barbares, without any sort or shew of civilitie. For the first sort, put straitly to execution the Lawes made alreadie by me against their Over-lords, and the chiefes of their Clannes, and it will be no difficultie to danton them. As for the other sort, follow forth the course that I have intended, in planting Colonies among them of answerable In-lands subiects, that within short time may reforme and civilize the best inclined among them; rooting out or transporting the barbarous and stubborne sort, and planting civilitie in their roomes.

Subtlety and discrimination are not terms normally associated with this passage, but it is important to acknowledge not only the distinction drawn between the mainland and the Isles, but also between redeemable and irredeemable inhabitants of the latter. The Isles were a special case, and the focus of this essay.

Recent treatments of James's Highland policies have been in the contexts of Stewart monarchy in Scotland, and Scottish state formation.[3] Stewart dynastic expansion in 1603 effects a seamless transition to British state formation, and the first attempt to create a British state in the modern era, with Stewart multiple monarchy at its core. The process is sometimes portrayed in terms of vanishing or closing frontiers within the British and Irish archipelago, as the contemporary hallmarks of 'civility' – the royal supremacy, law, Protestantism, trade and commerce based upon urban centres, and the English language – sought to penetrate the dark corners of the sceptred isles.[4] One construct which seems to gain life and legitimacy from such a vision is 'Greater Gaeldom': Gaelic Scotland and Ireland as one Gaelic world upon which nascent British imperium cut its teeth, and the adjacent, interconnected and equally recalcitrant regions of Ulster and the Isles as the hard centre of that world. Gaels were at the heart of an early modern domestic Other – a ragbag of anti-British elements also including Borderers, Catholics, witches, itinerants and the able-bodied unemployed – and the act of their suppression was a glue with which to bind the British project.[5] The validity of a 'Greater Gaeldom' will be questioned below, but it should be noted that in the Isles after 1603, policy was at points implemented by a quasi-British army and navy. Several expeditions were composite in character, while the forces headed by the Campbells in putting down the Islay Rising in 1615 drew upon the ships and soldiery of the Three Kingdoms.

A Gaelic world may also have existed in the minds of others insofar as it was the subject of a common lexicon of perception, a discourse founded on the basic distinction between civility and barbarity which could be used in turn as the wrapping in which policy was presented and legitimated. A genuinely Scoto-centric literature of colonisation originated in the 1620s with the Scottish schemes in Nova Scotia, and pamphlets such as Sir William Alexander's *Encouragement to Colonies*, and Sir Robert Gordon of Lochinvar's *Encouragements for New Galloway in America*.[6] By then, however, the great humanist scholar George Buchanan and others had established a Scottish dimension to European discourse on colonisation, while Scottish literati had long been engaging in a domestic version based on representations of their own aboriginals, the Gaelic Scots.[7] The main features of that characterisation, down to the point of the Reformation and James's accession, are well known.[8] What has not been done is to carry out a similar exercise for this reign, mining its very rich Highland materials such as the preambles to numerous acts of parliament and of the Privy Council. If in policy terms the reign saw little that was new, as will be argued below, was the same true of the attendant rhetoric? There was certainly nothing new in the image of Gaelic Scotland as a land of plenty inhabited by a benighted people who either did not know what to do with it, or did the wrong things with it. When and where did the concept of barbarity enter the lexicon, supplanting the older term *silvestres* or 'wood-dwellers'? Irreligiosity, which had barely featured in the older portrayal, now became pervasive. The Scottish Reformation may in itself have been enough to superimpose the image of the heathen, but crossfertilisation both with the broader European literature, and with English representations of Gaelic Ireland in the era of conquest, also needs to be taken into account.

Hand in hand with a literature of colonialism went a reordering of linguistic referentials within these islands. Conquest in Ireland and embryonic Britishness sharpened 'subordinate' ethnic and national divisions and labels. In 1590 a horrified Scottish Privy Council recorded the Clan Gregor acting 'in eithnik and barbarous manner'.[9] In 1603, in what was virtually James's last domestic act before heading south to take Elizabeth's throne, the very words 'MacGregor' and 'Gregor' were proscribed by the Privy Council.[10] Alteration of behaviour could be effected by alteration of language. In 1607 a charter of Kintyre to Gilleasbuig Gruamach, seventh earl of Argyll, granted in part for ethnic cleansing against the MacGregors, stipulated that no subinfeudations be made to those bearing the surnames MacDonald or MacLean.[11] In turn, when Sir Seumas MacDonald's escape from Edinburgh Castle precipitated the Islay Rising in 1615, the subsequent letters of this multilingual and highly literate individual resounded with his resolve never to surrender Islay to one of the name of Campbell, even unto death; 'seing my Race has bene tenne hundredth yeeris kyndlie Scottis men, vnder the Kinges of Scotland'.[12] In this same era Campbells and MacDonalds were pioneering the use of terminology involving the matter of Britain, such as 'North British' for Scottish, and 'British' for colonists settling in Ulster.[13] Within Ireland, Gaels and non-Gaels – the latter in the shape of the *sean-Ghaill* or 'old English' – were becoming conjoined on the basis of faith and territory as *Éireannaigh* or Irish, thereby creating a clear distinction between *Éire* or Ireland and *Saxain* or England to which the *nua-Ghaill* or 'new English' colonists belonged.[14] In Gaelic Scotland, although the evidence is much vaguer, similar adjustments in the Gaelic view of themselves and others may have been at work. The year 1600 seems to be the rough horizon at which *Goill* as a specific referent ceased to be able to mean English, and came to be equated exclusively and perhaps for the first time with Lowland Scots. Scottish Gaelic now drew a clear terminological distinction between Gaelic and non-Gaelic Scots, but nonetheless Lowlanders were recognised along with Scottish Gaels as *Albannaich* or Scots, inhabiting *Alba* or Scotland, and thus clearly distinguished from *Saxannaich* or *Sassanaich*, the inhabitants of England.[15]

If Ireland was a laboratory for Empire, then it is legitimate to ask if, at least as far as the king was concerned, Gaelic Scotland was a laboratory for Ulster. Particularly after 1603, it is not difficult to identify those for whom the Scottish Isles, Ulster and America were concentric circles in the one imperial ocean. When in 1610 Niall MacLeod, who had played such a prominent role in resistance to the colonisation of Lewis, intercepted a privateer laden with the illegal booty of empire, Sir Alexander Hay responded:[16]

> If his Maiestie would be pleased in regaird of the service done to direct Neill to the pairtes of Virginia and to direct a staite of inheritance to be gevin to him there, I think our country heir suld be best rid of him. There wald be no suche danger there as of his being in Iyireland, for albeit bothe the speiches be barbarous yit I hope he sall neid ane interpretour betuix him and the savaiges.

Andrew Stewart Lord Ochiltree was firstly active on the Anglo-Scottish border, and then expeditionary leader to the Isles in 1608, before becoming a prominent Ulster planter, latterly as Baron Castlestuart; while Andrew Knox was granted the bishopric of Raphoe in addition to that of the Isles.[17] Coming the other way was Sir Randal MacDonnell, a loyalist Gaelic planter in Ulster, but with ambitions to annexe the Scottish end of his kindred's historic patrimony. In Islay in 1613 this resulted in native accusations of Irish imperialism:[18]

> And that the said Sir Ronnald and his officiaris in his name intendis aganis the lawis of this realme to astrik and subject his Majesteis Tennentis forsaidis to the formes and lawis of Yreland and to compel thame to persew and defend in all thair actionis and caussis according to the forme and custome of Yreland. Quhilk is a mater of verie greit grief unto thame that they being his Majesteis native borne subjectis sould be reulit and governit be foreyne and strange lawis.

A crucial figure in advancing Campbell interests in the era of multiple monarchy was Gilleasbuig or Archibald Campbell of Glencarradale, nemesis of Andrew Knox and Sir Seumas MacDonald, and constantly on the move between London, Edinburgh, the west Highlands and Ireland.[19] More spectacular still was the career of John Mason of King's Lynn in Norfolk, naval captain and colonial pioneer. Mason led Bishop Knox's fleet on his Isles expedition of 1610, sailing Hebridean waters for 14 months in His Majesty's ship *The Golden Fleis*. His reward was a grant of the assize herring revenues of the North Isles, stretching from Buchan to Uist, from all folk, native or foreign. His attempt to collect this led to accusations of extortion by Fife fishing burghs whose boats were active in the Isles, and charges of piracy against him. In 1629 Mason was still trying to extract payment from the crown for his service in the Isles subduing 'rebellious redshankes'. By 1633, acknowledging his by now considerable experience in the region, he was chief factor of the Royal British Fisheries Company based in Stornoway in Lewis. Interwoven with his Hebridean life was another in Newfoundland, New England and the colony he founded, New Hampshire. Whether Ireland featured in his imperial portfolio awaits investigation.[20]

The statist perspective adopted thus far invites the drawing of parallels with Gaelic Scotland's role at subsequent key junctures in the evolution of an early modern British state. In the immediate aftermath of Anglo-Scottish parliamentary union in 1707, Scottish Gaels were again cast as the domestic 'Other', as Jacobites and inveterate enemies of the newly made and soon to be Hanoverian kingdom of Great Britain. One riposte from the private sector, although attracting state approval and sponsorship, was the religious and educational crusade in the Highlands led by the Society in Scotland for the Propagation of Christian Knowledge founded in 1709, and which included within its remit the spreading of the gospel among native peoples in the American colonies.[21] The failure of the '45 Jacobite rising unleashed a dramatic rip-tide of multi-pronged assimilation as the state confronted the issue of how to deal with a military capacity within its own borders which in the first instance was not its to command. The resolution saw the harnessing of Scottish Gaelic military potential in the interests of the blue-water British imperial mission.[22] Even more instructive, however, is a late-medieval and indeed medieval perspective, which reveals what happened in the west Highlands and Islands in the early seventeenth century as shaped by older structures and issues. The two main themes in what follows take their cue from this, and open a Scottish Gaelic window on the plantation of Ulster. They are the exercise of sovereignty in the west, where the sovereignties of Scotland, England and Ireland had long met and commingled in the Irish Sea world; and discussions of the Campbells and MacKenzies, the true gainers from Jacobean policy in the west Highlands and Islands, whose success defines the limits, frailties and fallacies of Stewart empire and early modern British state formation.

Once there was a fourth sovereignty distinct from those of Scotland, England and Ireland, exercised by the Norse over all the 'Scottish' Isles including Man, and, with the exception of Orkney and Shetland, formally extinguished in favour of the Scottish crown in 1266. Or was it? Just around the corner from 1266 lay two incontrovertible facts: the Kingdom of the Isles became the MacDonald Lordship of the Isles, forfeited to the Scottish crown in 1493, but whose ruling lineage survived in the person of Dòmhnall Dubh until 1545; and the outbreak of Scottish–English enmity, which prevailed until the Reformation. Some have built upon the first to argue that the spirit of Norse sovereignty lived on in a semi-independent MacDonald Lordship.[23] Others have coupled to the second both weakening English authority over Gaelic Ireland, and ineffectual or nominal Stewart control of Gaelic Scotland, to argue that the vacuum left by the Norse was filled by a 'Gaelic world' or 'Greater Gaeldom'.[24] A favoured metaphor is of a wedge first removed, allowing for a renewed Gaelic communion across the North Channel, with Clann Dòmhnaill, the Clan Donald, as the umbilical cord.[25] The plantation of Ulster was thus an Anglo-Scottish project partly or even mainly about restoring the wedge: the resources of the three kingdoms were trained upon Ulster and the Isles in order to streamline sovereignties, sunder 'Greater Gaeldom', and cauterize the wound.[26] Late medieval ecclesiastical politics in the Scottish west offered its own version of this narrative in the struggle for jurisdiction over the bishopric of the Isles, whose metropolitan see remained in Norway until 1472.[27] Locally, the MacDonald lords were patrons of the abbey of Iona, and the bishop lived in the shadow of the abbot, the heir of St Columba. Hence the attempt after 1493, only partially fulfilled, to relocate the bishop's see to Iona and make him commendator of the abbey. After the Reformation the bishop-commendator continued to struggle, this time vis à vis secular lords and their interest in the ecclesiastical lands of the Isles, especially the rich territorial inheritance of Iona.[28]

The *long durée* of Scottish–English enmity gave life and various manifestations to the fourth sovereignty. James VI was in a sense the heir of Scottish royal aspirations to rule Ireland which did not die with Edward Bruce at Faughart in 1318, and underlay the Irish ambitions of monarchs such as James IV and particularly James V.[29] A more clear-cut and convincing pedigree can be assembled for Scottish use of Ireland, and English use of the Isles, as second fronts from which to gain political advantage. Diplomatic bookends for this phenomenon are provided by the Treaty of Edinburgh in 1328 and the Treaties of Berwick in 1560 and 1586.[30] The most spectacular lighting up of the board came with Henry VIII's 'Rough Wooing' in the early 1540s. In pursuit of revived English claims to suzerainty over both Scotland and Ireland, Henry took hold of both maritime levers, attempting to harness the military power of the Isles against Ireland while simultaneously launching a seaborne invasion of Scotland in the west.[31]

It is when we seek to pursue this sovereignty on the inside that it becomes highly elusive and problematic, whether at the level of 'Greater Gaeldom', Gaelic Scotland or the Isles. The reality of pan-Gaelic sentiment for some can hardly be doubted, and indeed the climax of English conquest of Ireland threw up an outstanding instance in the reaction of one Scottish Gael to attempts to lure him and others to aid the English cause in early 1602, in the immediate aftermath of Kinsale:[32]

> one of his country whom he [Campbell of Glenorchy] meant to have employed had plainly protested though the King, Argyll and he should force them to go, yet they would not serve against that people they were come of and whose language was one with theirs, but be true to them against the Saxons (meaning English).

Nevertheless, such linguistic and ethnic fellow-feeling did not a unitary Gaelic polity, far less a sovereignty, make. Even before the Reformation added a confessional divide, major distinctions existed between Gaelic Scotland and Ireland in various spheres, not excepting areas of presumed uniformity such as language and culture.[33] *Ceannas nan Gàidheal*, 'the headship of the Gaels', was a high poetic concept first associated with the chief of the MacDonalds, and then contested by the chief of the Campbells in the sixteenth century. Neither could realistically lay claim to overlordship of the whole of Gaelic Scotland, while it has recently been stated that *ceannas nan Gàidheal*, albeit some of its Scottish poetic occurrences gave it an Irish dimension, was entirely absent from Irish classical poetry. To the Irish poets, Gaelic Scotland was an anomalous and inferior organism, Gaeldom was co-extensive with Ireland, and Gaelic sovereignty meant the high-kingship of Ireland.[34] By the later sixteenth century the yoke of constant billeting by Scottish mercenaries can have left equally little room for pan-Gaelic sentiment among the tenantry of Ulster.[35] On the Scottish side the main stumbling block to a 'Greater Gaeldom' was engagement with the matter of Scotland, especially the kingship of the Scots.[36] The Lordship of the Isles was unmistakably a polity possessing considerable autonomy, but its *raison d'être* was not the establishment of 'a separatist state in the west'.[37] Even in 1545 and the climactic attempt to restore the Lordship, Dòmhnall Dubh, who had little reason to regard the Scottish kingdom with affection after a lifetime largely spent in royal captivity, chose to make his oaths of allegiance to Henry VIII through the medium of the earl of Lennox, whose authority he and his supporters accepted 'as the true Regent and second person of the realm of Scotland'.[38] A Gaelic sovereignty was a chimera, a by-product of Scottish-English enmity existing largely in the fevered imaginations of Tudor and Stewart officialdom, and in no need of help from the Ulster plantation to obliterate itself.

What gave substance to metropolitan fears was the remarkable military capacity of Gaelic Scotland, specifically Argyll and the Isles, across the later middle ages. Its main outlet was permanent and then seasonal service in Ireland and as it waxed in tandem with the process of English conquest, so the struggle to control it intensified. *Ceannas nan Gàidheal* meant the right to command this resource. Irish poets played the Gaelic card and bade Gaelic Scots to come home and defend the motherland.[39] Henry, Elizabeth and James in turn sought to harness it to advance English or proto-British interests, and in James's case, as a bargaining chip over Elizabeth's succession. The impact of the redshanks on the north of Ireland at points in the later sixteenth century makes it worth asking to what extent a form of colonial society existed in Ulster in advance of the Plantation, even if the predominant English view was that they were part of the problem rather than the solution, and had to be removed.[40] No less profound were the domestic consequences of the end of the mercenary trade about 1595.[41] In the immediate aftermath there was an intensification of warfare in the Isles, as longer standing feuds between Clan Donald South and the MacLeans, and between the MacLeods of Dunvegan and MacDonalds of Sleat, erupted into a series of pitched battles. These contributed to the devastation of parts of Kintyre, Skye, Uist and Harris, and point to the existence of a highly militarised zone, and a climate of total war, on either side of the North Channel. Underlying the conventional discourse of civility and barbarity employed by James VI in *Basilikon Doron* was the reality that the Isles required large scale-demobilisation and fundamental social reform.

Historians have long been struck by the sheer range of policy initiatives applied to Gaelic Scotland across James's reign, but all of these were united by more than a common rhetorical framework.[42] James was happy to pursue whatever routes would result in answerability to God, crown, church, and law, and barbarity meant ignorance of these. Equally, the crown's financial expectations were throughout in inverse proportion to its outlay. Of those who acted on its behalf, some were favoured because of the money they brought to the table, some suffered because of expenditure deemed excessive or poorly accounted for, and some never received the promised rewards.[43] *Basilikon Doron* heralded a new phase, and one dominant policy strand between then and 1625 was the forcible expropriation of four kindreds from their patrimonies: the MacLeods from Lewis by 1610; Clann Eoin Mhòir or Clan Donald South from Kintyre and Islay by 1615; the MacIains from Ardnamurchan in 1624, and the MacGregors, proscribed in 1603, from Glenstrae by 1625. The ruling lineages of all these clans were vulnerable or at odds. Three of the four were in the west, and three times the Campbells made significant gains. Kintyre was granted them in 1607, in part for services rendered against the one kindred from outwith the west, the MacGregors.[44] The MacGregor proscription proved unattainable; Lewis was colonised not by the original grantees, the Fife Adventurers, but by the MacKenzies, and prominent voices expressed disquiet about the wisdom of granting Islay to the Campbells.[45] Furthermore, all these instances had antecedents in regional disputes concerning rival territorial claims, and both Ardnamurchan and Glenstrae saw the original possessors very heavily compensated by the Campbells for their loss of patrimony or *dùthchas*, a curious way to treat rebels subject to crown forfeiture.[46] Recent historians have been much more circumspect than their predecessors who echoed contemporary assertions that his Highland policies represented a great Jacobean success story, delivering universal peace in Gaelic Scotland by the end of the reign.[47] Closer investigation of the true winners in the west reveals instructive and uncomfortable perspectives on Stewart empire in its own backyard.

The staggering rise of the Campbells to power and influence in Argyll, much of the Isles, many other parts of Gaelic and non-Gaelic Scotland, Ireland,

and ultimately on the British stage across the late medieval and early modern eras, defies the textbooks and lays down a gauntlet to historians. They have been characterised as a 'semi-Lowland' family, or as a Trojan horse of Gaeldom.[48] John MacInnes, while emphasising that they were as Gaelic as anyone else, has stated memorably that 'if the Campbells had not existed we should have to invent them'.[49] Their success and longevity is often attributed to an utter lack of principle beyond self-aggrandisement and self-preservation, and an automatic and unquestioning crown loyalism. Recent invaluable studies have revealed Campbell power less as a cipher for that of Stewarts or Hanoverians than as an intelligence in its own right, discriminating in its behaviour, and ever-mindful of how far monarchical agendas dovetailed with its own conscience.[50] Down to the Reformation those at the helm of the kindred exhibited a deeply committed and consistent Scottish patriotism, uncompromising towards England. The Reformation saw the Campbells reborn as sincere Presbyterians with natural British sensibilities.[51] Adherence to these principles was a major cause of the many estrangements with their sovereigns before and after 1560, giving a history and pedigree to the deaths of two successive earls of Argyll on the scaffold in 1661 and 1685.[52] There were also times when doubts surfaced about over-reliance on the Campbells in the west.[53] James V's dramatic *volte face*, turning to Clan Donald South and imprisoning the fourth earl of Argyll, is precisely analogous to the strategy pursued by Andrew Knox as bishop of the Isles, fostering a close relationship with the same kindred, and through the Statutes of Iona seeking a Campbell-free future for the Isles.[54]

In the primary world in which they functioned, with Argyll secured, nothing was more natural than that the Campbells should aspire to lordship of the Isles and *ceannas nan Gàidheal*, or that they should follow the MacDonalds into Ulster. References to a state of outright enmity between the kindreds were made in 1545, 1596, and throughout the letters of Sir Seumas MacDonald.[55] A jostling for military supremacy was explicit in the offers made by Alasdair Cathanach MacDonald to James V in 1531, while in the thousands who rose first for Dòmhnall Dubh in 1545, and then for Gilleasbuig fourth earl of Argyll in the Pinkie campaign of 1547, we sense the passing of a baton.[56] The Reformation and Anglo-Scottish *rapprochement* then marked the start for the Campbells of a remarkable two hundred years of front-line Britishness. Coming from the avowed irreconcilable periphery, arguably no-one was more central to the making of the early modern British state. The kinsmen and successors of the fifth earl 'adopted his British dimension in their thinking and formulated their own British strategies. The marquis of Argyll in the mid-seventeenth and the dukes of Argyll in the eighteenth century conducted their careers on the British stage, at times dominating Britain's political world'.[57] As queen's commissioner the second duke steered the Union settlement through the Scottish parliament between 1705 and 1707; he came to the rescue of fledgling Hanoverian monarchy by thwarting the Jacobite rising of 1715; he and his brother were successively the first 'managers' of Scotland within the post-1707 British political system. In the aftermath of the '45 the third duke insulated the Highlands from the outraged instincts of a state initially determined to visit with extreme prejudice the Jacobean model of colonisation, transplantation and extirpation on the region.[58]

Whence did this Britishness come? Genealogy is the obvious starting point. It is well-known that of the three known versions of Campbell origins, the most plausible argues for descent from an early medieval North Briton called Arthur. As the kindred grew into a British political landscape, so this pedigree took centre stage, and the equation of Arthur with King Arthur, hitherto muted, became insistent.[59] A mid-seventeenth-century classical Gaelic panegyric to the first marquis of Argyll is soaked in Arthurianism,[60] but the process was already visible in *Dual ollamh do thriall le toisg* ('It is customary for a poet to travel on a diplomatic mission'), a poem of Irish authorship addressed to Gilleasbuig the seventh earl, probably in 1595 at a pivotal juncture in the Nine Years War. The poet coins for Gilleasbuig the striking neologism of *Breat-Ghaoidheal* or 'Brit-Gael', and urges him to succour Ireland not on the grounds that she had nurtured his kindred, as we find in parallel contemporary appeals to MacDonald chiefs, but because his progenitor King Arthur exacted tribute there.[61] Earlier still, the most important surviving literary manuscript of late medieval Gaelic Scotland, the Book of the Dean of Lismore, demonstrates a unique ability to navigate among the main linguistic cultures of pre-Reformation Britain and Ireland, except for Wales.[62] The dean in question was a Campbell protégé, and his book can be taken to represent a Campbell viewpoint. The same can be said of John Carswell, author of *Foirm na n-Urrnuidheadh*, a version of John Knox's *Book of Common Order*, the first Gaelic book to be printed in Scotland or Ireland, in 1567.[63] Carswell was a protégé of the fifth earl, and first post-Reformation bishop of the Isles; in the poem he dedicates to his book, we see a comparable 'British' vision, and willingness to approach or cross frontiers:[64]

> After that, travel each district
> throughout Scotland, gently, slowly,
> but, since they have no need of thee,
> do not take one step into Saxons' fields.
>
> After that, travel over each wave
> to the land of Ireland of liberal bounds,
> though the friars care little for thee,
> move westwards within their sight.

There was nothing intrinsically non-Gaelic about the Reformation. The Campbells provide *prima facie* evidence of this, and to secure Elizabethan support for Reformation in Scotland they could offer influence in Ireland: military assistance, the prospect of Campbell-led plantations, and the export of Reformation across the North Channel. The last could be achieved through plantation, but coupled to Carswell's 'Gaelic Protestantism' under the aegis of the 'godly chief', the Gaelic learned classes, the printing press and ministers preaching the Gaelic word from the Gaelic book.[65] The fifth earl pursued this pioneering and idealistic British policy between 1559 and 1565, before abandoning it in the face of English opposition to the presence of Gaelic Scots in Ireland that proved too deeply ingrained. Dawson concludes: 'instead of working for the integration and assimilation of Ireland into a united, Protestant British Isles, the fifth earl created the preconditions that made possible the Nine Years War'.[66] His post-1565 strategy of shoring up MacDonald power in Ulster delayed the Tudor reconquest for a generation and paved the way for the Plantation. The fifth earl was as capable of networking across cultures and frontiers as the Dean of Lismore and John Carswell. In his eyes, Ulster and the west Highlands were an axis not of evil but of opportunity, the starting-point for a Reformed theocracy embracing the Three Kingdoms. Gaeldom was the potential core of a true early modern Britain. He was perhaps the first but not the last Scottish politician to discover that God was an Englishman, and that his English counterparts either identified England with Britain, or regarded 'Britain' as useful only insofar as it was compatible with English interests and security.[67] Thus he helped lay down a road to Britishness that must needs be confrontational and colonial, with Gaelic Scotland and Gaelic Ireland cast as hostile peripheries to be conquered and suppressed in order to make good a very different imperial vision. It was a new order to which the Campbells readily adapted, and within which they were to prove serious rivals to the Stewarts.

The year 1609 saw both the start of plantation in Ulster and the Statutes of Iona, whose scope was all the Scottish Isles bar Lewis. I have argued elsewhere that Bishop Andrew Knox envisaged the Statutes as an alternative to colonisation on the Ulster and Lewis model; a new deal between central government and the Hebridean elite which deliberately excluded the Campbells. Knox had concluded that the supposed instruments of Stewart government and civility were a barrier to the delivery of peace, progress and justice, not to mention the authority of the bishop of the Isles. A key part of Knox's agenda was the reconstruction of his diocese on a sound fiscal footing so as to encourage the spread of Reformation.[68] In making the case for the Statutes as an antidote to Campbell imperialism in the west, there are two crucial pieces of evidence. The first was the Scottish Privy Council's repeal of a trade embargo between the Isles and Argyll on 28 September 1609, the very day on which Knox reappeared before the council to report on the mission which had delivered the Statutes of Iona:[69]

> quhairby as the makaris of that proclamatioun hes committit a very grite errour and ouersight in usurping upoun thame suche a soverane pouer and auctoritie, noway competent in the persone of a subject, and thairwithall hes defraudit and prejudgeit his Majesteis goode subjectis of the benefeit of thair lauchfull trade and intercourse of thair goodis, wairis, and mercheandice, so haif they verie far hinderit his Majstie in the tymous and thankfull payment of his dewyteis, in heich contempt of his Majesteis auctoritie and lawis…

The second was James's letter to the council on 8 May 1610, instructing it to ratify the Statutes, and to cancel the jurisdictions held or claimed by others in favour of those now granted to Knox, henceforth not only bishop but also sole lieutenant and steward of the Isles:[70]

> And, because some acclameing pouer of jurisdictioun over those Yllis in tyme past have not maid muche farder use of there offices then upoun that cullour sometymes to oppres the inhabitantis, nochtwithstanding all these Illis, at leist most pairt, be of oure propertie, over whiche no uther officer bot oure awne stewart sould have jurisdictioun and power, you ar thairfore to inhibite and dischairge all and quhatsumevir officeris other than oure Stewart and his deputis foirsaidis to use, exerce, and usurp any jurisdictioun in any of these Illis, upoun quhatsumevir pretext or challenge to be maid by thame that the same do lye within schirefdome quhairof thay haif the heritable graunte.

The anonymous usurpers of Stewart sovereignty were, almost certainly, the Campbells.[71] To have dared to speak their name would have been tantamount to condemning longstanding Stewart policy in the west Highlands.

The Statutes of Iona enjoyed an Indian summer before the Islay crisis of 1614–15 restored and tightened the Campbells' stranglehold over the west. Not only did they annex Islay and then Ardnamurchan, but by 1634 there was a Campbell bishop of the Isles with effective control of the extensive lands historically pertaining to the abbey of Iona. In the same decade, adroit exploitation of the growing financial pressures upon the elite of the west Highlands and Islands saw significant expansion of the feudal superiority of the house of Argyll. The Campbells had long held the heritable justiciarship of Argyll and the Isles as well as the lieutenancy and the sheriffdom of Argyll, the last now expanded to cover Kintyre and all the Isles south of Ardnamurchan. As part of Charles I's Revocation Scheme in 1628 their chief was effectively granted 'a virtual legal monopoly which afforded no appeal to royal courts' in respect of all civil and criminal cases throughout his lands and jurisdictions, except in instances of treason.[72]

For non-Campbell kindreds the dilemma was clear. Answerability to the law as a touchstone of civility became problematic when the law meant Campbell justice. Resorting to the law meant the prospect that the Campbells would outmanoeuvre them, and resorting to arms would bring the law down upon them. The Statutes of Iona, whose signatories became the local legal agents of the crown, represented one way out. Another was direct access to the monarch, and a notable theme is how difficult this proved for those kindreds which suffered expropriation, especially with the advent of absentee monarchy in 1603. The outstanding example was the Clan Donald South, whose leaders made numerous offers which fell on deaf ears. The Campbells played the system brilliantly, exploiting their victims' weaknesses or failings, gaining new legal title or reactivating claims long dormant, stirring the waters to induce confrontation, and then acting as the military agents of legally sanctioned private expropriation.

The crisis over Islay brought forth biting analysis from Knox and others who saw the Campbells as the problem not the solution, and laid bare the reality of the Stewart settlement of the west.[73] The Campbells were suspected of fomenting it. If Sir Seumas MacDonald escaped from jail to lead the rising, the seventh earl of Argyll was recalled from a debtor's exile to put it down, and Campbell of Auchinbreck was released from jail and given royal commission until the earl's arrival. What followed was the government-sanctioned resolution of a private war in which it openly took sides. Beneath the veneer of Stewart imperial success in the west lay something very rotten indeed. It is hard to agree with Goodare in seeing the triumph of the Campbells as representing any form of progress for 'the state'.[74] Rather, 'the state' threw in the towel. The virtually absolute power the Campbells aggregated unto themselves in the west in the Plantation era precisely demarcated the limitations of Stewart empire, and set the stage for the 1640s and the second Campbell attempt to forge a true Britishness based on religion.

That the Isle of Lewis should be the object of the first Scottish colonial scheme is easy to treat as a farce initiating the grand tradition of Scottish disasters abroad which peaked at Darien. As with Darien, however, the commercial logic was sound. The background was one of disputed succession within the ruling kindred, the powerful and important Sìol Torcail, consistent backers of the MacDonald cause after 1493, and thus at loggerheads with the crown at several points across the sixteenth century. The same era saw the growth of lordship of the MacKenzies in the neighbouring mainland region of Ross as the ostensible local representatives of the crown, the Campbells of the north. The groundbreaking research of Aonghas MacCoinnich allows what ensued to be presented afresh as an iconoclastic tale of four plantations.[75]

Under legislation of 1597 and 1598 the MacLeods lost title to Lewis whose main settlement of Stornoway was mooted as one of three new burghs in the west, and the island was granted to a consortium of gentlemen adventurers. This was classic colonisation. *Basilikon Doron* was the template and the mission statement, and the 'Fife Adventurers' were the responsible inland subjects who would plant the appropriate infrastructure of burgh, kirks and inns, civilising or displacing the MacLeods in the process. Economic motives featured prominently in the accompanying rhetoric, focusing first and somewhat surreally on a presumed 'incredibill fertilitie' of corn, shifting thereafter to the fruits of the sea.[76] The presence of east coast fishing boats in the northwest Highlands and Isles had long predated 1597, and bred tensions which could be readily packaged as ethnic, but which rather reflected the desire of the MacLeods and others to challenge the stranglehold of the royal burghs, and to maximise their own economic return in recognition of their lordship.[77]

The project had failed completely by 1610, and classic colonisation gave way to the native version. Both the Fife Adventurers and the MacLeods had been prompted and hindered in equal measure by the MacKenzie chief Coinneach Cam who played the king and council as surely as did Knox and the Campbells, knowing exactly which linguistic and psychological buttons to press. With both sides worn out and the king in a corner, MacKenzie now came forward at the head of his kindred to implement a swift and ruthless military conquest, replacing unanswerable with answerable Gaels. The subsequent grant of the earldom of Seaforth was no more than tokenistic royal acknowledgement of what had come to pass on the ground.

Phase three was one of enlightened colonisation. The MacKenzies first cut the royal burghs (with the exception of Aberdeen) out of the fishing trade, and introduced skilled English workers to Wester Ross to accelerate development of the indigenous iron industry. Then, in the late 1620s, they established the Company of Lewis: a partnership with members of the commercial elite of Zeeland to exploit the northwest fishery on the basis of a shared Calvinism, military connections, and commercial ambition. Cailean, earl of Seaforth, established Stornoway as a free burgh of barony, and brought in 12 Dutchmen and their families to man four Dutch busses which were fishing out of Stornoway by 1630. In stark contrast to what had gone before, this was fully consensual colonialism, suggesting that Seaforth was *au fait* with the latest thinking and practice. He offered the Zeelanders the monopoly of a fishery ripe for exploitation, the protective umbrella of his lordship on both sides of the Minch, and a fort to be built to protect Stornoway itself. The Zeelanders were fully committed to a project offering an opportunity to shore up their status vis à vis Holland and Spain. In Lewis they saw no dark corner, but a logical strategic and commercial hub for their western, northern and eastern trades, complementing the existing Dutch fishery in Shetland, and offering an ideal staging post on the homeward journey for shipping on the Dutch Atlantic and East Indian routes. As one contemporary noted, Stornoway would give them a base from which to trade with 'any part of Christendome in Barbary, Asia, Affrica, East and West Indies, and Newfoundlande'.[78] The Dutch undertakers were 'settlers rather than colonists'.[79] They would remain Dutch citizens with the right to bring in their own clergy and schoolmasters. At the same time they would be subject to the civil magistrate and the laws of Scotland, paying taxes to the crown, but equally enjoying the personal and commercial privileges afforded to Scottish subjects and burgesses.

The enterprise was wrecked in the first instance by the conservatism and opposition of the Scottish Convention of Royal Burghs and allied Scottish interests. These pulled monarchy in their wake, leading to the creation of the Association of the Fishing of Great Britain and Ireland, or the Royal British Fisheries Company, licensed by Charles I in 1632. Where the MacKenzies had led the Stewarts thus followed, but in a very different manner reminiscent of the Anglo-centric mindset that had so disillusioned the fifth earl of Argyll in Ireland. The company sought not to cooperate with the Dutch but to eclipse them in the inshore and deep-sea waters around the British Isles. The Scottish burghs, and their favoured way forward of planting Lowland Scots instead of natives or the Dutch, were likewise sidelined by English interests. The Royal British Fisheries Company was an essentially English enterprise, a 'cabal of London-based merchants and nobles' with its hub at Deptford and a projected fleet of 200 vessels to be based in Lewis.[80] The blueprint set out by Sir William Monson, vice admiral of the English fleet, even envisaged that the MacKenzies might lose Lewis. Seaforth would resign it to the king, who would grant it out to English and possibly Lowland Scottish interests to exploit the fishery. The Dutch were to be completely excluded, and dialogue between them and the Islesmen should cease. A governor would be appointed and a garrison built, not to protect the ships and harbour but to subdue the natives, who would be taught English through a rigorous educational programme. The pendulum had swung back to overt colonialism, but by English rather than Fife Adventurers. English fishermen were active in Lewis between at least 1631–36, with John Mason as chief factor of the Company, but the scheme foundered on entrenched MacKenzie power, grass-roots hostility, and complications arising from the status of the English colonists in Scots law. The tensions generated by the Royal British Fisheries Company were one source of the Scottish disaffection that precipitated the crisis of the late 1630s and the Covenanting Movement.[81]

To draw conclusions, what stands out is the intensity of the struggle in the Isles in the early seventeenth century and the number of competing parties and visions, of which Stewart monarchy was one. For the historian of Gaelic Scotland, the natural vantage point from which to observe and understand the process is not the contemporary one afforded by colonisation, the matter of Britain, and Stewart empire, but the fracturing of the legacy of the MacDonalds, and the destiny of the three titles to which they had laid claim in the fifteenth century: lords of the Isles, earls of Ross, and *ceannas nan Gàidheal*. The MacDonalds over-reached themselves in their attempts to realise the right they inherited to the earldom of Ross, and Ross helped to bring them down. The opposition of the MacKenzies, at that stage a prominent kindred within the earldom, was significant here, while in the same period the Campbells were a coming power within the ambit of the Lordship, having effectively made good title to the lordship of the subordinate province of Argyll. What we see in the early seventeenth century is confirmation of the disaggregation of that greater MacDonald empire of Ross and the Isles, with the MacKenzies annexing the former as earls of Seaforth. Campbell annexation of the Isles, however, was less convincing, while the disaggregation also cast doubt over the legitimacy of any Campbell claim to *ceannas nan Gàidheal*.

Michael Lynch has posited the recrudescence of Gaelic regional lordship as a counter-narrative to (and integral part of) early modern British state formation, and the Campbells and MacKenzies are prime examples.[82] Indeed, in uniting Lewis with mainland Ross as earls of Seaforth, the MacKenzies restored the earldom of Ross to its original medieval contours before Lewis was absorbed into the MacDonald Lordship. The MacKenzies and Campbells were classic cases of lesser kindreds originating within larger lordships which made it to the top, squeezing out their rivals and supporters as they expanded outwards and downwards. Those who suffered were the MacGregors, erstwhile Campbell clients; branches of the Clan Donald, confirming the disintegration of the kindred; and the MacLeods of Lewis, steadfast MacDonald allies from 1493 to 1615.

James III and James IV respectively had claimed the titles of earl of Ross and lord of the Isles for their dynasty through forfeiture in 1475 and 1493, as milestones in a first era of Stewart imperialism in its original Scottish sense of the exercise by monarchy of absolute sovereignty over its own dominion and subjects.[83] The early seventeenth century revealed that the Stewarts had failed to make these titles their own, and had no alternative but to restore them to regional dynasts. The retention by the heir to the throne of the honorary designation of Prince or Lord of the Isles offered no palliative to that hard fact, which forms one essential link between two phases of Stewart imperialism. Another is the uncanny parallelism to be drawn at virtually every turn between royal strategy in the Isles under James IV and James VI and I.[84] If the Stewarts had set their faces against various overmighty Scottish subjects in the fifteenth century, this imperial thrust hit the buffers in the Isles in the sixteenth century, with profound long-term consequences. Despite the best efforts of James IV and James V, Stewart monarchy failed the resource test for what it took to exercise lordship over the Isles, and continued to struggle even in an early seventeenth-century British context, leaving it with no choice but to allow older structures to win through, and to revert to that older model of reliance upon major regional magnates. But as the separation of the titles of Ross and the Isles demonstrated, neither the crown nor anyone else proved capable of being what the MacDonald Lords of the Isles had been in their heyday.

This does point towards one silver lining for the Stewart empire. The conquest of Ireland meant the withering of the military power of the Isles. It became clear very soon after 1600 that the great age of the Hebridean mercenary, galley and sea-girt castle – the very pillars of Clan Donald power – was gone. Insofar as the Campbells possessed lordship over the Isles, it was in a judicial sense. From a central perspective a powerful 'internal sovereignty' had been fatally reduced. Even so, the Scottish Gaelic west in the Plantation era revealed at its starkest Stewart frailty and misrule, and the economic and military limitations of Stewart sovereignty. There we see a Stewart chrysalis still tied to a late-medieval *modus operandi* as 'first among equals'; and the Campbells and MacKenzies as native imperialists at turns more successful, enterprising and even more truly British than the multiple monarchy itself. It is a deeply unedifying tale, a dreadful abnegation of monarchical responsibility which left its subjects in the Isles at a remove from royal justice, with the Campbells as an apparently impermeable barrier. The failure of Stewart monarchy in the region, epitomised by acquiescence in Campbell domination in the west, contributed to the collapse of tripartite monarchy in the 1640s, with the Campbells in the van of opposition. At the same time, and by an irony of ironies, it helped to cement the anti-Campbell coalition which would seek to come to the Stewarts' rescue.[85]

NOTES

1 Donald Gregory, *The History of the Western Highlands and Isles of Scotland, from A.D. 1493 to A.D. 1625, with a brief introductory sketch, from A.D. 80 to A.D. 1493* (1836; Edinburgh, 2008).
2 *King James VI and I: Political Writings*, ed. Johann P. Sommerville (Cambridge, 1994), p. 22. For an anticipatory royal proclamation in 1596, see Michael Lynch, 'James VI and the "Highland problem"', in Julian Goodare and Michael Lynch (eds), *The Reign of James VI* (East Linton, 2000), pp. 208–27, at p. 211, n. 1.
3 Goodare and Lynch (eds), *The Reign of James VI*; Julian Goodare, *State and Society in Early Modern Scotland* (Oxford, 1999); Julian Goodare, *The Government of Scotland 1560–1625* (Oxford, 2004).
4 Gordon Donaldson, 'James VI and vanishing frontiers', in Gordon Menzies (ed.), *The Scottish Nation* (London, 1972), pp. 103–17; Jane H. Ohlmeyer, '"Civilizinge of those Rude Partes": colonization within Britain and Ireland, 1580s–1640s', in Nicholas Canny (ed.), *The Oxford History of the British Empire, Volume I: The Origins of Empire: British Overseas Enterprise to the Close of the Seventeenth Century* (Oxford, 1998), pp. 124–47.
5 Michael Lynch, *Scotland: A New History* (London, 1991), p. 241; Goodare, *State and Society*, pp. 293–4.
6 *Royal Letters Charters and Tracts Relating to the Colonisation of New Scotland* (Bannatyne Club, Edinburgh, 1847); Edward J. Cowan, 'The myth of Scotch Canada', in Marjory Harper and Michael E. Vance (eds), *Myth, Migration and*

the Making of Memory: Scotia and Nova Scotia, c.1700–1990 (Edinburgh, 1999), pp. 49–72.

7 Arthur H. Williamson, 'Scots, Indians and empire: the Scottish politics of civilisation 1519–1609', *Past and Present*, 150 (1996), 46–83.
8 Martin MacGregor, 'Gaelic barbarity and Scottish identity in the later middle ages', in Dauvit Broun and Martin MacGregor (eds), *Mìorun Mòr nan Gall, The Great Ill-will of the Lowlander? Lowland Perceptions of the Highlands, Medieval and Modern* (Glasgow, 2009), pp. 6–44.
9 J. H. Burton et al. (eds), *The Register of the Privy Council of Scotland* (Edinburgh, 1877) (hereafter *RPC*), iv, 453–6.
10 T. Thomson and C. Innes (eds), *The Acts of the Parliament of Scotland* (12 vols, Edinburgh, 1814–75) (hereafter *APS*) iv, 550b.
11 J. M. Thomson and J. B. Paul (eds), *Registrum Magni Sigilli Regum Scotorum* (11 vols, Edinburgh, 1882–1914) (hereafter *RMS*) vi, no. 1911; *RPC*, vii, 749–50.
12 J. R. N. MacPhail (ed.), *Highland Papers* (Scottish History Society, Edinburgh, 1914–34), iii, pp. 222, 224–5, 263–9, 279; Gregory, *History*, p. 375.
13 Allan Macinnes, *Clanship, Commerce and the House of Stuart, 1603–1788* (East Linton, 1996), p. 60; *Highland Papers*, iii, 87.
14 Brendan Bradshaw, 'Native reaction to the Westward Enterprise: a case-study in Gaelic ideology', in K. R. Andrews, N. P. Canny and P. E. H. Hair (eds), *The Westward Enterprise: English Activities in Ireland, the Atlantic and America, 1480–1650* (Liverpool, 1978), p. 77, n. 5; Mícheál Mag Craith, 'Gaelic Ireland and the Renaissance', in Glanmor Williams and Robert Owen Jones (eds), *The Celts and the Renaissance: Tradition and Innovation* (Cardiff, 1990), pp. 57–89, at p. 81.
15 John MacInnes, 'Gaelic poetry and historical tradition', in Loraine Maclean (ed.), *The Middle Ages in the Highlands* (Inverness, 1981), pp. 142–63, at p. 144; idem, 'The Gaelic perception of the Lowlands', in William Gillies (ed.), *Gaelic and Scotland: Alba agus a' Ghàidhlig* (Edinburgh, 1989), pp. 89–100, at pp. 92–3; Wilson McLeod, *Divided Gaels: Gaelic Cultural Identities in Scotland and Ireland c.1200–c.1650* (Oxford, 2002), pp. 21–9; Màrtainn MacGriogair, '"Ar sliochd Gaodhal ó Ghort Gréag": An Dàn "Flodden" ann an Leabhar Deadhan Liosmòr', in Gillian Munro and Richard Cox (eds), *Rannsachadh na Gàidhlig 4* (Edinburgh, 2010), pp. 23–35.
16 *Highland Papers*, iii, 121–2.
17 Macinnes, *Clanship*, pp. 60, 85. For the involvement of Andrew Stewart's successor as Lord Ochiltree, James Stewart, in Scottish colonial initiatives in Nova Scotia in the 1620s, see J. Balfour Paul (ed.), *The Scots Peerage* (9 vols, Edinburgh, 1904–14) vi, 517–18.
18 The Iona Club (ed.), *Collectanea de Rebus Albanicis* (Edinburgh, 1847), p. 161.
19 Martin MacGregor, 'The Statutes of Iona: text and context', *Innes Review*, 57 (2006), 111–81, at 168.
20 Ibid., pp. 151, 162, 169; C. E. Clark, 'Mason, John (1586–1635)', in H. C. G. Matthew and Brian Harrison (eds), *Oxford Dictionary of National Biography* (60 vols, Oxford, 2004) xxviii, pp. 181–2; D. Laing (ed.), *Royal Letters Charters and Tracts Relating to the Colonisation of New Scotland, and the Institution of the Order of*

Knights Baronet of Nova Scotia. 1621–1638 (Bannatyne Club, Edinburgh, 1867), pp. 4–5, n. 3; Aonghas MacCoinnich, 'Tùs gu Iarlachd: Eachdraidh Chlann Choinnich, c.1466–1637' (Ph.D. dissertation, University of Aberdeen, 2004), p. 335, n. 1128; Aonghas MacCoinnich, 'Native, stranger and the fishing of the Isles, 1611–1637'. I am very grateful to Dr MacCoinnich for allowing me to consult this unpublished paper.

21 M. G. Jones, *The Charity School Movement: A Study of Eighteenth Century Puritanism in Action* (London, 1964), pp. 178, 182.

22 Macinnes, *Clanship*, pp. 210–21; Andrew MacKillop, *'More Fruitful than the Soil': Army, Empire and the Scottish Highlands, 1715–1815* (East Linton, 2000).

23 Alexander Grant, 'Scotland's "Celtic fringe" in the late middle ages: the MacDonald lords of the Isles and the kingdom of Scotland', in R. R. Davies (ed.), *The British Isles 1100–1500* (Edinburgh, 1988), pp. 118–41; Norman Macdougall, 'Achilles heel? The earldom of Ross, the lordship of the Isles, and the Stewart kings, 1449–1507', in Edward J. Cowan and R. Andrew MacDonald (ed.), *Alba: Celtic Scotland in the Middle Ages* (East Linton, 2000), pp. 248–75.

24 Steven G. Ellis, 'The collapse of the Gaelic world, 1450–1650', *Irish Historical Studies*, 31 (1999), 449–69, at 453; Jane H. Ohlmeyer, *Civil War and Restoration in the Three Stuart Kingdoms: The Career of Randal MacDonnell, Marquis of Antrim, 1609–1683* (Cambridge, 1993), p. 7.

25 T. C. Smout, *A History of the Scottish People, 1560–1830* (London, 1969), p. 40.

26 Macinnes, *Clanship*, p. 58; Ohlmeyer, '"Civilizinge of those Rude Partes"', p. 127.

27 Sarah Elizabeth Thomas, 'From Rome to "the ends of the habitable world": the provision of clergy and church buildings in the Hebrides, 1266 to 1472' (Ph.D. dissertation, University of Glasgow, 2008), pp. 1–5.

28 K. A. Steer and J. W. M. Bannerman, *Late Medieval Monumental Sculpture in the West Highlands* (Royal Commission on the Ancient and Historical Monuments of Scotland, Edinburgh, 1977), pp. 116–18, 208–9; MacGregor, 'Statutes of Iona', pp. 152–3; M. Dilworth, 'Iona Abbey and the Reformation', *Scottish Gaelic Studies*, 12 (1971), pp. 77–109. For more on 'greater Gaeldom' and religion, see Fiona A. MacDonald, *Missions to the Gaels: Reformation and Counter-Reformation in Ulster and the Highlands and Islands of Scotland* (Edinburgh, 2006).

29 Stephen Boardman, *The Campbells 1250–1513* (Edinburgh, 2007), pp. 330–2; Alison Cathcart, 'James V King of Scotland – and Ireland?', in Seán Duffy, *The World of the Galloglass* (Dublin, 2007), pp. 124–43.

30 G. W. S. Barrow, *Robert Bruce and the Community of the Realm of Scotland* (3rd edition: Edinburgh, 1988), p. 259; Jane Dawson, *The Politics of Religion in the Age of Mary Queen of Scots: The Earl of Argyll and the Struggle for Britain and Ireland* (Cambridge 2002), pp. 1–2; David Edwards, 'Securing the Jacobean succession: the secret career of James Fullerton of Trinity College, Dublin', in Duffy (ed.), *The World of the Galloglass*, p. 203.

31 Gregory, *History*, pp. 151–7, 163–79.

32 J. Bain *et al.* (eds), *Calendar of the State Papers Relating to Scotland and Mary, Queen of Scots, 1547–1603* (Edinburgh, 1898–) (hereafter *CSPS*), xiii pt 2, p. 937, no. 762.

33 McLeod, *Divided Gaels*, pp. 6–7; Martin MacGregor, *Gaelic Scotland in the Later Middle Ages* (forthcoming).

34 Pía Dewar, 'Kingship imagery in classical Gaelic panegyric for Scottish chiefs', in Wilson McLeod, James E. Fraser and Anja Gunderloch (eds), *Cànan & Cultar/ Language & Culture: Rannsachadh na Gàidhlig 3* (Edinburgh, 2006), pp. 39–56, at pp. 41–3, 46–7; McLeod, *Divided Gaels*, ch. 3, esp. pp. 173–93.

35 Steven G. Ellis, *Ireland in the Age of the Tudors 1447–1603: English Expansion and the End of Gaelic Rule* (London, 1998), p. 247; Dawson, *Politics of Religion*, p. 204.

36 MacInnes, 'Gaelic poetry and historical tradition', pp. 146–7; Dewar, 'Kingship imagery', pp. 42–50.

37 John Bannerman, 'The lordship of the Isles', in J. M. Brown (ed.), *Scottish Society in the Fifteenth Century* (London, 1977), pp. 209–40, at pp. 214–15. For other views see above, n. 23.

38 Gregory, *History*, pp. 170–1.

39 McLeod, *Divided Gaels*, pp. 173–91.

40 Dawson, *Politics of Religion*, p. 203; Michael Lynch, 'National identity in Ireland and Scotland, 1500–1640', in Claus Bjørn, Alexander Grant, and Keith J. Stringer (eds), *Nations, Nationalism and Patriotism in the European Past* (Copenhagen, 1994), pp. 109–36, at p. 112.

41 Macinnes, *Clanship*, p. 82; Edwards, 'Securing the Jacobean succession', pp. 205–7.

42 Julian Goodare, 'The Statutes of Iona in context', *Scottish Historical Review*, 77 (1998), 31–57, at 54–7; Lynch, 'James VI and the "Highland problem"', pp. 216, 227.

43 MacGregor, 'Statutes of Iona', pp. 168–9.

44 *RMS*, vi, no. 1911; *RPC*, vii, 749–50.

45 Gregory, *History*, pp. 289–90, 355–6.

46 Ibid., pp. 406, 411–12; M. D. W. MacGregor, 'A political history of the MacGregors before 1571' (Ph.D. dissertation, Edinburgh, 1989), pp. 272–3; Cosmo Innes (ed.), *The Black Book of Taymouth* (Bannatyne Club, Edinburgh, 1855), p. 61.

47 Goodare, *State and Society*, pp. 280–5; Lynch, 'James VI and the "Highland problem"', pp. 225–7; Ohlmeyer, '"Civilizinge of those Rude Partes"', pp. 143–5. For optimistic counterpoints see Gordon Donaldson, *Scotland: James V–James VII* (1965: Edinburgh, 1990), pp. 228–32, and the contemporary evidence cited in Gregory, *History*, p. 402.

48 Donaldson, *James V–James VII*, pp. 3–4; Lynch, *Scotland: A New History*, p. 242.

49 John MacInnes, 'The panegyric code in Gaelic poetry and its historical background', *Transactions of the Gaelic Society of Inverness*, 50 (1976–8), 435–98, at 442.

50 Boardman, *Campbells*; Dawson, *Politics of Religion*.

51 Boardman, *Campbells*, pp. 9, 13, 84, 121, 169–71, 202–3, 213–4, 249–50, 330–35; Gregory, *History*, p. 154; Dawson, *Politics of Religion*, pp. 8–10.

52 Boardman, *Campbells*, p. 203; Jane E. A. Dawson, 'Two kingdoms or three?: Ireland in Anglo-Scottish relations in the middle of the sixteenth century', in Roger A. Mason (ed.), *Scotland and England 1286–1815* (Edinburgh, 1987), pp. 113–38, at p. 125.

53 Boardman, *Campbells*, pp. 278–83, 317–19.

54 Gregory, *History*, pp. 129–43; MacGregor, 'Statutes of Iona', pp. 163–6.

55 James Gairdner and R. H. Brodie *et al.* (eds), *Letters and Papers, Foreign and Domestic, of the Reign of Henry VIII* (21 vols, London, 1862–1932), xx, pt 1, p. 345, no. 665, p. 435, no. 865; *CSPS*, xii, 202, no. 173; cf. *RPC*, vii, 749–50.
56 Gregory, *History*, pp. 171–4; Boardman, *Campbells*, p. 333.
57 Dawson, *Politics of Religion*, p. 219.
58 Macinnes, *Clanship*, pp. 211–17, 235–6; Alex Murdoch, 'Scottish sovereignty in the 18th century', in H. T. Dickinson and Michael Lynch (eds), *The Challenge to Westminster: Sovereignty, Devolution and Independence* (East Linton, 2000), pp. 45–6.
59 W. D. H. Sellar, 'The earliest Campbells: Norman, Briton or Gael?', *Scottish Studies*, 17 (1973), 109–25; William Gillies, 'Some aspects of Campbell history', *Transactions of the Gaelic Society of Inverness*, 50 (1976–8), 256–95; idem, 'Arthur in Gaelic tradition part II: romances and learned lore', *Cambrian Medieval Celtic Studies*, 3 (1982), 41–75; idem, 'The invention of tradition, Highland-style', in A. A. MacDonald, M. Lynch and I. B. Cowan (eds), *The Renaissance in Scotland: Studies in Literature, Religion, History and Culture Offered to John Durkan* (Leiden and New York, 1994), pp. 144–56; idem, 'The "British" genealogy of the Campbells', *Celtica*, 23 (1999), 82–95; McLeod, *Divided Gaels*, pp. 122–3; Martin MacGregor, 'Writing the history of Gaelic Scotland: a provisional checklist of "Gaelic" genealogical histories', in Colm Ó Baoill and N. R. McGuire (eds), *Caindel Alban: Fèill-sgrìobhainn do Dhòmhnaill E. Meek/Scottish Gaelic Studies* 24 (2008), pp. 357–79, at pp. 362–3.
60 W. J. Watson, 'Unpublished Gaelic poetry IV, V', *Scottish Gaelic Studies*, 3 (1929–31), 138–59, at 138–51.
61 Gillies, 'Some aspects of Campbell history', p. 260; McLeod, *Divided Gaels*, pp. 99, 123, 191–2, 214–15; Wilson McLeod, 'The galloglass in Irish classical poetry', in Duffy, *The World of the Galloglass*, pp. 169–87, at p. 178; Dewar, 'Kingship imagery', p. 46. The poem refers to Margaret Douglas, whom the seventh earl married before October 1594, and who died in 1607. For 1595 as offering a likely context, either before or possibly after the start of Ó Néill's rising and his victory at Clontibret, see McLeod, 'Galloglass', p. 178; Edwards, 'Securing the Jacobean succession', pp. 206–7; Macinnes, *Clanship*, p. 82.
62 Martin MacGregor, 'Creation and compilation: *The Book of the Dean of Lismore* and literary culture in late medieval Gaelic Scotland', in Ian Brown, Thomas Owen Clancy and Murray Pittock (eds), *The Edinburgh History of Scottish Literature, Volume One: From Columba to the Union (until 1707)* (Edinburgh, 2007), pp. 209–18, at p. 210.
63 R. L. Thomson (ed.), *Foirm na n-Urrnuidheadh: John Carswell's Gaelic Translation of the Book of Common Order* (Edinburgh, 1970).
64 Ibid., pp. 13, 181.
65 Donald E. Meek, 'The Reformation and Gaelic culture: perspectives on patronage, language and literature in John Carswell's translation of "The Book of Common Order"', in James Kirk (ed.), *The Church in the Highlands* (Edinburgh, 1998), pp. 37–62. Cf. Brían Ó Cuív, 'The Irish language in the early modern period', in T. W. Moody, F. X. Martin and F. J. Byrne *The New History of Ireland*, vol. iii, *Early Modern Ireland 1534–1691* (Oxford, 1976), pp. 509–42, at p. 513.

66 Dawson, *Politics of Religion*, p. 143.
67 Cf. Lynch, *Scotland: A New History*, p. xix.
68 MacGregor, 'Statutes of Iona', pp. 150–70.
69 *RPC*, viii, 757.
70 Ibid., ix, 17.
71 MacGregor, 'Statutes of Iona', pp. 164–5.
72 Macinnes, *Clanship*, pp. 46–7, 74–6, 79–80.
73 Gregory, *History*, pp. 289–90, 355–6.
74 Goodare, *State and Society*, p. 284.
75 What follows relies heavily upon MacCoinnich, 'Native, Stranger and the Fishing of the Isles'. See also MacCoinnich, 'Tùs gu Iarlachd', ch. 5; Esther Mijers, 'A natural partnership? Scotland and Zeeland in the early seventeenth century', in Allan I. Macinnes and Arthur H. Williamson (eds), *Shaping the Stuart World, 1603–1714: The Atlantic Connection* (Leiden, 2006), pp. 233–60.
76 *APS*, iv, 160–1, 248–51.
77 *Collectanea de Rebus Albanicis*, pp. 99–105.
78 MacCoinnich, 'Native, stranger and the fishing of the Isles'.
79 Ibid.
80 Ibid.
81 Allan Macinnes, *Charles I and the Making of the Covenanting Movement, 1625–1641* (Edinburgh, 1991), pp. 108–13.
82 Lynch, 'National identity', pp. 110–12.
83 Roger A. Mason, *Kingship and the Commonweal: Political Thought in Renaissance and Reformation Scotland* (East Linton, 1998), pp. 126–37.
84 Boardman, *Campbells*, chs 10, 12, esp. pp. 322–3, 327–8; MacGregor, 'Statutes of Iona'.
85 I am deeply indebted to the editors, to participants at the conference who asked or prompted questions, and to Dauvit Broun, Alison Cathcart, Jane Dawson, Colin Kidd and Aonghas MacCoinnich, for their comments on a draft of this paper. For an overlapping discussion which appeared too late to take account of here, see Alison Cathcart, 'The Statutes of Iona: the archipelagic context', *Journal of British Studies*, 49 (2010), 4–27.

4

Plantation and civil society

PHIL WITHINGTON

The starting point for this essay is a paradox surrounding what has been called the Elizabethan 'monarchical republic'.[1] First coined in a series of essays by Patrick Collinson, this builds on the insight that although Tudor and early Stuart England was ostensibly a monarchy, the nature of governance was such that enormous power and responsibility increasingly devolved across the social spectrum after the death of Henry VIII – not merely to the Privy Council and parliament at the political 'summit', but to the governors of counties, cities and boroughs, and parishes of provincial England. The result was a participatory 'commonwealth' (as contemporaries described it) in which renaissance ideals of personal and public government were increasingly disseminated and valorised by not merely nobles and gentry but also the urban and rural 'middling sort'. At both the local and metropolitan levels Elizabethan England became, in effect, 'a republic which also happened to be a monarchy: or vice versa'.[2] The paradigm has caught the attention of literary and intellectual historians interested in the social transmission, extent, and appropriation of political ideas, in particular forms of civic humanism.[3] It also resonates with social and political historians interested in the discursive developments which imbued social action and social relations with their contemporaneous meanings.[4] The cumulative impression is that, in their dealings with and miscomprehension of this variegated and devolved political culture, the Stuarts repeatedly ran into serious – indeed fatal – problems. Yet this implicitly positive account of an English commonwealth capable of governing itself and policing its rulers sits uncomfortably alongside what we know about the English in Ireland over the same period. For English poets and statesmen who in many respects epitomised the ideal of 'monarchical republicanism' it was a shibboleth that Irish society could and should be 'civil' like England.[5] To this end, however, the same political culture justified remorseless critiques of 'degeneracy' against the 'Old English', policies of expropriation and forcible reformation against the barbarous 'mere Irish', war and plantation.[6] The defence of 'liberty' in one context becomes the imposition of slavery in another.

The paradox has been explained in various ways. Long before the notion of monarchical republicanism was even coined Brendan Bradshaw blamed the deterioration of Anglo-Irish relations on militant Protestantism. Over the course of the sixteenth century the cosmopolitan and encompassing humanism practised by the generation of Thomas More was corrupted, and narrowed, by the exigencies of state-formation and pathologies of Reformation.[7] The development of the Elizabethan 'monarchical republic' clearly challenges this diagnosis, with Markku Peltonen convincingly charting not so much the withering of civic humanism under Elizabeth as the proliferation of classical and republican templates and their application to the problems of modern governance. English towns offered, for example, 'favourable circumstances in which humanist and republican vocabulary could be applied': by the 1570s, even an obscure town clerk like John Barston of Tewkesbury could appropriate Cicero to envisage 'a community where everyone would be willing to disregard his private good and to practice his civic virtues in pursuit of the common good'. In this way 'the main aim of the urban commonwealth was to maintain liberty'.[8] It was the misfortune of Ireland to provide, in contrast, 'the context for a thorough usage of Machiavellian republicanism'. Most notoriously, the colonist Richard Beacon adapted the theories of the Florentine to argue that, by ruthlessly conquering and reforming the 'servile' Irish, the 'free' English would hone their 'warlike discipline' and 'promote the greatness of England'.[9] This was merely one example of how the colonial experience exposed the dark and dangerous underbelly of humanist conceits.[10] A third explanation for the paradox has recently been suggested by Rory Rapple, who argues not that civic humanism was eclipsed by Protestantism, nor that 'monarchical republicanism' provided a range of potentially contradictory positions, but that Elizabethan Ireland, as a land of opportunity for ambitious English soldiers, became the antithesis of monarchical republican values. Rapple suggests that by the later sixteenth century 'the godly civic republicanism espoused throughout England in both local and national government, with its abhorrence of pre-eminence, left a fault-line in Elizabethan political culture which arose out of the disintegration of Henrician legacy'. Because 'English martial men' could no longer be 'chivalrous' or 'pre-eminent' in England they went to Ireland instead.[11]

While there is no doubt some truth in all these explanations, the concern of this essay is with a dimension of 'monarchical republicanism' which is often neglected by historians but which is nevertheless vital for understanding plantation. This is simply its pronounced *corporatism*, by which is meant the increasing importance from around 1570 of theories and practices of association – or what contemporaries described as 'company' and 'society' – which were not simply manifestations of 'the family' or 'the state'. More to the point, the essay suggests that the lineaments of plantation society were neither the result of some ongoing colonial experiment nor antithetical to monarchical republicanism. On the contrary, plantation involved the transposition of English corporate practices – most notably institutions associated with incorporated boroughs and cities – into an (albeit) hostile environment. The resulting tensions between newcomers and inhabitants replicated tensions which already characterised English provincial life, though in Ulster these were massively amplified by religious and ethnic differences.

In making this argument the paper also addresses the assumption that British involvement in Ireland in general – and the plantation of Ulster in particular – should be understood purely as a process of early modern state-formation. The importance of 'the state' (however defined) is an entrenched position among historians of colonialism; it clearly reflects a basic truth. As Michael J. Braddick puts it, the Americas and Ireland were peculiar in terms of the general expansion of the British state in that, for various reasons, 'local elites did not provide the basis of civil government'. Instead the Flight of the Earls created the vacuum into which the Ulster plantations could be inserted: 'Gaelic legal codes were discountenanced and titles of land were only recognized if they were good in common law. Alongside the religious and legal instruments of civility went administrative and political regularization – county divisions, sheriffs, justices of the peace and assizes. And, of course, garrisons.'[12] For William J. Smyth, the logic of this process was centralisation, the 'monarchical state' utilising 'the modernizing segments in the society to override and subordinate local and regional governments and powers to the goals of the centre'.[13] The concern here is not to belittle these developments, deny the inherent violence of colonialism, nor the need for organised military force to enable and sustain plantation; far from it. Rather it is to highlight some of the corporate vocabulary and institutions that also shaped the colonial experience. Corporate discourse framed how servitors, undertakers, planters, and their tenants presented and organised themselves. It also influenced how the indigenous cultures and communities under threat were conceived and represented. The Ulster plantations in general – and those orchestrated by the London companies in particular – marked the quintessence of this corporate dimension. This corporatism was, in turn, closely related to the culture of civic humanism that characterised the thinking of English elites in the later sixteenth and seventeenth centuries, and which first encouraged Collinson to coin the slogan 'monarchical republicanism'.

The essay explores this corporatism in three stages. The first section traces its presence in the work of humanist writers looking to resolve the problem of Ireland. It suggests that the 'agenda' for plantation was set not by Edmund Spenser in the 1590s, as is often assumed, but by Spenser's mentor Sir Thomas Smith's in the 1570s.[14] Viewed in these terms Spenser's *A View of the Present State of Ireland* (1596; pub. 1633) served as a kind of ideological bridge between Smith's *A Letter from I. B.* (1571) and Thomas Blennerhasset's *A Direction for the Plantation of Ireland* (1610), which successfully marketed the project to potential investors back home.[15] The corporatism of these humanists is then placed in the context of a more general discursive development – the assimilation of the words 'company' and 'society' into the printed English vernacular from the 1570s. The final section discusses the types of urban 'society' which served as a template for plantation, unpacking the implications for notions of citizenship, freedom, and slavery not merely among the New English but within the monarchical republic more generally.

Sir Thomas Smith was one of the foremost humanists of his day and a latter day hero of the 'monarchical republic'. His *De Republica Anglorum*, or *The Commonwealth of England*, went through eleven editions between 1583 and 1640 and provided both the title and subject matter for Collinson's lecture on the Elizabethan polity.[16] His biography is well known and might be taken as the archetypal Tudor success story. Born into a yeoman family in the borough of Saffron Walden in Essex his precocity as a child saw him moved to Cambridge at the age of eleven where he became the 'flower' of the university – a prominent scholar and celebrated humanist; university and ecclesiastical office-holder; royal secretary, ambassador, and MP. He was not only knowledgeable in Greek and Roman but sought to apply his classical learning to contemporary problems and issues: this was his defining characteristic as a humanist. This he did personally, through his work as a pedagogue, politician, and writer, and indirectly, through his advice to friends and patrons, like William Cecil, and eager mentors, like Gabriel Harvey.[17] Harvey was, in turn, a great friend of Edmund Spenser, another Cambridge graduate, and through Harvey we see a direct link between Smith and the mid-century network of humanists who so dominated government to the writers of the *fin-de-siècle* Renaissance, Spenser foremost among them.[18] The common denominator in these respects was the University of Cambridge, the hot-house of Tudor humanism: it was here Smith and Spenser both made their most lasting friendships and learned the skills and outlook which characterised their contributions to the English humanist project.

Although Thomas Blenerhasset is a much less well-known figure his biography is similar.[19] The younger son of the lower gentry/'middling sort', Blenerhasset attended Cambridge in the 1570s, where he translated Ovid's *De Remidio Amoris* and became embroiled in the same arguments about linguistic reform that enflamed Smith, Harvey and Spenser. He also looked to make his personal fortune through service to his commonwealth and was drawn to Ireland as an obvious sphere of activity. His choice of public service was martial; while garrisoned in Guernsey in 1578 Blenerhasset composed a new edition of *The Mirror for Magistrates*, a text which, according to Scott Lucas, 'presented to decades of readers a compelling intellectual endorsement of monarchical-republican governance'.[20] Blenerhasset acquired property in Ireland through the patronage of the Denny family in the 1590s and would have known Spencer and the circle of influential New English poets in Dublin. More obliquely, although Blenerhasset's early schooling is unknown his brother-in-laws were products of the Merchant Taylor's School, where they would have been taught by the great English humanist and pedagogue, Richard Mulcaster, who was also Spenser's teacher. He was, in short, a 'soldier-poet' in the manner of Philip Sidney and John Harington, though hardly recognisable as Rapple's stereotype of a 'martial man'.[21] Given this background, it is hardly surprising that in his 1610 pamphlet on *A Direction for the Plantation of Ireland* Blenerhasset sang from the same civic sheet which also characterised Smith's modest but extraordinary influential corpus of work.[22]

This corpus includes *A Letter Sent by I. B.*, the pamphlet which publicised Smith's plan for the ultimately unsuccessful plantation of east Ulster in 1571. As is well known, Smith's *Letter Sent by I. B.* offered the solution to settling in Ireland without direct fiscal and military aid from the crown. As he put it:

> There be many that not considering what facility it is by good order and willing means to bring great things to pass, but wondering rather at the greatness of the sum, which must furnish so many soldiers, carry them over, and maintain them there for a year or there about (that must necessarily be supplied from England) are of the opinion, that it cannot be done without the Princes pay. But I will inform you an easy way, to bring this without her majesties expenses to pass.[23]

Historians have usually characterised the 'facility' by which 'to bring great things to pass' as 'private'. In the 1880s Richard Bagwell described the resulting 'Enterprise of Ulster' as an attempt to found 'private principalities in Ireland'.[24] A hundred years later Hiram Morgan repeatedly typified Smith's 'private undertaking' as an early experiment in 'private colonization'.[25] Morgan is hardly unusual in this – it is an orthodox view of colonial historiography that 'English colonization, although conducted under royal patents, was always pursued by private companies'. As a result, 'the success or failure of the enterprises rested entirely on the ability of private interests to raise capital and personnel'.[26]

There are, however, serious problems with this characterisation, not least because for Smith and other humanists 'private' was the antithesis of the project he had in mind. The word carried two connotations. In the first instance, 'private' referred to anything not 'public' in the political (*res publican*) sense of that term. As Smith put it in *De Republica Anglorum*, 'the division of these which be participant of the commonwealth is one way of them that bear office, and which bear none, the one be called magistrates, the other private men'.[27] 'Magistrate' here meant lowly 'subalterns' such as jurors or constables as well as sheriffs, aldermen, and privy councillors: it was not a term of distinction so much as the participatory extent of civil government.[28] Second, 'private' described not merely abstention or exclusion from public office but the very opposite of 'civil society'. Smith explained that:

> if one man had as some of the old Romans had (if it be true that is written) v. thousand or x. thousand bondmen whom he ruled well, though they dwelled all in one city, or were distributed into diverse villages, yet that were no commonwealth: for the bondman hath no communion with his master, the wealth of the Lord is only sought for, not the profit of the slave or bondman.

The relationship was equivalent to 'instruments of the husbandman' like 'the plough, the cart, the horse, ox or ass': 'though one husbandman had a great number of all those and looked well to them, yet that made no commonwealth nor could not so be called'. This was because 'the private wealth of the husbandman is only looked for and there is no mutual society or pact no law or pleading between the one and the other'. In the same way 'the bondman or slave which is bought for money' was 'not otherwise admitted to the society civil or commonwealth, but is part of the possession and goods of his Lord'.[29] For Smith as for other humanists 'private' was antithetical to the principles upon which commonwealth and society rested and which colonialism, as acts of society, was designed to extend. Indeed, if 'private' described anything it was Gaelic Ulster, where men lived not in 'society' but feudal *septs* (clans) and in which, as Smith was informed by his Dublin correspondent Roland White, the warrior aristocracy enjoyed effective possession of their 'churls'.[30]

In this instance, then, 'private' as a category of historical analysis discounts the word's early modern meaning; it also obscures what Smith envisaged by plantation and the vocabulary through which he did so. It quickly transpires that the 'facility' he has in mind is 'company'. Organised as a joint-stock, the enterprise was to be funded initially through 'common charges', 'the common stock', and 'the company's stock', with 'every man putting in a share'.[31] This was intended to 'furnish' in Ulster

> a company of Gentlemen, and others that will live friendly in fellowship together, rejoicing in the fruit and commodity of their former travails, which (through noble courage) for estimation sake, and the love of their own country the first enterprised [sic], deserving if I may speak it, that am resolved one of the same company, to be crowned with garlands of honour and everlasting fame.

Readers were encouraged 'to be a partaker with him [Smith] in person'; they would find him 'vigilant and careful, coveting more the well doing hereof, and the safety of his company, than the glory of victory in any rash attempt, more desiring to please and profit every man, than looking for ceremonious courtesy and reverence'. Rules and orders were nevertheless needed to ameliorate those petty feuds that inevitably 'disturb the whole company' and these would 'be drawn by the advice of the best captains, and shall be read unto the whole company'.[32] Hiram Morgan has rightly observed that in its basic approach, *A Letter* was conceived very much in the spirit of Smith's *A Discourse of the Commonweal of the Realm of England*.[33] This argued that personal profit and the common good were not irreconcilable, as medieval and mid-Tudor moralists claimed, but reciprocal. What Morgan fails to stress is that for Smith 'profit' should be pursued and organised not through 'private' possession but rather corporate association. It was through this 'facility' that the settlers were 'motioners and ring leaders of so many English families, to be planted forever in the *Ardes*'.[34]

A Letter initiated, in effect, an extenuated process of corporate association. Smith instructed 'all men willing to adventure in this most honourable and profitable voyage' to 'resort into Paul's churchyard to the Sign of the Sun'. There he could view 'the Indentures of Covenants' authorising the project, 'pay such money as he is disposed to adventure', and 'receive his assurance from Thomas Smith the son'. Draft proposals for the institution of colonial society show that in Ards itself one-half of the colony was to be organised along classically military lines, with 'a Chieftain', 'Deputy Colonel', 'Captains', 'footmen' and 'horsemen' led by a 'Centurion' and 'Decurion', and other military officers.[35] These officers and 'their companies' were represented by a 'Privy Council of Martial Affairs', which took decisions of military significance on a consultative basis. The colony, however, also consisted of 'adventurers' and their tenants: the 'artificers to work in the town', 'husbandmen in the field', 'merchants to travel into fairs and markets', and the 'Fathers of the Colony' – 'the first founders of the Colony from which honour the poorest and meanest of all adventurers is not to be excluded'. These were represented by a common counsel, a body which formed the political centre of the colony: it made laws, set taxes, chose officers, and was responsible for 'all weighty affairs'. The chieftain required 'the consent of the more part of them' before doing anything 'of consequence' and a select group of 12 common councillors were responsible for advising the Privy Council about 'making of war, or concluding of peace with the enemy'. Historians have noted the debts to the classical Roman colony.[36] In terms of its civil organisation, however, the plantation was, as is shown below, a faithful transplantation of English incorporated cities and boroughs.

The Ards venture failed and Smith's son died trying to make it work. Its imprint, however, can be found on the much larger projects to colonise Munster from 1583 and the Ulster plantations in the 1610s. Edmund Spenser was a direct beneficiary of the Munster initiative in terms of office and lands; the Nine Years War, which provided the backdrop to *A View of the Present State of Ireland*, imperilled these benefits (Spenser's household was destroyed by insurgents in 1598). Written in the midst of this interminable and occasionally ferocious conflict, *A View* was clearly not a promotional tract like *A Letter*. Rather, it used the dialogic form in order to debate and resolve the practical problem of what Spenser perceived to be Irish 'incivility'. This format allowed Spenser to develop the thesis that Irish incivility was manifest socially and politically; afflicted both the 'mere Irish' (the Gaelic inhabitants) and 'Old English' (the medieval settlers); and was deeply embedded in the 'manners', 'customs', and 'habits' of the people. The result was a prognosis which advised, among other things, the destruction of the institutions of Gaelic and Old English association, be they military, sociable, legal.[37] These should be replaced with corporate structures that embedded English civility in Irish locales.[38]

The destruction of septs and other kinds of customary association went hand-in-hand, therefore, with the creation of 'societies civil'; and for Spenser as for Smith the 'facility' was urban corporatism. Just as strategically placed garrisons were the ultimate guarantors of order in colonial Ireland, so 'at every of these forts I would have the seat of a Town laid forth and encompassed, in the which I would wish that there should inhabitants of all sorts as Merchants, Artificers, and Husbandmen, be placed, to whom there should Charter, and Franchises granted to incorporate them'.[39] The obvious prospect of 'much profit' and 'great commodity' meant 'it will be no matter of difficulty to draw out of England, persons which would very gladly be so placed'. The establishment of these and other urban communities would galvanise the country as a whole. Irenius explained that 'there is nothing doth sooner cause civility in any country than many market towns, by reason that people repairing often thither for their needs, will daily see and learn manners of the better sort'. He argued that 'there is nothing doth more stay and strengthen the country, than such corporate towns, as by proof in many Rebellions hath appeared, in which when all the countries have swerved, the towns have stood fast'. As importantly, 'there is nothing doth more enrich any country or Realm, than many towns, for to them will all the people draw and bring the fruits of their trades, as well to make money of them, as to supply their needful uses'.[40] Networks of incorporated market towns meant that 'everyone that is not able to live of his free-hold' could be apprenticed in 'a certain trade of life, to which he shall find himself fittest'. They meant that 'keeping of cattle' ('very barbarous and uncivil') could be superseded by 'tillage and husbandry' ('peace and civility'). They also allowed the building of parish and grammar schools, whereby the 'youth...in short space grow up to that civil conversation, that both the children will learn their former rudeness in which they were bred, and also the parents will even by the example of their young children perceive the foulness of their own behaviour'.[41] It was, moreover, symptomatic of Irish barbarism and degeneracy that most of the towns originally 'seated' in Ireland were 'utterly wasted and defaced' or in the thrall of 'lords and gentry'. Irenius now opined that 'as I wished many corporate towns to be erected, so would I again wish them to be free, not depending on the service, nor under the command of any but the Governor'. There could be no better example than 'all those free-boroughs, in the Low Countries, which are now all the strength thereof'.[42]

Thomas Blenerhasset, whom Canny describes as 'one of the most vigorous propagandists for the plantation', knew there were plenty of examples closer to home. Like Smith, Blennerhasset used cheap print to publicise the undertaking; like Smith, he envisaged the formation of a joint-stock company (centred on the patronage of the Earl of Shrewsbury); and, like Smith, he understood the potential for personal profit to be dependent on the creation of effective and mediating corporate structures, the more so in a hostile environment like Ulster. As Blennerhasset noted at one point in *A Direction for the Plantation of Ulster*, 'Oh this word Myne is a strong warrior, every man for his own will adventure far', though only if 'he will rather increase then decrease his number': 'Therefore in this our undertaking, let all the people be such as shall enjoy every man more or less of his own.'[43] More specifically, isolated garrisons as in Lifford in Donegal were expensive and ineffective. However, if 'Lifford and the lands adjoining...were undertaken by many, their many helping hands (everyman respecting his own profit) they would not regard charge, nor be weary with labour and pains to frame a perpetual security, and good success to their business.'[44] Blenerhasset's conception of how to structure this collectivism was also drawn from Smith: as Blenerhasset explained, the only possible means of success in Ulster was to construct 'so many goodly corporations, as it would be a wonder to behold'. Indeed, the only substantial differences between the 1571 and 1610 pamphlets were the *language* of association they deployed and a much greater emphasis by Blenerhasset on religion. Whereas Smith's organising term was 'company', Blenerhasset preferred 'society' or, like Spenser, 'corporation'. He insisted that by replacing garrisons with 'corporations' the undertakers and 'old worthy soldiers' alike would find 'their security...much better, and the society far excel'. With corporations 'there would be instead of popery true religion; & a comfortable society'.[45] In 'a scattered plantation, for many undertakers to be dispersed three score miles in compass, alas they shall be now at the first like the unbound sticks of a brush faggot, easy to be gathered in, neither shall there be true Religion, sweet society, nor any comfortable security amongst them'. In contrast, with corporations 'Ulster which hath been hitherto the receptacle and very den of Rebels and devouring creatures, shall far excel Munster, and the civilest part of that country... in civility and sincere religion, equal even fair England herself, with a Christian and comfortable society, of neighbourhood'.[46]

The onus placed on the words 'company' and 'society' by Smith in 1571 and Blenerhasset forty years later reflected their familiarity with, and predilection for, a form of association that has often eluded the attention of historians. This neglect stems, perhaps, from the apparent obviousness of the concept. Yet as the plantation of Ulster demonstrates, this is a mistake: for Smith, Spenser and Blennerhasset – not to mention the London Companies and the Irish Society – 'company' and 'society' were both the means and the end of the colonial process. What did the words mean? Derived from the Latin *societas*, 'society' was a synonym for the Anglo-Saxon word 'fellowship' and Romance term 'company'; its primary meaning between the sixteenth and eighteenth centuries was *purposeful and voluntary association*.[47] The initial translation of *societas* into the vernacular was the achievement of English humanists like Thomas More, Thomas Elyot and Smith, who took the communicative skills and mutual reciprocities requisite of 'society' to be central to the human condition.[48] Thereafter the term 'society' was ideological in a way that is difficult to appreciate today. This idealism was palpable in the first printed text explicitly devoted to the concept, *The Safeguard of Societie*, which was written by John Barston, the Tewkesbury town clerk, in 1576. Peltonen takes Barston's treatise to be specific to England. However, insofar as English urbanism became a template for plantation then *Safeguard of Society* was also an instrument of colonisation. This is the more so because Barston – yet another product of the University of Cambridge – shared the same patron as Spenser (the Earl of Leicester) and the same publisher as Smith (Henry Bynneman).

There are a number of points worth emphasising about Barston's concept of 'society'. First, its overarching purpose was always collective and ethical: 'every societie of people is established for [their] common weal' and 'felicity unto all' and that such ends are achieved through 'lawful government'; the promulgation of 'virtue', 'prudence' and 'wisdom'; and 'civil behaviour'.[49] Second, Barston did not think in terms of single and monolithic entity called *society* so much as an amalgamation of inter-related associations: *societies*. These ranged from 'that special kind of societie and fellowship of one people, gathered together in one town, which resembles the beginning of all civility' – towns like Tewkesbury – to 'the first society, and the very beginning of all other': the association of 'marriage and household'. In between were a host of institutions – networks of kin, neighbours, and friends; guild and corporations – which were societies in their own right.⁵⁰ Third, Barston echoed Smith in assuming that even the most intimate societies were always public in important respects. He explained that he 'sequestered societies, into sundry kinds, to the only intent, that the necessity thereof appearing in private causes, the public society of all may be duly honoured'. Indeed, 'All peculiar fellowship be as it were members of civil society': it was in their everyday forms of 'societie' – the 'practice [of] common weal by their private constitution' – that people learnt virtue, civility, and 'the last end of common weal', and so reconciled the needs of 'public utility' with the pursuit of 'private lucre'.⁵¹ It followed that, fourthly, the qualities and skills of 'civility' inculcated in smaller interactions could only strengthen and improve the larger social body: it was 'by observing the decorum and comely behaviour in particular degrees of fellowship, the common preferment of all may be more easily perfected'.⁵² Finally, magistrates and others responsible for 'common provision' were bound to ensure the 'profit, preferment, safeguard and estimation of all societies' through equitable and just governance. The alternative, Barston argued, was 'that the simple are quelled with extremities and the best of all shall possess neither life nor goods in safety, to the great dishonour of magistrates and the utter disparagement of the common weal'.⁵³

In the decades following Barston's initial discussion 'society' became a more regular fixture on English title-pages. Indeed the appearance of the word and its synonyms on the front of printed texts provides a rough and ready method of tracing its assimilation into more general usage. Such appearances can be measured using the *English Short Title Catalogue* (ESTC) in conjunction with the *Early English Books Online* (EEBO), both of which are now available electronically. The resulting database suggests a basic chronology which, within the parameters of the sample, is relatively systematic and complements the more familiar techniques of cultural and literary analysis: for example, the collation on anecdotal evidence (and the serendipity this entails), focus on one or two writers in their canonical contexts, or the close analysis of a text or genre of texts.⁵⁴ The basic contour of discursive change is outlined in Figure 1, which shows the chronology of appearances of society or company in printed title-pages between 1500 and 1700 and compares the number of *first editions* in which the words and their variants were used with their appearance in *all editions* (i.e. later editions of existing works). The chart shows a sudden upsurge in the use of the words from the 1570s onwards – the decade in which experiments in plantation, like Smith's, began in earnest, as well as when Barston published his treatise. It also shows another peak around the time of the Ulster plantations in the 1610s. Over the period as whole, neither term appeared on vernacular printed title-pages in the 1490s; by the 1590s, 23 first-editions and 35 editions deployed either company or society; and by the 1690s the figure was 511 and 705.

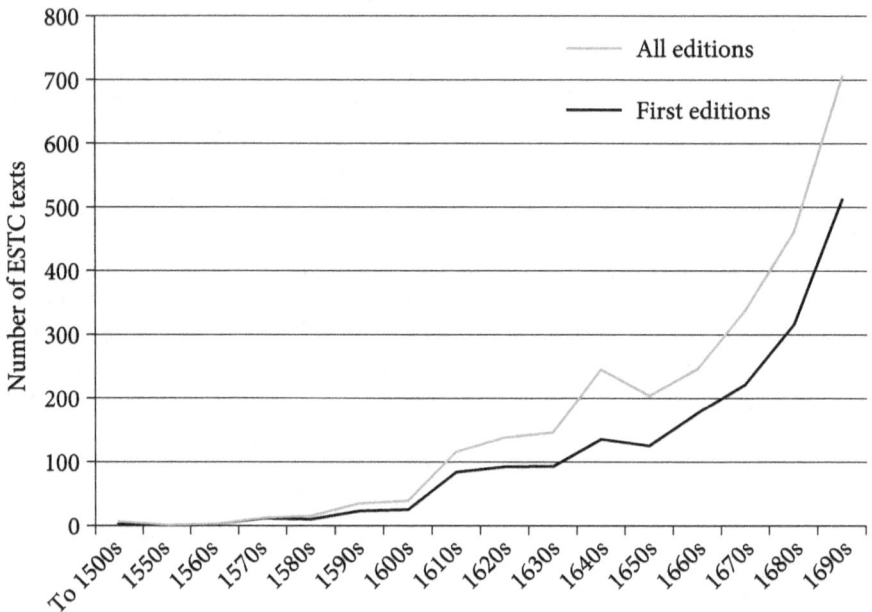

Figure 1 The number of ESTC texts with 'society' or 'company' in the title, 1500–1700

The same period witnessed, of course, a sustained increase in the production of printed material. In such circumstances it might well be expected that the number of texts that happened to have 'society' or 'company' emblazoned on their title-page should also rise. It is significant, therefore, that when the increase in print production is taken into account then the shape of the graph outlined in Figure 1 remains broadly the same. Figure 2 shows the same data as a percentage of all texts catalogued on the ESTC between 1500 and 1700. This demonstrates the general increase in the terms' visibility. It also emphasises the rise of the 1570s (2% for first-editions and 2.5% for all editions) and the peak of the 1610s. However, as might be expected the trajectory of 'company' and 'society' were by no means identical. Figure 3 shows that company predated society on title-pages by 60 years and that, after its first appearances in the 1570s, it was not until the 1590s that 'society' appeared with any regularity. It also shows that society enjoyed a moment of comparative prominence at precisely the moment Blenerhasset was writing his pamphlet.

Just to give some context to these trends, Figure 4 compares appearances of 'society' or 'company' with the words 'state' and 'family'. It shows that 'state', which derived its many early modern meanings and applications from the core concept of 'condition', was fairly ubiquitous throughout the period.

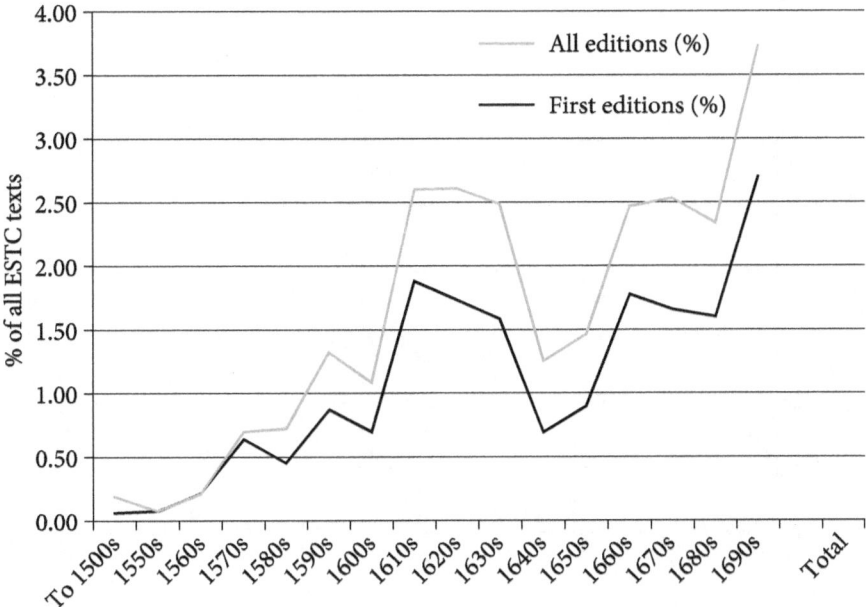

Figure 2 The % of ESTC texts with 'society' or 'company' in the title, 1500–1700

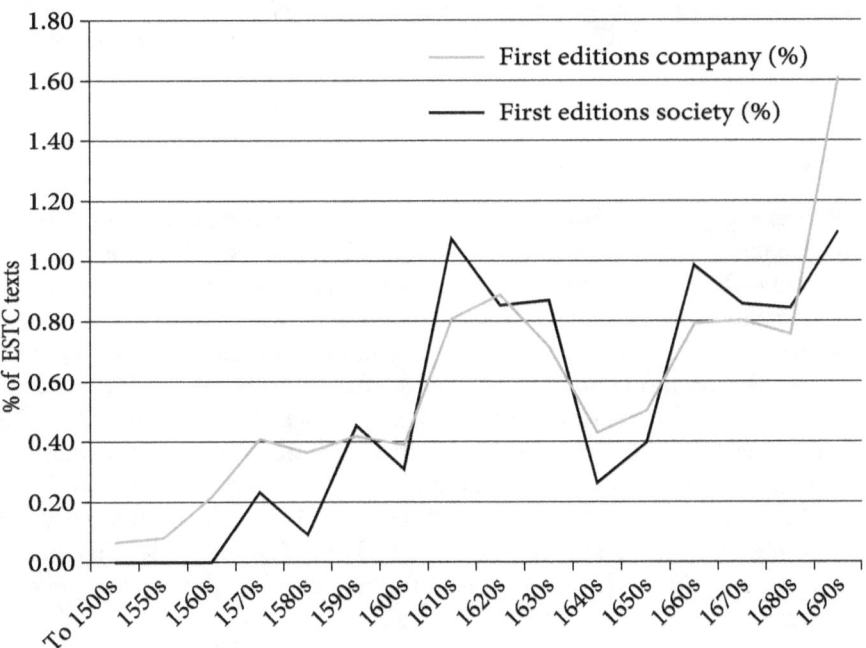

Figure 3 The % of ESTC texts with 'society' and 'company' compared, 1500–1700

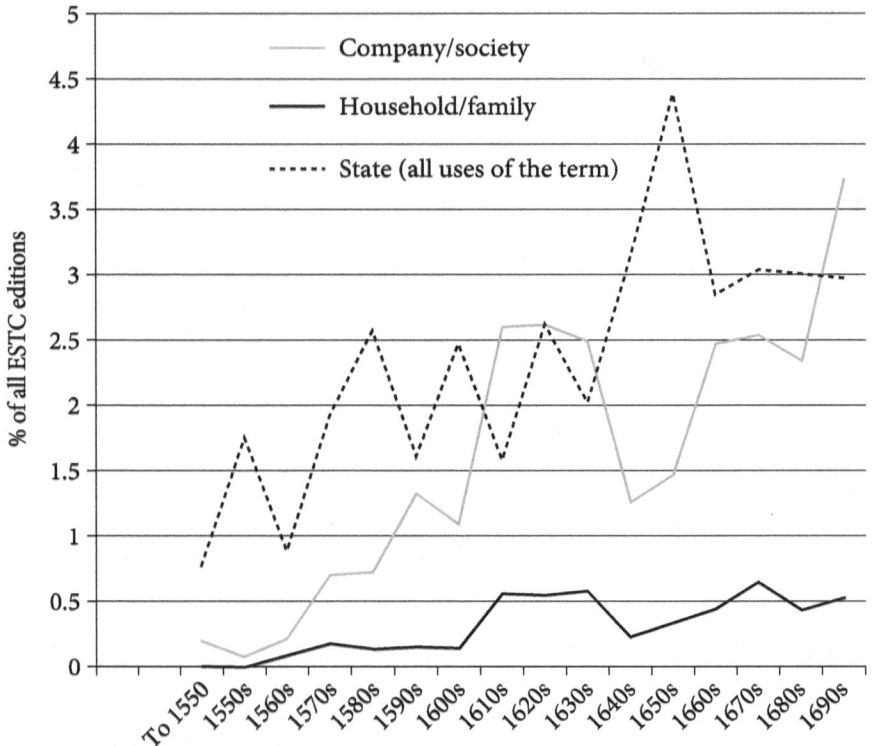

Figure 4 State, society/company and household/family, 1500–1700

This was in contrast to 'family', with society/company somewhere in-between. The Jacobean plantation of Ulster, however, coincided with one of those moments when company/society overtook 'state' in terms of print visibility.

The figures, therefore, show the establishment of the concept of purposeful and voluntary association in the print-consciousness of English readers at precisely the moment 'company' and 'society' were being elucidated as instruments of colonialism. In the context of Ulster the terms described the institutional means by which undertakers organised and funded themselves (through the joint-stock companies and urban communities); they were the paradigm through which existing Irish and Old English communities were negatively evaluated; and they described the kind of communities that would ideally be created in their stead. It is to this aspect of the process that we can now turn.

Smith, Spenser and Blenerhasset all recognised that the most likely means of turning an associational ideal into practical reality was the creation of incorporated cities and boroughs. As such, it is important to stress that Barston's conflation of humanist and urban life was widespread. Angel Daye echoed the general sentiment when he observed in 1586 that urbanity in English meant 'civil, courteous, gentle, modest, or well-ruled, as men commonly are in cities and places of good government'.[55] For Henry Manship, cities and boroughs embodied 'a certain community or Society, both of life and goods, which makes a civil body, formed and made of divers members, to live under one power, as it were under one Head and Spirit, and more profitably to live together in this mortal life, that they may the more easily attain unto life eternal life forever'.[56] We have already seen the enormous faith that Blenerhasset placed in this conceit; his preference for corporations over other models of settlement, like garrisons and fortified manors, was deliberate and reasoned. The prominent role accorded London citizens in the plantation of Ulster indicates that the preference was widely shared. Certainly Smith liked cities. This is clear from the way he read books. On his copy of Marguerite de Navarre's *L'Heptameron* (1560), for example, he copied the names of Tarbes, Narbonne, Barcelona, Marseilles, and Aigues-Mortes in the margin and drew urban skylines underneath. It is also clear from his template for colonial governance, which in terms of its classicism, corporatism, and civic institutions bore all the hallmarks of English incorporated cities and boroughs. The major difference was, of course, the accentuated military presence: Smith warned potential settlers that 'it must be understood, that you are either wholly in the war, or half in war and half in peace, And because you must begin as you were all in war, these [military] offices be necessary'.[57] This should not deflect how both soldiers and adventurers were portrayed as citizens.

This use of urbanity for colonial purposes was not the mere product of over-ripe imaginations. Rather, it was borne of experience and practice. Just as corporations were a crucial dynamic in the plantation of Ulster after 1610, so they had figured prominently in the wide-ranging social and economic reforms initiated in England since the 1540s.[58] The origins of these reforms were many, complex, and varied.[59] However, in terms of sanction by central government, the driving force – including urban incorporation – was Smith, Cecil and other members of their sprawling Cambridge mafia who dominated the higher echelons of royal power for much of the Edwardian and Elizabethan eras.[60] More to the point, one of the outcomes by the turn of the seventeenth century was a discernible 'corporate system' by which cities and boroughs – or 'little commonwealths', as contemporaries described them – had filled the topography of provincial England.[61] As Figures 5 and 6 indicate, viewed in this context Ulster appears not so much a 'laboratory' of empire as a red-hot crucible for precisely the kind of 'civil society' that already characterised much of provincial England.[62] Indeed Ulster's rate of incorporation in the

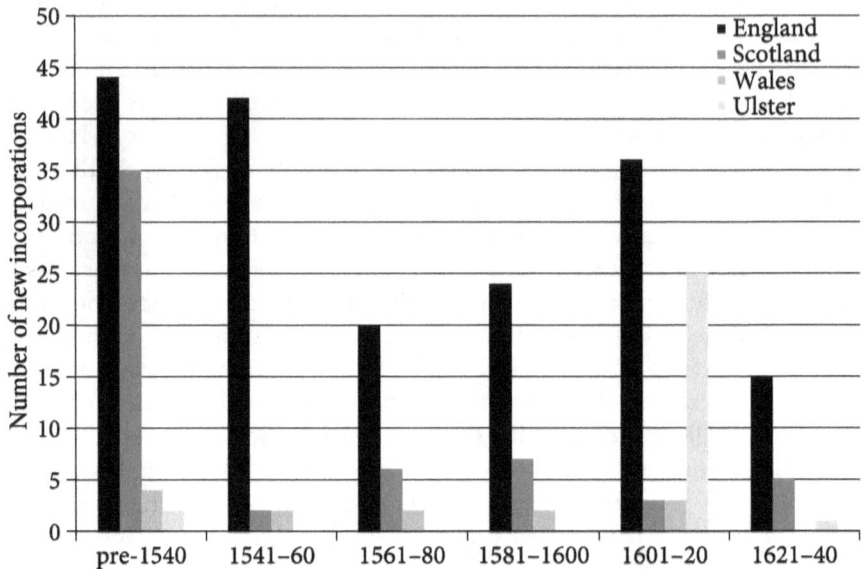

Figure 5 Rates of incorporation in England, Wales, Scotland and Ulster, 1540–1640

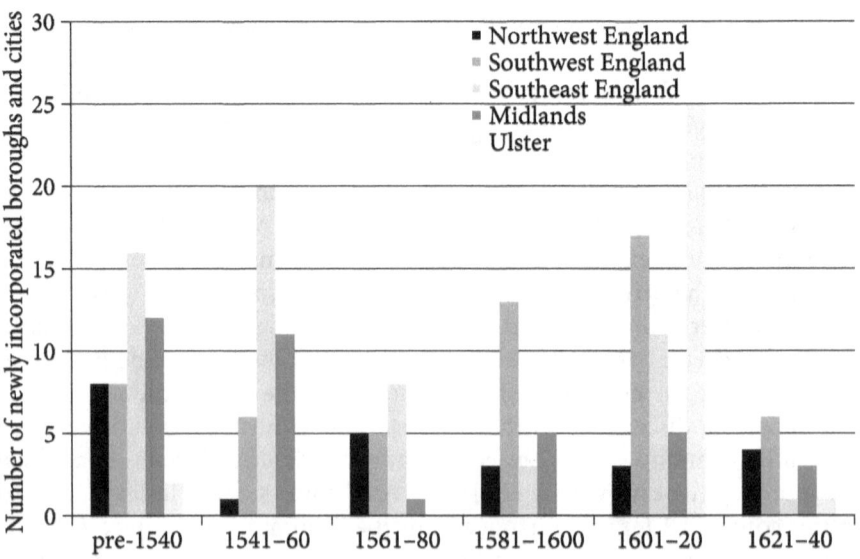

Figure 6 Rates of urban incorporation in Ulster and English regions, 1540–1640

Jacobean era was second only to England; and compared with individual English regions, levels of urbanisation in Ulster (in terms of the provision of urban institutions) lurched from lowest to highest almost overnight.[63]

The incorporated communities of early modern England gave their enfranchised inhabitants – burgesses, freemen and citizens – extensive public powers, responsibilities and autonomy. There is every indication that the same culture of Protestant corporate citizenship was intended for Ulster – indeed how could it not when a key colonial body, the Irish Society, charged with managing the Londonderry plantation, was an association of citizens drawn from the most powerful urban community on the islands.[64] Urban corporations were the practical and symbolic foundations of civility, Protestantism, law and commerce in the province. As Jonathan Barry has shown, they also endowed their inhabitants with a collective freedom and civility that transcended the sum of their individual parts.[65] While this was good news for companies of undertakers and settlers it was less so for an indigenous population unable or unwilling to partake in 'commonwealth'. Indeed the flipside of Ulster's 'societies civil' – and of civic humanism generally – was a powerful set of discourses that not only stigmatised the excluded but also perpetuated the likelihood of exclusion.

If Smith was the key theorist of civil society then he was also pivotal in enunciating its alternatives. In his 1571 pamphlet Smith distinguished between the Irish 'enemy', by whom he meant the indigenous warrior aristocracy, and the Irish 'husbandman', 'which they call churls'.[66] Smith did not doubt that these churls would, in 'great numbers', 'come and offer to live under us, and to farm our grounds: both such as are of the country birth, and others, both out of the wild Irish and the English pale'. This was in part because 'the churl of Ireland is a very simple and toilsome man' and in part because the exactions and oppressions of 'coin and livery' made legal tenancy, waged labour, and so the possibility of being 'Master of his own' eminently preferable. We have already seen that, as things stood, the mere Irish were the 'private' possession of men and so living in what Smith took to be a state of slavery, whereby 'a bondman or a slave is as it were (saving life and some reason) but the instrument of his Lord, as the axe, the saw, the chisel and gauge of the carpenter'. In this condition the Irish did not constitute a commonwealth or society because a commonwealth was, as he noted in *De Republica Anglorum*, 'a society of common doing of a multitude of free men collected together and united by common accord and covenants among themselves'. What Smith and subsequently Spenser and Sir John Davies offered the churl was, as Deborah Shuger puts it, the possibility of 'passing from feudal bondage to the status of a "free subject" with full legal rights and protection'.[67] This scenario depended, of course, on a particular conception of Gaelic lordship, the accuracy of which is less important than its perception through the lens of 'society'. What matters is that as early as 1571 Smith was offering the Irish the chance (as he saw it) of being 'admitted to the societie civil or commonwealth' rather than remaining 'the possession and goods of his Lord'; Smith's confidence was such that he feared 'the sweetness which the owners shall find in the Irish churl...will hinder the country much in the peopling of it with the English Nation'.[68] Yet by the time Blenerhasset published *A Direction* forty years later it seemed clear that many of the 'mere Irish' – churls or otherwise – were either unwilling or (perhaps) unable to make that transition. Instead Blennerhasset describes an indigenous population of 'Rebels', 'wood-kern' (i.e. outlaws), and 'many other (who now have put on the smiling countenance of contentment) [but who] do threaten every hour, if opportunity of time and place doth serve, to burn and steal whatsoever'.[69] The logic of monarchical republicanism suggested that, whether they liked it or not, the Irish remained 'slaves'.

The choice between feudal bondage and civil society so implied was not, however, as straightforward as Smith intimated. To appreciate this fully we need to return, finally, to the moment of Smith's initial rise to public prominence in the later 1540s under Lord Protector Somerset. Smith sat in the 1547 parliament and has been identified – speculatively but convincingly – as one of the authors of the infamous Vagrants Act of 1547.[70] This observed that 'idle and vagabond persons, being unprofitable members, or rather enemies of the commonwealth, had been suffered to remain and increase'. It opined that 'if they should be punished by death, whipping, imprisonment, and with other corporal pain, it were not without their deserts, for the example of others, and to the benefit of the commonwealth'. The statute nevertheless conceded that 'if they could be brought to be made profitable, and do service it were much to be wished and desired'. It was accordingly ruled that if any 'master' offered 'such idle person service, and labour, and that by him refused', and could 'prove the idle living of the person' with 'two honest witnesses' before a Justice of the Peace, then that person 'was to be marked with a hot Iron in the breast, the mark of V and to be his slave' for two years. The master was to give 'the said slave bread and water, or small drink, and such refuse of meat as he shall think meet [and] cause the said slave to work by beating'. 'Slaves' who absconded were to be marked on the forehead and enslaved in perpetuity; if they ran away again they were to be killed.[71]

Hiram Morgan has noted the importance of Sir Thomas More's *Utopia* on Smith's colonial theory and C. S. L. Davies has done the same for the Vagrancy Act.[72] Utopian citizens enjoyed communality and equality not because it was their right but because they were civil enough to enjoy it. Because they were inherently rational and industrious people they welcomed and internalised rules based on reason. As in ancient Athens and Rome, however, Utopian citizenship also depended on a system of slavery, which was both a punitive institution for those who lapsed in their civility and a source of labour.[73] Transposed by Smith onto Edwardian England, enforced slavery became an institution promising to combat vagrancy and idleness and to compel men, women and children into 'service'. Moreover, the breakdown of the English manorial system and concurrent development of a market economy meant that slavery was an incipient feature of social relations in any case. As Smith noted in *De Republica Anglorum*, 'necessity and want of bondmen hath made men to use free men as bondmen to all servile services: but yet more liberally and freely, and with more equality and moderation, than in time of gentility slaves and bondmen were want to be used'.[74] The position of the English commons as articulated in the sixteenth article of 'Kett's Demands' (issued during the Norfolk rebellion of 1549) was unequivocal: 'We pray that all bond men may be made free for God made all free with his precious blood shedding.'[75] Perhaps because of popular resistance the Vagrancy Act was subsequently repealed, though Smith and other members of a nominated committee recommended it be 'revived and duly put in execution' in 1558.[76] In the context of Ulster, the sentiments that shaped the legislation raised the prospect of three kinds of social stigmatisation for the 'mere Irish' after 1570 – as the 'very slaves and villeins' of tyrannical lords; as the 'idle' and 'unprofitable enemies of the commonwealth'; and as the 'servile' workers upon whose 'simplicity' and 'toil' the plantation economy ultimately depended.

The plantation of Ulster was, among other things, a corporate process. It involved men forming 'companies' and 'societies' for the good of both their commonwealth and their own profit. Sanctioned by royal authority and facilitated by military conquest, these companies looked to form civil societies in the specific sense of urban communities and in the more general notion of commonwealth in Ulster. Soldiers, undertakers, and urban inhabitants all presented themselves as civil citizens in the civic humanist sense of those terms. This was largely at the expense of the indigenous populace, who were conversely perceived as slavish in at least three respects: as bondsmen, vagrants, or servile workers. The fairly rapid marginalisation of Irish tenants to the poorer areas of the planted regions suggests that it was the latter notion of slavery that ultimately proved most pervasive, and that religious and ethnic conflict in Ulster need to be considered in relation to the kinds of social exclusion and economic inequality that characterised the simultaneous incorporation of England.[77] Corporations would create 'sweet society', as Blenerhasset promised; they would also ensure 'the generation of the Irish...be sufficiently bridled'.[78] All of which suggests that the paradox of 'monarchical republicanism' lay not simply in the insemination of confessional hatred, the potential for Machiavellianism, nor the capacity of 'martial men' to slip the civic leash. It was inherent to the concept of 'civil society' itself.

NOTES

1 Patrick Collinson. 'The monarchical republic of Queen Elizabeth I', *Bulletin of the John Rylands University Library of Manchester*, 69 (1987), 394–424, reprinted in his *Elizabethan Essays* (London, 1994), pp. 31–57 (text cited). Patrick Collinson, *De Republica Anglorum, or History with the Politics Put Back* (Cambridge, 1990).
2 Collinson, 'Monarchical republic', p. 43.
3 Markku Peltonen, *Classical Humanism and Republicanism in English Political Thought 1570–1640* (Cambridge, 1995); Cathy Shrank, *Writing the Nation in Reformation England, 1530–1580* (Oxford, 2003); Jennifer Richards, *Rhetoric and Courtliness in Early Modern England* (Cambridge, 2003); Andrew Fitzmaurice, *Humanism and America: An Intellectual History of English Colonisation 1500–1625* (Cambridge, 2003); John F. McDiarmid (ed.), *The Monarchical Republic of Early Modern England: Essays in Response to Patrick Collinson* (Aldershot, 2007).
4 Stephen Alford, *The Early Elizabethan Polity: William Cecil and the British Succession Crisis 1558–1569* (Cambridge, 1998); Mark Goldie, 'The unacknowledged republic: office-holding in early modern England', in Tim Harris (ed.), *The Politics of the Excluded, c.1500–1850* (Basingstoke, 2001); Phil Withington, *The Politics of Commonwealth: Citizens and Freemen in Early Modern England* (Cambridge, 2005); idem, 'Public discourse, corporate citizenship and state formation in early modern England', *American Historical Review*, 112:4 (2007) 1016–38; Richard Cust, 'Reading for magistracy: the mental world of Sir John Newdigate', in McDiarmid (ed.), *Monarchical Republic*, pp. 181–200; idem, 'The "public man" in later Tudor and early Stuart England', in Peter Lake and Steve Pincus (eds), *The Politics of the Public Sphere in Early Modern England* (Manchester, 2007), pp. 116–43.
5 Deborah Shuger, 'Irishmen, aristocrats, and other white barbarians', *Renaissance Quarterly*, 50:2 (1994), 494–525.
6 Ciaran Brady, 'Spenser's Irish crisis: humanism and experience in the 1590s', *Past & Present*, 111 (1986), 17–49; Nicholas Canny, *Making Ireland British 1580–1650* (Oxford, 2003), 1–55; Peltonen, *Classical Humanism*, 54–118.
7 Brendan Bradshaw, 'Review: the Tudor commonwealth: reform and revision', *Historical Journal*, 22:2 (1979), 455–76; idem, 'Transalpine humanism', in J. H. Burns and Mark Goldie (eds), *The Cambridge History of Political Thought 1450–1700* (Cambridge, 1991), pp. 95–113.
8 Peltonen, *Classical Humanism*, p. 72; John Barston, *The safegarde of societie: describing the institution of lavves and policies, to preserue euery felowship of people by degrees of ciuill gouernement: gathered of the moralles and policies of philosophie* (1576).
9 Peltonen, *Classical Humanism*, p. 74; Richard Beacon, *Solon his follie, or a politique discourse, touching the reformation of common-weales conquered, declined or corrupted* (1594).
10 Nicholas Canny, 'Spenser's Irish crisis: humanism and experience in the 1590s', *Past & Present*, 120 (1988), 201–9; Andrew Hadfield, *Edmund Spenser's Irish Experience: Wilde Fruit and Salvage Soyle* (Oxford, 1997).
11 Rory Rapple, *Martial Power and Elizabethan Military Culture: Military men in England and Ireland, 1558–1594* (Cambridge, 2009), p. 84.

12 Michael J. Braddick, *State Formation in Early Modern England, c.1550–1700* (Cambridge, 2000), pp. 379, 388.
13 William J. Smyth, *Map-Making, Landscape and Memory: A Geography of Colonial and Early Modern Ireland* (Cork, 2006), p. 15.
14 Canny, *Making Ireland British*, pp. 55–8.
15 Edmund Spenser, *A view of the state of Ireland, written dialogue-wise betweene Eudoxus and Irenæus, by Edmund Spenser Esq. in the yeare 1596. Whereunto is added the history of Ireland by Edmund Campion, sometime fellow of St Iohn's Colledge in Oxford*, ed. James Ware (Dublin, 1633); Thomas Smith, *Letter sent by I. B. Gentleman vnto his very frende Maystet [sic] R. C. Esquire, vvherin is conteined a large discourse of the peopling & inhabiting the cuntrie called the Ardes, and other adiacent in the north of Ireland, and taken in hand by Sir Thomas Smith one of the Queenes Maiesties priuie Counsel, and Thomas Smith Esquire, his sonne* (1571); Thomas Blenerhasset, *A direction for the plantation in Vlster* (1610).
16 Thomas Smith, *De Republica Anglorum. The maner of gouernement or policie of the realme of England* (1583), ed. Mary Dewar (Cambridge, 1982); Collinson, *De Republica Anglorum*, passim.
17 Shrank, *Writing the Nation*, pp. 143–181; Richards, *Rhetoric*, pp. 78–86, 101–6; Ian W. Archer, 'Smith, Sir Thomas (1513–1577)', in *Oxford Dictionary of National Biography*, online edition, ed. Lawrence Goldman (Oxford, 2004) (*ODNB*) www.oxforddnb.com/view/article/25906 (accessed 30 September 2011).
18 Richards, *Rhetoric*, pp. 113–38; Canny, *Making Ireland British*, pp. 1–9; Andrew Hadfield, *Edmund Spenser's Irish Experience*; idem, 'Spenser, Edmund (1552?–1599)', in *ODNB*, www.oxforddnb.com/view/article/26145 (accessed 30 September 2011).
19 Sidney Lee, 'Blenerhasset, Thomas (c.1550–1624)', rev. Andrew Hadfield, in *ODNB*, www.oxforddnb.com/view/article/2636 (accessed 30 September 2011).
20 Thomas Blenerhasset, *The seconde part of the Mirrour for magistrates, conteining the falles of the infortunate princes of this lande. From the conquest of Cæsar, vnto the commyng of Duke William the Conquerour* (1578); Scott Lucas, '"Let none such office take, save he that can for right his prince forsake": *A Mirror for Magistrates*, Resistance Theory and the Elizabethan Monarchical Republic', in McDiarmid, *Monarchical Republic*, p. 107.
21 D. J. B. Trim, 'The art of war: martial poetics from Henry Howard to Philip Sidney', in Mike Pincombe and Cathy Shrank (eds), *The Oxford Handbook of Tudor Literature 1485–1603* (Oxford, 2009), pp. 587–605; Rapple, *Martial Power*, pp. 56–8.
22 Phil Withington, '"For This is True or Els I do Lye": Thomas Smith, William Bullein and the mid-Tudor Dialogue' in Pincombe and Shrank, *Oxford Handbook*, pp. 455–71.
23 Smith, *Letter*, sig Eiii.
24 Richard Bagwell, *Ireland under the Tudors: With a Succinct Account of the Earlier History* (3 vols, London, 1885), ii, p. 211.
25 Hiram Morgan, 'The colonial venture of Sir Thomas Smith in Ulster, 1571–1575', *Historical Journal*, 28:2 (1985), 263, 265, 266, 267.

26 Karen Ordahl Kupperman, *Providence Island 1630–1641: The Other Puritan Colony* (Cambridge, 1993), p. 19; Fitzmaurice, *Humanism and America*, p. 7.
27 Smith, *De Republica Anglorum*, p. 65.
28 For 'subaltern' see Richard Mulcaster, *The first part of the elementarie vvhich entreateth chefelie of the right writing of our English tung* (1582), pp. 12–14.
29 Smith, *De Republica Anglorum*, p. 57.
30 Shuger, 'Irishmen, aristocrats', p. 494; Morgan, 'Colonial venture', pp. 262, 274–5.
31 Smith, *Letter*, sigs Eiii, Fi.
32 Ibid., Fiv, Gi, Gii.
33 Morgan, 'Colonial venture', pp. 269–70.
34 Thomas Smith, *A Discourse of the Commonweal of the Realm of England* (written c.1549), ed. Mary Dewar (Charlottesville, 1969); Keith Wrightson, *Earthly Necessities: Economic Lives in Early Modern Britain* (New Haven, 2000), pp. 154–6.
35 Essex Record Office (ERO), D/Sh/01/7, 'Offices Necessarie in the Colony of the Ardes and Orders Agreed Upon' (1572).
36 Canny, *Making Ireland British*, pp. 121–3.
37 Spenser, *A view of the state of Ireland*, pp. 4, 48–50, 53–6, 101.
38 Ibid., pp. 103, 106, 109.
39 Spenser, *A view of the state of Ireland*, pp. 88–9.
40 Ibid., p. 116.
41 Ibid., pp. 109–10, 11.
42 Ibid., pp. 116–17.
43 Blenerhasset, *A direction*, C2.
44 Ibid., B2ii.
45 Ibid., B2i–ii.
46 Ibid., C, D.
47 Phil Withington, *Society in Early Modern England: The Vernacular Origins of Some Powerful Ideas* (Cambridge, 2010), ch. 5.
48 Thomas More, *The supplycacyon of soulys. Made by syr Thomas More knyght councellour to our souerayn lorde the Kynge and chauncellour of hys Duchy of Lancaster. Agaynst the supplycacyon of beggars* (1529), ii, p. 19; Thomas Elyot, *The boke named the Gouernour* (1531), iii, sig. yiv.
49 Barston, *Safeguard*, Avr, p. 111r.
50 Ibid., B, Bi, pp. 60–6.
51 Ibid., p. 29.
52 Ibid., Bi.
53 Ibid., Bii, p. 112.
54 For a fuller discussion of the method see Withington, *Society in Early Modern England*.
55 Angel Daye, *The English secretorie. VVherein is contayned, a perfect method, for the inditing of all manner of epistles and familiar letters, together with their diuersities, enlarged by examples vnder their seuerall tytles* (1586).
56 Henry Manship, *The History of Great Yarmouth*, ed. Charles John Palmer (London, 1854), p. 11.
57 ERO, D/DSh/1/7.

58 Robert Tittler, *The Reformation and the Towns in England: Politics and Political Culture, c.1540-1640* (Oxford, 1998); Withington, *Politics of Commonwealth*, passim.
59 Paul Slack, *From Reformation to Improvement: Public Welfare in Early Modern England* (Oxford, 1999), pp. 6-28; Keith Wrightson, *Earthly Necessities*, pp. 132-58.
60 Steve Alford, *The Early Elizabethan Polity: William Cecil and the British Succession Crisis, 1558-1569* (Cambridge, 1998).
61 Withington, *Politics of Commonwealth*, pp. 16-50.
62 Jane Ohlmeyer, '"Civilisinge of those Rude Parts": Colonization within Britain and Ireland, 1580s-1640s' in Nicholas Canny (ed.), *The Oxford History of the British Empire*, vol. I, *The Origins of Empire: British Overseas Enterprise to the Close of the Seventeenth Century* (Oxford, 2001), p. 146.
63 For urbanisation as structural and cultural as well as simply demographic enlargement see Jan de Vries, *European Urbanization 1500-1800* (London, 1984), pp. 10-17.
64 T. W. Moody, *The Londonderry Plantation 1609-1641: The City of London and the Plantation of Ulster* (Belfast, 1939); Philip Robinson, *The Plantation of Ulster* (Belfast, 1984), pp. 80-2.
65 Jonathan Barry, 'Civility and civic culture in early modern England: the meanings of urban freedom', in Peter Burke, Paul Harrison and Paul Slack (eds), *Civil Histories: Essays Presented to Sir Keith Thomas* (Oxford, 2000), pp. 181-97.
66 Smith, *Letter to I. B.*, sig. D1ii.
67 Shuger, 'Aristocrats', p. 515.
68 Smith, *De Republica Anglorum*, p. 57; *Letter to I. B.*, sig. D1iii.
69 Blenerhasset, *A Direction*, B.
70 C. L. S. Davies, 1966. 'Slavery and Protector Somerset: the Vagrancy Act of 1547', *Economic History Review*, 19:3 (1966), 533-49.
71 England and Wales, *Statutes made in the Parliament, begon at Westminster the fourthe daie of Nouember, in the firste yere of the reigne of our moste dread souereigne lord Edvvard the. VI* (1548).
72 Morgan, 'Colonial venture'; Davies, 'Slavery', pp. 541-2.
73 Thomas More, *Utopia*, ed. George M. Logan and Robert M. Adams (Cambridge, 1999), pp. 80, 83.
74 Smith, *De Republica Anglorum*, p. 142.
75 Cited in Anthony Fletcher and Diarmaid MacCulloch, *Tudor Rebellions* (5th edn, London, 2008), p. 158.
76 Davies, 'Slavery', p. 544.
77 Keith Wrightson, *English Society 1580-1680* (London, 1982), pp. 222-7.
78 Blenerhasset, *A Direction*, sig. B2ii.

5

The city of London and the Ulster plantation

IAN W. ARCHER

On Saturday 28 February 1635 the government of Charles I delivered a mighty humiliation to the magistrates of the city of London. They were found guilty in the Court of Star Chamber of a variety of offences relating to their management of the Londonderry plantation. The city was accused of having falsely procured the charter of incorporation of the Irish Society, 'many things being inserted which were not in the warrant'; of having failed to observe the conditions of the articles of plantation, principally 'by not building houses in number or quality according to their covenants', and by failing to provide adequate fortifications so that the king was put to unnecessary expense; of having destroyed woods of an inestimable value; and of failing to provide adequate glebe land for the churches on the plantation. The lords declared that the lands of the Londonderry plantation were forfeit and that the city should pay a swingeing fine of £70,000.[1] The case was a *cause célèbre*. As the Reverend George Garrard wrote to Thomas Wentworth, the king's lord deputy in Ireland, 'this business is of great consequence; the eyes of all men have been fixed upon it, never in our time hath a case been fourteen days handled in the star chamber before'.[2] The newsletter writers suggested that the Recorder Robert Mason who presented the city's case had the better of the argument, winning an 'immortal fame by the well managing of the city's defence'; 'he did please the auditory so well that they gave a loud "Hem" to the disturbance of the court'.[3]

The progress of the Londonderry business was closely watched by the king himself who had presided over a meeting of the Privy Council two days before the verdict at which the case was fully discussed.[4] Also watching closely was the city's great adversary, Sir Thomas Phillips, former governor of Coleraine and proprietor of Limavady in the heart of the plantation, who had fought the city every inch of the way over its alleged betrayal of the plantation project.

The day after the sentence was passed, Phillips wrote to Bishop Bramhall of Derry: 'I have after twenty-five years employment in this business accomplished my long and tedious labour against the great city of London.' Bramhall responded warmly: 'To surprise that great and rich city after so long a siege was a work worthy of yourself.'[5] But contemporaries knew that the city had been harshly treated. Wentworth had pressed Secretary Coke for the humbling of the city with a keen eye on the customs revenue of Londonderry and Coleraine currently enjoyed by the Irish Society, 'a feather...not fit to be worn in the round cap of a citizen of London'.[6] But even he warned Charles that the Londoners had laid out 'great sums upon the plantation, and that it were...very strict in their case...if the uttermost advantage were taken'.[7]

If contemporaries were well aware of the significance of the case, historians have been curiously neglectful of it. The story of the plantation has been recognised as a formative experience in the history of Ireland, but its significance for English political developments is rarely discussed.[8] The Ulster plantation figures in most livery company histories but it has not been properly incorporated in the narratives of city–crown relations.[9] It merits a very brief mention in Valerie Pearl's *London and the Outbreak of the Puritan Revolution*; it is only sketchily present in Keith Lindley's and Robert Brenner's accounts of popular and merchant politics respectively; only Robert Ashton gives it due weight.[10] Within the general accounts of the origins of the civil war it has very low visibility. It is not mentioned by either Conrad Russell in *The Fall of the British Monarchies*, or by John Adamson in *The Noble Revolt*, or even by Kevin Sharpe in his account of *The Personal Rule of Charles I*.[11]

This neglect by three door-stopping books is curious because the verdict and its aftermath seem to have played an important role in the destabilisation of Charles I's rule. Although as was customary in Star Chamber cases, the Londoners had their fine mitigated (it was reduced to £12,000), they had to endure the loss of their Irish estates which, but for the mounting attacks on their position from the mid-1620s onwards, might have been yielding a healthy profit after years of haemorrhaging capital expenditure. Doubtless the city exaggerated the financial impact of the confiscation and fine, but they were invoked in the closing years of the Personal Rule as a reason for not lending money to the crown. In February 1639 the earl of Cork was told that city's reluctance to lend money for the prosecution of the Bishops' Wars was 'in regard of the many taxes imposed upon them and [loss] of their land in Londonderry'.[12] In August 1640 when Charles turned in desperation to the livery companies for financial support, most of them claimed that they were unable to assist him because the plantation had 'consumed their stocks'.[13] In November Samuel Vassall, one of their MPs, said that 'this business of Londonderry had much exhausted the city, and that this sticks upon them to £160,000'.[14] They took the opportunity of the assault on the instruments and policies of the Personal Rule to launch their own campaign against the Star Chamber verdict, preparing a petition in January 1641 comprehensively refuting the charges brought against them, and declaring the proceedings 'illegal and irregular'.[15] They were vindicated when on 26 August 1641 a committee of the House of Commons upheld their position, ordering that the fine be repaid, the rents handed back, and the property restored.[16] The Londonderry plantation was at the heart of Charles I's attempted rapprochement with the city on 25 November 1641 when on the occasion of the superbly choreographed royal entry, the king declared that 'one thing I have thought of as a particular affection to you which is to give back to you freely that part of Londonderry which heretofore was evicted from you'. He had of course ruefully to admit that 'this, as that kingdom is now, is no great gift, but I hope to recover it first, and then to give it you whole and entirely'.[17] It was a gesture in the right direction, but as Clarendon suggested, the Londoners imputed the concessions to the 'power of Parliament and remembered how they had been taken from them rather than by whom it was restored'.[18]

The case had clearly done great damage to Charles's standing in his capital. But in Ulster the fall-out was, as Jane Ohlmeyer has shown, nothing short of disastrous. The king's commissioners had driven up rents and called in the leases of some of the greater landowners, among them Sir John Clotworthy, tenant to the Drapers' Company lands. The 'extreme and cruel usage' of these commissions was a key element in the remonstrance of the Irish House of Commons against Wentworth (now the earl of Strafford) on 7 November 1640, and Clotworthy 'not otherwise to be named but as a firebrand brought from Ireland to inflame this kingdom', was to become one of Strafford's leading opponents in the Long Parliament, orchestrating the impeachment case. As Ohlmeyer sums up, 'by sequestering the City's plantation in the first place, and then by trying to extract whatever profit he could from it Charles dragged issues, previously confined to the periphery, into the central arena, and in so doing inadvertently helped to destabilise royal government in Dublin, Edinburgh, and London at a critical point in his reign'.[19]

So, how had things reached the sorry pass of the Star Chamber suit? The answer lies in a mixture of factors: the city's initial reluctance to undertake the plantation; the unrealistic expectations placed upon the magistrates; the difficulties they faced in managing their affairs at a distance; the fact they were too tempting a target for a cash-strapped monarchy; and the tensions between aristocratic and civic values their involvement brought into focus. Nonetheless, the Londoners' achievement was in the circumstances by no means negligible, and they brought a distinctive approach to the central issue of urbanism, seen as crucial to the civilising mission in Ireland.

The city claimed repeatedly that it had never desired the plantation but had undertaken it out of deference to the crown's wishes, 'an endeavour of obedience not of contract'. It is not clear where the impetus for the city's involvement came from, for it was certainly not part of the original plan for the plantation, and not an element in the discussions over the shape of the plantation between Sir Arthur Chichester, Sir John Davies and Sir Francis Bacon.[20] It may well have been the lord treasurer, the earl of Salisbury, who saw the advantages of bringing in the Londoners, 'the ablest body to undergo so brave and great a work as the plantation of that country'.[21] Sir Thomas Phillips claimed that he gave Salisbury a sense of the profits that might redound to the Londoners, and in this period of gestation Phillips was certainly responsible for some of the more optimistic assessments of the prospective revenues.[22] Perhaps Salisbury grasped the commercial potential of the area around Londonderry, and saw the Londoners' involvement as the fittest means to realise it. Certainly the 'Motives and Reasons to induce the city of London to undertake the plantation in the north of Ireland', dated 25 March 1609 and soon to be circulated to all the livery companies, suggested that Londonderry and Coleraine would be suitable places for the establishment of corporate towns whose fortifications would be crucial to the security of the plantation but which would also sit at the centre of a flourishing commercial system. The text waxed lyrical about Ulster's agricultural and marine commodities: 'it yieldeth store of all things necessary for man's sustenance in such measure as may not only maintain thyself but also furnish the City of London yearly... the sea fishings of that coast are very plentiful of all manner of usual fishings'. The towns would lie at the centre of a commercial network linking Ireland with England, Scotland, Spain and Newfoundland. And, in an attempt to push all the buttons, it played to Londoners' anxieties about their own growth and overcrowding. By providing employment opportunities, the new towns would also 'reap a singular commodity... by easing of an insupportable burden which so overchargeth all the parts of the City that one tradesman can scant live by another which in all probability would be a means also to free and preserve the City from infection'.[23] It is an indication of the crown's seriousness of purpose about the economic development of the area that from the outset it was prepared to make major commercial concessions to the city of London, granting it the customs revenues from the ports, and the lucrative salmon fisheries and Admiralty jurisdiction for 21 years.[24]

The propagandists and panegyrists warmed to the theme. According to Thomas Blenerhasset, the Londoners, 'successors of high renowned Lud' would 'there re-edify a new Troy... They have O'Cahan's country and whatsoever Ireland's Eden can afford, and therefore even in respect of their own reputation, they of themselves will perform this the most honourable action that ever they attempted. Therefore let Coleraine rejoice for the heart of England (London herself) will no doubt make her more beautiful than many and furnish Lough Foyle with a goodly fleet'.[25] The citizens themselves were rather more sceptical. There may have been some enthusiasm for the project among aldermen with Irish business interests like the draper, John Jolles, and the skinner, William Cockayne, both of whom had profited from army victualling contracts, and to whom the government turned in the ensuing months; Cockayne was to be governor of the Irish Society for four years. But the perspective of the Londoner on the streets is captured in Barnabe Rich's *New Description of Ireland* (1610), dedicated to Salisbury and Cockayne, and claiming that within six days of his arrival in London from Dublin 'I was asked sixteen several times what I thought of this plantation in the north of Ireland, and whether it were possible that those labourers and workmen that are now sent over for the building could save their throats from cutting, or their heads from being taken from their shoulders before the work were finished'.[26]

When the embryo project was unveiled to the livery companies in July 1609, and individuals invited to adventure, there was a marked lack of enthusiasm.[27] The Mercers were perhaps the frankest. While thanking the king for his offer, they pointed out that 'they are for the most part men that live by merchandise and therefore are very inexperienced in managing business of that nature and withal want means and ability for the accomplishment thereof. [So] this company are not willing to have a hand or intermeddle in the same'.[28] The Ironmongers expressed their 'desire with our best means to help the state and commonwealth, but what we would we cannot in respect of weakness'.[29] When it came to attempts to generate subscriptions, members were curiously absent or unavailable because they were dwelling out of the city. Of the 46 men on the Ironmongers' subscription list, 9 were absent, 10 out of the city, and 2 allegedly 'not of ability'. The story was much the same elsewhere.[30]

It was quite clear that voluntary subscriptions were not going to work, and once the scale of the capital requirements became clear the aldermen and Common Council fell back on the expedient of compulsory levies on the 55 livery companies through which much of London's economic life was regulated.[31] The leading 12 companies which provided 80% of the requirement were dominated at their upper levels by wholesalers and prosperous retailers, though some like the Merchant Tailors and the Clothworkers had large numbers of restless artisans among the rank-and-file. It was not unusual, however, to turn to the companies for support for royal and civic projects. Prior to 1585 the city had used the companies as a mechanism for financing and levying troops; members were also expected to subscribe to the corn monies which financed grain purchases made by the city for resale at below market prices to the poor. It was to the quotas for corn money that the Common Council turned as a basis for the assessments on the companies to support the Londonderry plantation.[32]

What was unusual was the scale and frequency of the calls for money. On 15 December 1609 London's Common Council announced its intention of financing the plantation with a levy of £15,000 on the companies, but the Privy Council made it clear that this would be insufficient, and the city raised its offer to £20,000 a week later to be paid in two instalments by February 1611. But this proved inadequate to the task and additional levies were authorised, £10,000 on 10 July 1611, £10,000 on 30 April 1613, £5,000 on 17 December 1613, £7,500 on 11 January 1615, £5,000 on 18 October 1615, and £2,500 on 1 October 1616 – so that by 1616 the companies had coughed up £60,000.[33] We need to put these figures into perspective. They were small in comparison with the huge amounts invested in East India Company stock: £418,691 in the first joint stock of 1613 and £1.63 million in the second joint stock of 1617.[34] But they were large in relation to compulsory levies. The national yield of a parliamentary subsidy (the main direct tax) in the first decade of the seventeenth century was £67,000, while London's total direct tax burden (on a fairly inclusive definition) in the 1590s has been estimated at only £12,818 per annum.[35] So, it is not surprising that the levies were collected with some difficulty. The Mercers complained of the inequity in basing the assessments on the outdated corn quotas which did not reflect the shifting distribution of wealth between the companies (the Mercers were a shadow of their former selves). Rebuffed by the aldermen, they eventually took their case to the Privy Council. It eventually secured a revision in the quotas but only after their wardens endured a night's imprisonment.[36] There were regular cases of defaulters being sent to prison on the lord mayor's commandment until they paid up. In August 1610 the Grocers threatened all defaulters with the loss of the benefit of their freedom, an indication that the stakes were really high, as such a sanction debarred them from taking apprentices, making freemen, or engaging in retail trade.[37] As the scale of the demands escalated, companies began to pay the money from their general stock by depleting their reserves. In several cases the companies corporately took over individual subscriptions.[38] That they were able to do so is an indication of the relative financial health of the livery companies in the early seventeenth century, a result of their swelling property portfolios and rising rental income. They were taking on more charitable trusts and found their coffers replenished by entry fines (often as much as the rental income) which could be spent on general purposes, as trusts were discharged from basic rents.[39]

Not only were the companies reluctant to pay their assessments, they were also very hesitant about the request in early 1611 that they accept proportionate shares of the escheated estates in the plantation in lieu of their subscriptions. If they accepted they would be entitled to dividends on their investments in the general plantation at Londonderry and Coleraine, but if they did not, they would still be liable for all the charges for the development of the general plantation.[40] The Clothworkers, Haberdashers, Mercers, Merchant Taylors, and Skinners initially refused, the Clothworkers doubtless speaking for the rest when they declared that 'with such difficulty have payments in past been gathered and payment has been so unsavoury to everyone, and with so hard hand drawn from them that the master and wardens do not see any hope of collecting any more money'. In the end the resisting companies capitulated, although the Haberdashers immediately transferred their interests to two members, William Freeman and Adrian Moore.[41]

The propaganda accompanying the plantation stressed not only that the plantation would be to the city's profit, but that it would also redound to its honour. It was an action that would prove both 'honestum et utile', 'a matter tending to their present honour and future commodity'.[42] The city's honour was to be advanced through nothing less than a civilising mission. Londoners were urged to consider 'the place being in former times the nest of rebellion and now the parties become fugitive it [is] easy by plantation to civilise the same'.[43]

There was, of course, a wide gap between expectation and fulfilment in the civilising mission that informed the plantation, but it is worth dwelling upon the distinctive contribution of the Londoners, in particular their contribution to urban development. Towns were regarded as essential agents of civility; the corollary was that civility was a core element in urban identity. As the author of 'The Apology of London', appended to John Stow's *Survey* (1598) put it, 'men are congregated into cities and commonwealths for honesty and utility's sake...First, men by this nearness of conversation are withdrawn from barbarous ferity and force to a certain mildness of manners and to humanity and justice...Also the doctrine of God is more fitly delivered and the discipline thereof more aptly to be executed, in peopled towns than abroad, by reason of the facility of common and often assembling...And whereas commonwealths and kingdoms cannot have, next after God, any surer foundation than the love and good will of one man towards another, that also is closely bred and maintained in cities, where men by mutual society and companying together, do grow to alliances, commonalities, and corporations'.[44] In the specific Irish context, Edmund Spenser made the connection between urbanism, civility, security and prosperity. 'Nothing doth sooner cause civility in any country than many market towns by reason that the people repairing often thither for their needs will daily see and learn civil manners...besides there is nothing doth more stay and strengthen the country...than many towns.'[45]

The crown envisaged 25 corporate towns in the plantation; by 1613 only 14 had been established, and the scheme eventually produced 16.[46] Londonderry and Coleraine, the responsibility of the Londoners, stood in a league of their own. By the terms of the agreement with the crown, the city was required to provide 200 houses at Londonderry with room for 300 more, and 100 at Coleraine with room another 200. The number actually built was always a matter of contention: whereas Phillips reported only 102 houses at Londonderry in 1622, the city claimed a total of 219. The explanation may be that the London authorities had built small houses with single rooms and attics. The inhabitants united them to make more viable units, so that there were only 121 households within the walls. For its part, the city claimed that the houses had been amalgamated not because of their small size, but because of the allocations of land that went with them.[47]

By comparison with the mother metropolis, Londonderry and Coleraine were only anaemically urban, but in the Ulster context they encapsulated the aspirations to social engineering of the plantation's promoters. With 500 males at Londonderry and 300 in Coleraine they were the largest settlements in the plantation; the next largest Ulster town was Strabane at 208.[48] Elsewhere a major obstacle to urbanisation was that the settlers were expected to live upon their lands. As a 1619 commentator put it, 'all the tenants do dwell dispersedly... and cannot dwell together in a village because they are bound everyone to dwell upon his own land'.[49] The threshold for 'urbanism' could be extraordinarily low. George Canning, agent for the Ironmongers' plantation claimed that the six houses he had constructed at Athgeave were sufficient 'in this place to be called a town'![50] So, in spite of their modest size Londonderry and Coleraine were exceptional. Their occupational structures were more diversified, and although they did not become the thriving metropoloi envisaged by the propaganda of 1609-10, they did enjoy a significant mercantile presence. Merchants from Scotland, Chester and London were soon frequenting the two ports, while as early as 1614-15 a merchant fleet of seven ships accounted for 18.5% of Londonderry's exports.[51] Londonderry boasted urban amenities not available elsewhere. Its streets were paved; it had a town hall costing between £500 and £1,000; its school was founded by the London merchant Matthew Springham, its master receiving a salary of 20 marks per annum through the Irish Society; its cathedral church of St Columba, the first purposely built Protestant cathedral in the three kingdoms, costing at least £3,800 opened in 1633 with a capacity of 1,000 people. True, Londonderry lacked other key features found in English towns: there was still no bridge; a recommendation that a bridewell should be built was resisted; and there were no almshouses; indeed there was little sign of any charitable activity at all.[52] A key variable in determining the relative success of Londonderry and Coleraine was the fact that the landlord was directly involved in building whereas elsewhere in the plantation urban development was promoted through the granting of building leases. Urban settlements elsewhere were terribly under-capitalised.

The two towns were part of the so-called general plantation for which the Irish Society took responsibility, while in the rural areas the livery companies undertook to develop of the portions allocated to them. As major undertakers they were expected to build a castle and a bawn (that is a fortified wall) on their proportions, and to populate them with British tenants, displacing the natives. Quite apart from the doubtful feasibility of this social engineering, to which we shall return, the envisaged plantation posed formidable problems of management, and whether the companies could overcome them depended to a very considerable extent upon the quality of the agents appointed. Management of the plantation from a distance meant that issues of trust were paramount. Correspondence between the high command in London and the agents on the ground is full of assurances of loyal and disinterested service, reinforced by personal ties. Lieutenant Perkins, the Ironmongers' first agent, repeatedly stressed his duty to the company 'whereof I am a poor member' ... 'being a freeman of the company brought up by Mr Garton and Mr Huntley while they lived' (both of them assistants and company benefactors), but all this was part of a carefully crafted rhetoric designed to secure a permanent post with a salary of 100 marks per annum.[53] 'Whatsoever it shall please them to allow me more than I shall ask I will as well in all respects deserve it as any man that shall be employed. Otherwise let me lose the dignity of an ironmonger which cost me many years service.'[54] In the event Perkins was dropped, and the company turned to George Canning, a gentleman from Barton in Warwickshire with an income of £100 per annum, and the brother of one of the assistants, William Canning.[55] In managing their estates the Londoners fell back on the kind of kinship ties which inflected so many of their business dealings.

Canning, as we shall see, probably did his company good service, but he used the position to leverage a lucrative tenancy for himself. Other company agents were much less successful in balancing their own interests against those of the company. The Drapers employed John Rowley, a freeman of their company, but he was the discredited general agent whose frauds had been exposed by the city's delegation of Matthew Springham and George Smith in 1613.[56] Rowley was assisted by Robert Russell, a man full of bright ideas like that of making a canal from Lough Neagh to the Drapers' proportion, but never able to bring them to fruition. His brewhouse project depended on paying his workers' wages in barrels of beer and maintaining three taphouses which seemed to reduce the villagers to a state of permanent inebriation.[57]

There is no doubt, however, that the companies were being set extraordinarily demanding tasks in fulfilling the articles of the plantation, and we can see something of the difficulties through an examination of one of the proportions where the survival of correspondence between the agent and the London Company is very good. The Ironmongers' proportion comprised 19,540 acres in the barony of Coleraine, but it was fragmented, being split up by church lands, a 'great maim' as Lieutenant Perkins had put it. The church lands, he had warned, 'will be as suckers at the root of a tree and will waste much of the sap if they be not pruned off in time'.⁵⁸ One worry was that as the natives were removed they would take up residence on the bishop's land, and 'having the best knowledge may eat out some part upon the same'.⁵⁹ The resolution of boundaries was important but held up by the death of the bishop of Derry, John Tanner, and by the truculence of the ubiquitous John Rowley, tenant of the bishop's land.⁶⁰

Perkins had identified a suitable site for the main settlement with the castle and bawn complex required by the articles. Athgeave was considered suitable 'in respect of the goodness of the soil, the levelness of the land about it, and the portableness of the rivers which will help much for transportation of materials in time of building and yield great relief to the castle in time of trouble or rebellion'.⁶¹ But the main problems were a shortage of suitable building materials, the high costs of transport, and the unreliability of local suppliers. Timber was plentiful but in 'no way suitable for carriage'. There was no locally available source of stone or lime, and these had to be brought from Coleraine. The poor quality of local lime had been the undoing of the Mercers, whose first castle collapsed almost immediately on completion.⁶² In November 1614 Canning contracted with Edward Heyward of Coleraine for timber, slate, limestone and lath, but the merchant was unable to fulfil the terms of the contract because of the 'extremity of the weather...the waters so extreme with the abundance of snow that it was impossible to pass anything upon the bar' (at Coleraine).⁶³ When the stone did arrive it was of poor quality. Of the castle, Canning wrote that 'it doth almost stand of necessity to make your building round because the stone is so bad it will not make quoynes for the corners'. Better stone was available on the Fishmongers' proportion, but the cost of carriage was too high. As for bricks, 'I find they prove not so good in this country as about London and in all men's judgement will not be half so strong as that which is made of stone.'⁶⁴ It was eventually agreed that the lower levels of the castle should be of stone, and the upper of brick. Good workmen were also in short supply. Canning looked forward to employing Peter Benson, the bricklayer who had worked on the original walls of Londonderry, and expressed satisfaction with his choice of carpenter, Simon Mortimer, who would work for 9s. per week 'whereas other workmen will not work under 12s. per week and not so good neither'.⁶⁵ But he recognised the crucial importance of promptly paying wages to secure loyal service: 'It is a hateful thing to set poor men to work and not to pay them their hire in due time, but let them cry and complain their wages when they have earned it. It hath been too much practised in these parts since the beginning of the plantation.'⁶⁶ Linked to the issue of payment of wages was the scarcity of money. Canning took up money by bill of exchange, but he was often left waiting for instructions from the company, and one had to move quickly. Taking up £100 from Tristram Beresford in February 1616 (money which was to be repaid by exchange to Alderman Bennett), he remarked that 'if I had not spoke of it when I did, the Fishmongers' agent had had it; I am very beholding to Mr Beresford for he was very willing to supply my wants'.[67]

Money was an ever present anxiety as Canning sought to balance the obligation to provide good quality buildings against the demands of economy. He was concerned about the cost of his own house (between £60 and 100 marks, comprising two large rooms above and two below and measuring 38' × 18'),[68] and the company warned him that the lesser companies in partnership with the Ironmongers might question the expense, 'he being a brother of the Ironmongers'.[69] In seeking more money he was keen to emphasise his pursuit of economy. 'Let me have supply as shall be fitting for the plantation of so worthy a society for it is not expected that the Londoners' buildings shall be performed like unto private undertakers, yet I will presume and pawn my credit to you forever that your plantation shall be performed with as little expense of money (the remoteness of the materials considered) as any of the twelve proportions and as well to the content of those that shall have view of the plantation.'[70] The company wanted to pursue economy but it also wanted buildings that would stand it in good credit, as there was a certain degree of rivalry among the Great Twelve, which encouraged compliance with the articles of plantation.[71] As Perkins had put it, 'it behoves our Company to be forward that they be not behind other in their plantation both for building and inhabiting their lands'.[72] In 1614 the Merchant Taylors explained to their agent George Costerdyne, customer of Coleraine, that they believed that 'no other company shall go before us either in good husbandry as well managing these affairs'.[73]

In spite of all the obstacles Canning was reasonably satisfied with the progress of the building works. By the end of 1616 the castle was complete at least in its outward works; it measured 50' by 30'; its walls were 31' high and 4' thick. It was, he claimed, 'held to be the best house that is or will be built upon the city's plantation, yet I know some far costlier'. There were 6 small houses of timber, 3 covered with slates, 1 with straw for want of slates, and 2 which 'must of necessity remain uncovered until next spring'. Meanwhile 13 houses had been built elsewhere on the proportion by the tenants.[74]

The finding of tenants was Canning's other major task. The government had envisaged the almost immediate removal of the native Irish to the church lands or to the proportions allocated to the 'deserving Irish' to create a segregated native society, but to have implemented this straightaway would have led to economic collapse and probably dearth, so extensions were granted, and leases offered to the natives on 6-month terms.[75] It was later claimed that proclamations had been made in the city of London that land would be available in the plantation at 4d. an acre, but there is a lack of contemporary evidence for this; in any event there was minimal effort to recruit colonists from London.[76] In 1616 a dozen children were sent from London's Christ's Hospital orphanage and apprenticed with tradesmen in Londonderry and Coleraine, but the scheme was never systematised.[77] On the Ironmongers' proportion Canning seems to have made no effort to recruit directly from England, but he took very seriously his obligation to find English and Scottish tenants. His solution was to recruit artisans from Londonderry and Coleraine, most of them employed upon the plantation, and to offer them leases of 31 years on condition that they take only English and Scottish undertenants, expel the native Irish, erect specific buildings, including houses of brick, stone or timber after the English manner, and enclose a garden and orchard. They were also required to have subdivided and enclosed their entire allocation with quickset hedges within 3 years.[78] The English, he complained, were unwilling to pay the present rents; while the Scots were willing to pay, he thought they would not perform 'so good building'. Getting suitable tenants was a real challenge. 'Here is such catching after tenants that I think it not fit to put any away that will condescend to indifferent conditions.'[79] Recruitment was also hampered by the unsettled state of the plantation in the wake of the rebellion of 1615.[80] Canning's letters are filled with lurid accounts of the violence of the kerns still at large in the woods. 'I pray God in his mercy spare us from these bloody minded villains and put into their minds which are in authority to take some speedy course to cut them off, else I fear it will be the overthrow of the whole plantation'.[81] Canning's efforts paid dividends. In December 1615 he was able to report that 29 of the townlands had been let for 31 years and that the tenants were required to build 20 houses.[82] But he remained anxious as to whether the tenants would perform the terms of their agreements. In March 1616 he reported that 'I can hardly hold your tenants to keep their bargains'.[83] In May 1616 he was able to report that some tenants had begun building, but 'others have done little yet and one or two I fear will give me the slip', and once they did begin building they became reluctant to pay their rents, and arrears soon mounted.[84]

In addition to the leaseholders, the companies were supposed to establish at least six freeholders on their proportions, and the Irish Society in London issued a stream of orders to get them to comply. Freeholders were an essential element in the attempt to clone English conditions and ensure that the expensive burdens of jury service were met.[85] The Ironmongers were very reluctant to offer good terms to the freeholders, instructing Canning to place them in the remoter parts of the proportion and not to offer them more than one townland. Canning urged more generous terms: 'most men here do think that the freeholders that shall be on the companies' lands are worthy to pay little or no rent for it in regards of the continual charge and trouble which they shall be subject to for attendance at assizes and sessions'.[86] But his labours paid off and by December 1616 six freeholders had been identified.[87]

What was the overall balance sheet? Canning looks like a pretty conscientious agent. Nicholas Pynner's survey of 1618 found that there were 56 British males on the Ironmongers' plantation, more than any other, and it would seem that superficially the articles had been complied with. The problem, however, was that the 'fundamental ground' of the plantation was the avoiding of natives. Pynner noted that 'here is an infinite number of Irish upon the land which give such rents that the English cannot get any land'.[88] The fact of the matter was that everywhere on the plantation the British tenants sublet to native Irish, so that there was little change in the actual occupancy of the land.

And not all other companies were as successful as the Ironmongers. The articles of plantation required that every 1,000 of the notional acreage should be populated with at least 24 British males. Each company had a proportion of 3,210 acres, making a total of 38,520 acres, so there should have been 912 British males. Pynner's survey suggests that there were actually only around 500; the figure for 1622 was 617. There may have been improvements over the 1620s as the 1628 survey recorded 947 and the 1630 muster 894.[89]

It was widely suspected that the Londoners' non-compliance was due to the 'covetousness' of the merchants.[90] Sir Thomas Phillips produced some rather fantastic figures purporting to show the enormous profits they had made from the plantation, and his wilful distortions were adopted by the Star Chamber prosecutors in 1635.[91] They have been carefully dissected by Theo Moody and it is difficult to improve on his analysis. He demonstrates that the companies spent £22,000 on building works. Their rental income can be estimated at £37,500 in the 21 year period November 1613 to November 1634 (about £1,785 per annum), so there was a modest profit of £15,500. But against that has to be set their failure to gain any return on their investment in the general plantation. They had raised a total of £62,000 between 1610 and 1635, but only received a dividend between 1620 and 1626. The total dividend was only £5,940, so their losses on the general plantation would have been £56,000 and their overall losses about £40,500.[92] Quite why the Irish Society paid such low dividends, and none at all after 1626, is less clear. We can attempt to reconstruct its finances. The rentals from Londonderry and Coleraine were £650 and £410 per annum respectively (total £1,060 per annum); the fisheries originally leased for £600 per annum in 1613 were yielding about £1,000 per annum in the 1620s, and the Londonderry and Coleraine customs about £700 per annum. So the annual income of the Irish Society would have been about £2,760; from this had to be deducted the king's rent of £205 per annum, wages and running costs. It is possible that the dividend of £1,000 per annum in the 1620s represents the actual surplus.[93] But the dividends were temporary as the Society's finances collapsed in the latter part of the decade. Rental arrears were very high, but the nail in the coffin was the successive sequestrations of

September 1625 to July 1627 and from May 1628. There were also the expenses connected with the building of the new cathedral and the mounting costs of litigation, together estimated at £7,000.[94]

There is thus considerable evidence that the Londoners tried reasonably conscientiously to fulfil the terms of the articles of plantation, and delays in implementing the programme were due to forces largely beyond their control, though they were vulnerable, as were all the planters, on the question of the removal of natives. They were the victims of James's intermittent interest in the plantation which had the effect of giving their shortcomings a high profile, and then of the financial embarrassments of Charles I's government, which made them too tempting a target.[95] That they received such hostile attention also owes something to the prejudices of the crown and key elements of the Irish establishment about their involvement in the first place. The Londoners' participation ran counter to the principles of aristocratic society, for it rested on an extraordinary degree of collective decision making. The monies for it were voted by the Common Council, a body of 212 men, which was the largest representative body in the kingdom after the House of Commons. The general plantation was administered by the Irish Society whose ruling body comprised a governor, deputy governor, five aldermen assistants, and 21 common councillor assistants.[96] Each of the Great Twelve established committees to run their proportions, and handle the correspondence with their agents; these committees drew in representatives from the lesser companies associated with each of the Great Twelve. It is also clear that livery company involvement resulted in a high degree of political discussion, as the making of assessments required considerable negotiation.[97] The management of the Irish business is a superb example of corporate structures at work, engaging in the 'structured conversations' that Phil Withington sees as central to civic sociability and the politics of the commonwealth.[98] The Privy Council was rather more sceptical about their value, seeing the city's failures as being due to the fact that 'as usually falleth out when a business concerneth a generality [it] did put the managing of it over to such as knew not otherwise how to employ themselves here and the handling of their private interests could not promise any great hope of advancing a public service'.[99] The servitor class in Ulster resented the presence of the Londoners. Sir Arthur Chichester sneered at their involvement from the very beginning of the project hoping that 'they prove not like the London women who sometimes long today and loathe tomorrow'.[100] The Ironmongers recognised that the surveys of the plantation were prejudicial to the city, 'all urged in the worst sense by the servitors who envy that which the city do'.[101] Canning warned them that the judges at Londonderry might be biased: 'I verily suppose that the servitors who are the chiefest in that court at the Derry would gladly yield advantage to any gentleman against any of the city's agents.[102]

The city was particularly unfortunate in acquiring an enemy as tenacious as Sir Thomas Phillips, the displaced governor of Coleraine, now removed to Limavady, a servitor enclave in the heart of their plantation. Phillips had fallen out with them as early as 1611 when the deputy governor and assistants of the Irish Society complained of his intention to 'domineer over their jurisdiction'; he seems to have blamed the city for the delay in establishing his title to his land.[103] Thereafter he dogged them every inch of the way. In 1617 he denounced the city for failing to remove natives and colonise with British settlers; in 1622 with Richard Hadsor, he undertook the survey of the plantation for a royal commission of inquiry. It was to this inquiry that we owe the collection of maps by Thomas Raven, the purpose of which was to point out 'the many defects and omissions in the Londoners' plantation, a place principally designed by Your Majesty for the future and continual settlement of the whole province of Ulster which I have not manifested out of malice to the Londoners as they unjustly charge me, but out of my zeal to Your Majesty's service and the safety of the commonwealth'.[104] Phillips assisted another set of investigating commissioners in 1627.[105] His subsequent collection of materials against the city was assembled in part because he thought that they had taken too soft a line with the Londoners who had allegedly shown their contempt for the whole process by entertaining the commissioners with a performance of *Much Ado About Nothing*.[106] Phillips' magnificent collection, a mixture of official documentation with his own fantastic calculations on the Londoners' profit and loss, provided ammunition for Secretary Coke who was a prime mover in the development of the Star Chamber case.[107]

Phillips' malice undoubtedly weakened the city's position but it has to be said that the Londoners made some key errors in the 1620s. The least plausible element in their defence in 1635 was the notion that they were not bound by the articles of the plantation, and that these had been superseded by the terms of the agreement between the city and the crown which did not make specific reference to them.[108] There was it is true some ambiguity over this, for Adrian Moore and William Freeman who took on the Haberdashers' portion explained to Tristram Beresford in 1614 that 'although the City be not tied by any condition by their charter to any building, yet it will be expected they should perform in some ample manner according to the king's book though not with the strictness in all points as is therein contained'.[109] But when they argued for the priority of housing over the castle, they were forced to give way to Beresford's insistence on the castle, and a year later Moore and Freeman noted the 'great complaint made of several companies in the slackness of their buildings'.[110] So, the city's claim that it was not bound by the articles was somewhat disingenuous, as the companies acted through the 1610s as though they considered themselves bound.[111] It was only in 1623 that the claim crystallised that the city was not bound by the articles.[112]

Whereas elsewhere in Ulster the crown moved from 1619 to the policy of fining the undertakers for allowing natives to reside, and from 1628 granting new patents to the undertakers, doubling their rents and allowing them to settle natives legally on one quarter of their lands, the Londoners stood their ground, arguing that these provisions did not apply to them. They were in fact leaving themselves very dangerously exposed.[113]

The city was desperate not to admit its guilt, offering Charles a cash payment for his grace but without acknowledging its fault, but the crown's ministers were determined to press home their advantage.[114] During the trial the city was able to address many elements of the prosecution case: the claim that they had secured their charter by subterfuge was specious; the woods may have been spoiled but this was the fault of the farmers not of themselves; they had put up more buildings at Londonderry and Coleraine than required; the fortifications at Coleraine only looked inadequate because they had so exceeded their brief at Londonderry; and so on.[115] But on the central question, the removal of the natives, the city's record indeed looked poor, and it was this that gave their critics the crucial lever to secure their humiliation.

NOTES

1 British Library, Stowe MS 397, fols 67–133v; The National Archives (TNA), SP 63/255/7; T. W. Moody, *The Londonderry Plantation 1609–1641: The City of London and the Plantation in Ulster* (Belfast, 1939), pp. 259, 264, 355–73.
2 W. Knowler (ed.), *The Earl of Strafford's Letters and Dispatches* (2 vols, London, 1739), i, 374.
3 TNA, C115/106/8450, 8451, 8452.
4 *Strafford's Letters*, i, 374.
5 Moody, *Londonderry Plantation*, p. 373.
6 *Strafford's Letters*, i, 200.
7 Ibid., ii, 25.
8 T. W. Moody, F. X. Martin, and F. J. Byrne (eds), *A New History of Ireland*, vol. iii: *Early Modern Ireland, 1534–1691* (Oxford, 1976), pp. 197–222; P. Robinson, *The Plantation of Ulster. British Settlement in an Irish Landscape, 1600–1700* (2nd edn, Belfast, 1994); M. Perceval-Maxwell, *The Scottish Migration to Ulster in the Reign of James I* (Belfast, 1990); N. Canny, *Making Ireland British 1580–1650* (Oxford, 2001), pp. 187–242; S. Connolly, *Contested Island. Ireland 1460–1630* (Oxford, 2007), pp. 289–308.
9 A. Crawford, *The History of the Vintners' Company* (London, 1977), pp. 134–52; A. H. Johnson, *The History of the Worshipful Company of Drapers* (4 vols, Oxford, 1914–22), iii, pp. 34–50, 124–37; I. Doolittle, *The Mercers' Company, 1579–1959* (London, 1994), pp. 56–60; R. J. Hunter, 'The Fishmongers' Company of London and the Londonderry plantation, 1609–1641', in G. O'Brien (ed.), *Derry and Londonderry: History and Society* (Dublin, 1999), pp. 205–59. An immense amount can be excavated from J. S. Curl, *The Londonderry Plantation*

1609–1641: *The History, Architecture, and Planning of the Estates of the City of London and its Livery Companies in Ulster* (Chichester, 1986); idem, *The Honourable the Irish Society and the Plantation of Ulster, 1609–2000* (Chichester, 2000).
10 V. Pearl, *London and the Outbreak of the Puritan Revolution: City Government and National Politics, 1625–1643* (Oxford, 1961); R. Brenner, *Merchants and Revolution: Commercial Change, Political Conflict, and London's Overseas Traders, 1550–1653* (Cambridge, 1993); K. Lindley, *Popular Politics and Religion in Civil War London* (Aldershot, 1997).
11 C. S. R. Russell, *The Fall of the British Monarchies, 1637–42* (Oxford, 1991); J. S. A. Adamson, *The Noble Revolt: The Overthrow of Charles I* (London, 2007); K. Sharpe, *The Personal Rule of Charles I* (London, 1992).
12 *Lismore Papers, by Richard Boyle, Earl of Cork*, (ed.) A. B. Grosart (10 vols, London, 1886–88), second series, iv, 16.
13 Moody, *Londonderry Plantation*, p. 408.
14 Ibid. p. 409.
15 London Metropolitan Archives (LMA), Journal of Common Council (JCC) 29, fols 164–6v; Johnson, *Drapers*, iv, 591–6.
16 *The Journal of the House of Commons of England* (5 vols, London, 1742), ii, 172.
17 *His Royal Maiesties Speech Spoken in...Parliament...on December the 5 1641. With the Love which His Majesty Hath Shown to the City of London by Knighting Five Aldermen...and Royally Giving Them Again into Their Lands in London-Derrie* (London, 1641).
18 Edward Hyde, earl of Clarendon, *The History of the Rebellion and Civil Wars in England*, (ed.) W. D. Macray (6 vols, Oxford, 1888), book iv, section 180.
19 J. Ohlmeyer, 'Strafford, the "Londonderry Business", and the "New British History"', in J. F. Merritt (ed.), *The Political World of Thomas Wentworth, Earl of Strafford, 1621–1641* (Cambridge, 1996), pp. 209–29.
20 Moody, *Londonderry Plantation* p. 360; J. Spedding (ed.) *The Life and Letters of Sir Francis Bacon* (London 1868), iv, pp. 116–26; T. W. Moody (ed.), 'Ulster plantation papers', *Analecta Hibernica*, 8 (1938), 281–6; G. Hill, *An Historical Account of the Plantation of Ulster at the Commencement of the Seventeenth Century* (Belfast, 1877; repr. Shannon, 1970).
21 *Londonderry and the London Companies Being a Survey and Other Documents Submitted to Charles I by Sir Thomas Phillips* (Belfast, 1928), pp. 11–12.
22 *Calendar of the Carew Manuscripts* (6 vols, London, 1867–73), 1603–23, pp. 148–53.
23 Guildhall Library, London [GL], MS 11588/2, pp. 536–40; Hill, *Plantation of Ulster*, pp. 360–3.
24 Hill, *Plantation of Ulster*, pp. 379–83; *Attorney General v. Irish Society (Ireland), Appendix One* (London, 1899), pp. 103–8. This book contains invaluable transcriptions of key texts from the city archives relating to the early years of the plantation.
25 T. Blenerhasset, *A Direction for the Plantation in Ulster* (London, 1610).
26 B. Rich, *A New Description of Ireland* (London, 1610).
27 Moody, *Londonderry Plantation*, pp. 70–2.

28 *Attorney General v. Irish Society*, p. 60.
29 GL, MS 17278/1, fol. 6.
30 Ibid. fols 7-8; GL, MSS 34010/4, pp. 396-9; 11588/2, pp. 544-5, 547.
31 LMA, JCC 28, fols 19-20v.
32 I. W. Archer, 'The burden of taxation on sixteenth-century London', *Historical Journal*, 44 (2001), 616-17.
33 LMA, JCC 28, fols 19-20v, 24r-v, 239v-40; 29, fos 49r-v, 186, 299, 345; 30, fol. 139; GL, MSS 34010/4, pp. 421, 468, 509-10, 532-3; 34010/5, pp. 13, 52, 82, 83, 110, 149-50, 213, 360; 11588/2, pp. 570-1, 643, 689.
34 W. R. Scott, *The Constitution and Finance of English, Scottish, and Irish Joint Stock Companies to 1720* (3 vols, Cambridge, 1910-12), ii, 123-8.
35 Archer, 'Burden of taxation', pp. 623-4.
36 *Attorney General v. Irish Society*, pp. 112, 114, 121, 123-4, 132-3.
37 GL, MS 11588/2, pp. 572-3, 588, 596-7, 599. Cf. GL, MS 34010/4, pp. 554-5.
38 Moody, *Londonderry Plantation*, pp. 90, 91, 96.
39 I. W. Archer, 'The livery companies and charity in the sixteenth and seventeenth centuries', in I. Gadd and P. Wallis (eds), *Guilds, Society, and Economy in London, 1450-1800* (London, 2002), pp. 19-20.
40 Moody, *Londonderry Plantation*, pp. 154-6; idem, 'Ulster plantation papers', pp. 299-311; Hill, *Plantation of Ulster*, pp. 431-4.
41 Moody, *Londonderry Plantation*, pp. 89-90; for the Haberdashers, see National Library of Scotland, MS RH91/33; Canny, *Making Ireland British*, pp. 222-9.
42 *Londonderry and the London Companies*, p. 2.
43 GL, MS 17278/1, pp. 7-8; Hill, *Ulster Plantation*, p. 366.
44 *John Stow's Survey of London*, (ed.) C. L. Kingsford (2 vols, Oxford, 1908), ii, 197-8.
45 E. Spenser, *A View of the Present State of Ireland*, (ed.) W. L. Renwick (Oxford, 1971), p. 165.
46 A. J. Horning, '"Dwelling houses in the old Irish barbarous manner": archaeological evidence for Gaelic architecture in an Ulster plantation village', in P. J. Duffy, D. Edwards, and E. Fitzpatrick (ed.), *Gaelic Ireland, c.1250-c.1650: Land, Lordship, and Settlement* (Dublin, 2001), pp. 375-96.
47 Moody, *Londonderry Plantation*, p. 197.
48 R. J. Hunter, 'Towns in the Ulster plantation', *Studia Hibernica*, 11 (1971), 40-56. See also R. Gillespie, 'Small towns in Ulster, 1600-1700', *Ulster Folklife*, 36 (1996), 23-30; idem, 'The origins and development of an Ulster urban network', *Irish Historical Studies*, 24 (1984), 15-29.
49 *Calendar of Carew Manuscripts*, 1603-1625, pp. 410-11.
50 GL, MS 17278/1, fol. Xx.
51 Hunter, 'Towns in the Ulster plantation'.
52 Moody, *Londonderry Plantation*, pp. 274-80; Avril Thomas, 'Londonderry and Coleraine, walled towns: epitome or exception?', in G. O'Brien (ed.), *Derry and Londonderry: History and Society* (Dublin, 1999), pp. 259-301.
53 GL, MS 17278/1, fols 28, 30v.
54 Ibid., fol. 35.
55 Ibid., fols 58v, 91, 120.

56 *Attorney General v. Irish Society*, pp. 369–78, 388–90.
57 Johnson, *Drapers*, iii, 31–6.
58 GL, MS 17278/1, fols 28, 38, 39v–40.
59 Ibid., fol. 37v.
60 Ibid., fol. 74.
61 Ibid., fol. 31v.
62 Ibid., fol. 107.
63 Ibid., fols 46v–7, 47v–8, 107v, 109v.
64 Ibid., fol. 74.
65 Ibid., fols 52, 106v.
66 Ibid., fols 107v, 109.
67 Ibid., fols 109, 117, 152v.
68 Ibid., fols 68v, 69, 106r–v.
69 Ibid., fol. 66v.
70 Ibid., fol. 107v.
71 Ibid., fol. 21v.
72 Ibid., fol. 35v.
73 GL, MS 34010/5, p. 175.
74 GL, MS 17278/1, fols 152v, 153v–4v, 155r–v.
75 Moody, *Londonderry Plantation*, pp. 105–11.
76 *Londonderry and the London Companies*, pp. 107, 114.
77 GL, MS 17278/1, fol. 118v; *Attorney General v. Irish Society*, pp. 423–43.
78 GL, MS 17278/1, fols 76–9, 84–100v.
79 Ibid., fol. 52.
80 Moody, *Londonderry Plantation*, pp. 165–7.
81 GL, MS 17278/1, fols 109v–110, 132v–133, 135, 140r–v.
82 Ibid., fol. 108v.
83 Ibid., fol. 122v.
84 Ibid., fols. 132v, 135v.
85 Ibid., fols 75v, 139v–40, 161v.
86 Ibid., fols 110v, 111, 122v–3, 133.
87 Ibid., fol. 162.
88 *Attorney General v. Irish Society*, pp. 466ff.
89 Robinson, *Plantation of Ulster*, pp. 212–24. For the survey by Sir Josias Bodley of 1614, see *Report on the Manuscripts of the late Reginald Rawdon Hastings* (HMC, 4 vols, London, 1928–47), i, 159–82; for Pynner's survey see Hill, *Ulster Plantation*, pp. 449–590; for the 1622 survey, *Calendar of State Papers Relating to Ireland, 1615–25* (London, 1880), pp. 364–78; Curl, *Londonderry Plantation*.
90 *Londonderry and the London Companies*, p. 5.
91 Ibid., pp. 140–5; Historical Manuscripts Commission, *Twelfth Report* (London, 1888–89) appendix ii, pp. 91–2.
92 Moody, *Londonderry Plantation*, pp. 335–9.
93 Ibid., pp. 270–4.
94 Ibid., pp. 228–36, 238–42.
95 GL, MS 17278/1, fols 53–4; 'Ulster plantation papers', pp. 277–9; Moody, *Londonderry Plantation*, pp. 164–5, 192, 355–6.

96 LMA, JCC 28, fols 46–9; Moody, *Londonderry Plantation*, pp. 81–3.
97 GL, MSS 17278/1, fol. 15; 34010/4, pp. 396–9, 421, 509–10.
98 P. Withington, *The Politics of Commonwealth: Citizens and Freemen in Early Modern England* (Cambridge, 2005), pp. 124–55.
99 *Attorney General v. Irish Society*, p. 268.
100 Hill, *Ulster Plantation*, p. 372.
101 GL, MS 17278/1, fol. 43.
102 Ibid., fol. 70r–v.
103 *Calendar of State Papers relating to Ireland, 1611–14* (London, 1877), pp. 29–30; Hill, *Ulster Plantation*, pp. 392–5; Moody, *Londonderry Plantation*, pp. 114–18.
104 Johnson, *Drapers*, iv, 531; Moody, *Londonderry Plantation*, pp. 191–204, 211–22.
105 Moody, *Londonderry Plantation*, pp. 237–9.
106 *Londonderry and London Companies*, p. 4; Moody, *Londonderry Plantation*, pp. 238–40.
107 *The Manuscripts of the Earl Cowper, K. G., Preserved at Melbourne Hall, Derbyshire* (HMC, 3 vols, London, 1888–9), ii, 92; Moody, *Londonderry Plantation*, pp. 259–61.
108 *A Collection of Such Orders and Conditions as are to be Observed by the Undertakers of the Escheated Lands in Ulster* (London, 1608); Hill, *Ulster Plantation*, pp. 80–8.
109 NLS, RH91/33, Freeman and Moore to Beresford, 3 March 1614.
110 Ibid., Freeman and Moore to Beresford, 17 April 1615, 10 May 1615.
111 Cf. GL, MS 17278/1, fol. 39v.
112 Moody, *Londonderry Plantation*, pp. 217–18.
113 Ibid., pp. 236–7.
114 Ibid., p. 265; TNA, C115/106/8445.
115 Ibid., pp. 359–65.

6

Success and failure in the Ulster plantation

RAYMOND GILLESPIE

In much of the historical writing about early modern Ireland the problem of plantation tends to dominate the discussion on social and political change. The formulation of the idea of plantation, however, serves to obscure rather than clarify thinking about how change happened and how successful it was. At one level historians tend to link plantations into a coherent policy, driven by central government to effect political change across Ireland. Thus plantations have been interpreted as all embracing, top-down engines of change accounting for developments in political control as well as economic and social anglicisation.[1] As a result, plantations are seen as moments of disjuncture in which one world transmuted into another by the will of central planners. Some attempts have been made to smooth the transition and to depict some contemporaries as viewing plantation as an evolutionary process, stressing continuity and forgetting violence, but to date these have found little favour.[2] Political disjuncture naturally linked plantation to the alienation of the native Irish from the Dublin administration but, as with the link between economic commercialisation and plantation, a neat correlation is difficult to sustain. The rising of 1641, as has been pointed out, was a conservative affair initiated by those who benefited from the plantation scheme and, in the short term at least, had little interest in overturning that world.[3] This does not mean that all those involved in the rising were loyal subjects but it does suggest that the insurrection was a more complex affair than a simple reaction to plantation with diverse levels of loyalty and alienation among those who participated.[4] Thus, when understood from the world of Dublin, the idea of plantation forms a convenient organising principle within which to describe the development of early modern Irish society and particularly the failure of settlement policy in the upsurge of violence in 1641.

This view of plantation is not the only way of describing the phenomenon. Reconstructions from a more local perspective reveal a different aspect to the process. In the case of Ulster a series of studies of the actual workings of the settlement through the surveys of 1611, 1613, 1619 and 1622 atomised the distribution of land and its settlement. Beginning with George Hill's 1877 study, a number of theses and articles in local journals reveal a rather different plantation to that depicted in Dublin-focused surveys, highlighting a scheme plagued by a shortage of leadership, settlers and capital.[5] These local reconstructions depict a world much too fragile to support the weight of the powerful instrument of change urged onto the historiographical stage by those who write about plantations as agents of social change at a national level. Those historians who attempt to resolve this divergence between the image of plantation as a transformative process and its reality as a tenuous toehold that could easily fragment have tried to do so by differentiating between the theory and practice of plantation or between the desires of the government and the realities of profit that attracted the settler.[6]

This dichotomy between different ways of looking at plantations is not a problem created by historians since at times contemporaries also took diametrically opposed views. Some regarded the process as an unmitigated disaster. Even from the earliest stages of the scheme the Irish lord deputy, Sir Arthur Chichester, was convinced of its failure. In 1611 he noted 'the king's intention of bringing colonies out of Great Britain does not go forward as is to be wished' and highlighted the settlers' lack of success in fulfilling the terms of the plantation.[7] Chichester's voice was hardly an impartial one as he had objected to the form of the scheme since its inception. Many powerful people, however, supported Chichester's criticisms, especially after the four major plantation surveys between 1611 and 1622. In the wake of the last of these, Francis Blundell, a Wexford settler and Irish vice treasurer, who often disagreed with Chichester, confirmed his views. Calling Ulster 'a poor ragged quarter plantation', he condemned those settlers who had failed to carry out the terms of the scheme.[8] Others were inclined to be more sanguine about the Ulster experience. At the same time as Blundell penned his denunciations of the scheme another settler, probably William Parsons, although the author is identified only by the initials W.P., wrote a defence of the process of plantation. This did not focus on the accountancy of settlement projects but rather considered their importance in 'the new framing of commonwealths'. In this regard he adjudged them a success in breaking an older order and 'by this means shall that people [the Irish] now grow into a body compounded and a commonwealth'.[9] W.P.'s propositions had support from others. One tract written in the early 1640s described how Ulster had been 'divided into competent portions amongst lords and gentlemen of English birth who brought...tenants out of England and Scotland who built and planted those rude and unmanured lands...in such part as it was now become as comfortable for the life of man and as commodious for all that had occasion to travaille or commerce as any part of Ireland and little inferior to many parts of England'.[10] While the Old English Catholic Richard Bellings did not specifically mention the outcome of the Ulster plantation in his history of the 1640s he probably had the scheme in mind when he wrote that in 1641 'the colonies (setting aside their different tenets in matters of religion) were as perfectly incorporated, and as firmly knit together as frequent marriages, daily ties of hospitality and the mutual bond between lord and tenant could unite any people'.[11] Clearly contemporary judgements on the plantation process could vary considerably, based on the assumptions of the individual about its aims.

Assessing these different viewpoints is a difficult matter. There are, for instance, problems in knowing what the planners of plantation had in mind when they drew up their schemes for Ulster. Much of the final planning happened not in Ireland but in London as matters relating to Ireland were ultimately the concern of the king. Unfortunately, the Privy Council records, which contained at least some clues to the debates that shaped Ulster plantation scheme, are missing for the crucial years. In trying to resolve this problem it is important to acknowledge that those who planned the plantation did not live in ivory academic towers. They were practical politicians who knew the limits of government from a local level through to the centralised factional world of court and parliament. Moreover, they did not lack advisers with direct experience of plantation settlements, who in 1607 and 1608 wrote tracts urging a range of solutions for the province of Ulster. Richard Spert, a veteran of the Munster plantation, recommended that the king adopt a similar course to that followed in the southern province but said nothing about what this might mean for the native population. John Bell, vicar of Christ Church, Newgate Street, London, offered the king advice that he claimed could solve not only the problem of Ulster but also help to alleviate the problem of poverty in England.[12] Another tract grasped the bull firmly by the horns and advocated the wholesale clearance of the Irish and their cattle from Ulster to create a totally new settlement.[13] From within Ireland too those with vested interests lobbied the government, while historical precedents existed to guide the planners, as proposals for the colonisation of Ulster had been compiled in the late sixteenth century and were certainly on hand to be used.[14] More obviously, the example of the Munster plantation might serve as precedent and warning. Towards the end of 1607 someone in London made a list of available documents on the Munster plantation 'necessary at this time'.[15] King James VI and I also had practical experience of these problems. In 1597, 1605 and 1607 plans by Fife merchants to settle the Isle of Lewis as part of the strategy of solving the Highland problem in Scotland had come to nothing.

To try and understand something of the priorities of the planners it is worth following the fate of one piece of advice they received. Within weeks of the Flight of the Earls in September 1607 the Irish lord deputy, Sir Arthur Chichester, made proposals for what should happen to Ulster. On 17 September he informed the king that the best way to deal with the situation was to create freeholders in Ulster in order to sever the relationships between the former lords and their followers. He also advised the king to seize ecclesiastical lands. The new freeholders would be granted the lands of the earls, 'every man of note or good desert [being given] as much as he can conveniently stock or manure for five years to come', with the balance going to servitors who would 'bring in colonies of civil people of England and Scotland at his majesty's pleasure with condition to build castles or storehouses on their lands'. Soldiers would also be introduced, to be paid for initially by the crown and after five years by the settlers. If the king did not approve of that course 'which is the best of all others that he can think of' then it would be necessary to clear the inhabitants from the lands, leaving only garrisons. The chief beneficiaries were to be Chichester's former colleagues in the army who had been disappointed by the settlement after the defeat of O'Neill in 1603 since, as victors, they had expected the lion's share of the spoils of war.[16] This scheme appeared to meet with royal approval and the Privy Council informed Chichester on 29 September that 'for the plantation which is to follow upon attainder [of O'Neill] the king in general approves of his project'.[17]

In March 1608 Chichester worked the plan up in a more formal way, drafting his 'Notes of remembrance'. This proposed a much more sophisticated division of the lands, informed by intelligence gathered about the holdings of the earls, arguing that most of the settlement should be created 'as near to the form of that of Monaghan as we may'. As with the settlement in late sixteenth-century Monaghan significant constraints on action already existed. Those gentry of the Pale who had purchased lands in Cavan, where 'they have begun a civil plantation', and former soldiers, such as Sir Thomas Phillips, who had established colonies on land they acquired after the war, were to be left alone. Large parts of Tyrone and Fermanagh would remain in the hands of the older inhabitants. In addition to grants to the Church of Ireland, settlers would be introduced from England and Scotland but Chichester made explicit the implication from his earlier letter that former soldiers were best placed to act as planters: 'And for that the parties, who are in my opinion most fit to undertake this plantation are the captains and officers who have served in these parts.' These soldiers could act as a resident garrison to deal with any residual threats to the scheme, after which they would 'be left to their portion of lands as the rest of the undertakers and their wards and entertainments to cease'. Some thirty or forty undertakers were to be introduced. More optimistically Chichester argued that 'no great care need to be taken for the inferior natives of those three counties for they will all settle themselves and their dependencies either upon the bishops, undertakers or the Irish that shall be established by his majesties gracious favour for most of them are by nature inclined rather to be followers and tenants to others than lords and freeholders of themselves'. Accordingly, they would settle into towns and villages, giving up their traditional lifestyle.[18]

Chichester continued to press home his proposals with a text revised in October 1608, based on new information but with the same argument.[19] The scheme was based on a well-tried precedent. The settlement of Monaghan in 1596 created a series of freeholds for both settlers and natives held directly from the crown thus eliminating the problem of over mighty subjects, while ensuring that trustworthy natives obtained a place in the new order.[20] The use of this technique for balancing power and establishing authority in the Irish regions continued into the early seventeenth century. At the same time as the planners considered the Ulster plantation scheme for the six escheated counties the government carried out a similar freeholding scheme in the Magennis lordship of Iveagh, in the non-escheated county of Down.[21]

Despite all the historical precedent and contemporary practice that Chichester drew on, the final planners of the plantation did not accept his scheme. In October 1608 the Irish Privy Council sent the Irish chief justice, Sir James Ley, and the attorney general, Sir John Davies, with a draft of its own proposals, probably heavily influenced by Chichester, to the English Privy Council.[22] The latter body redrafted the scheme to produce the 'Project' of the Ulster plantation, which seems to have been printed in January 1609.[23] Chichester reacted furiously and wrote to the Privy Council on 10 March voicing his objections to the plan. He believed the allocation of land to individuals was insufficient to attract men of worth to hold the plantation together, the tenure of common soccage was too light, the time scale for carrying out the scheme was too short and too many Irish were to be allowed lands so that 'very few here will bear any part in this intended plantation'. He frequently returned to these themes.[24] The missing Privy Council registers mean that no definitive answer can be given as to why it diluted the focus of Chichester's early plan for the servitors by the inclusion of more undertakers and native Irish and why additional requirements to build and introduce settlers were imposed on the undertakers. A significant part of the explanation may lie in the way that the problem of Ulster was reformulated by the Privy Council in late 1608. Chichester sought to deal with the presence of over-mighty 'uncivil' lords in Ulster yet, as the king knew, plantation as understood by the Lord Deputy would not deal with that problem. In his *Basilicon doron* of 1599 King James related how he had tried to tackle a similar problem in the Isles through legislation and 'following the course that I have begun, in planting colonies among them of answerable inland subjects, that within a short time may root them out and plant civility in their rooms'.[25]

Yet this strategy had limited effect and by 1609 the emphasis had shifted towards reordering the locus of power within society as indicated in the Statutes of Iona. These statutes proved more in line with the sort of thinking that informed the making of the Ulster plantation scheme.

The shift in government policy on Ulster in late 1608 is a difficult problem to deal with given the loss of important evidence. This has been compounded by the way historians thought the plantation would work, informed by a rather narrow corpus of texts, mainly those canonised in George Hill's 1877 collection of material. The key texts, and the standards which they set down for plantation, were thus held to be the 'Project' of 1608 or early 1609 and the 'Orders and Conditions' of 1609, revised in 1612. Most attention has focused on what this limited range of documents said about the infrastructure of plantation rather than trying to understand the assumptions behind them, and in particular the language they used to articulate those assumptions. A case in point is the 'Project' of 1608 which set out the detailed divison of lands in the escheated counties.[26] In seventeenth-century terms a 'project' involved a practical scheme for exploiting material things, such as the introduction of new crops or developing new manufacturing processes in response to rising consumer demand.[27] More ambitious projects centred on monopolies, intended to allow their originators and the crown to make a profit. Some of those who considered settling in Ulster certainly had this sort of model in mind. When Lord Audley, who knew something of plantations since his estates lay in Munster, applied to be part of the scheme in 1609 he asked for a licence to erect iron mills, to make iron and glass and to sow the new consumer crop of woad for 41 years. He also sought to control the appointment to livings on his land and to monopolise certain areas of justice.[28] Audley's offer was not taken up but the parallel with monopolies in the form of land grants that could be exploited was a real one, payment for these monopolies over lands being in the form of subsequent investment by the grantees. Legal structures set out by the government had to guarantee the monopoly as happened in the production of goods. Thus the primary function of government was not to settle Ulster – that was a matter for the holder of the monopoly or land grant – but to provide the property and legal infrastructure that made such a settlement possible. Thus the 'Project' was not so much a document about the distribution of land as a framework for the organisation of settlers.

A second instance of how contemporary language can provide an insight into the unarticulated assumptions of the plantation planners is provided by considering how they chose to describe the settlers. The earliest draft of the plantation proposals refers to settlers as English or Scottish but in June 1609 a note of business to be transacted for the plantation referred to undertakers as 'servitors and Britons', a phrase subsequently used in other government documents.[29] The use of the term 'British' in each of the official plantation surveys up to 1622 was clearly an artificial construct. In the depositions taken in the early months of the 1641 rising the Ulster settlers are commonly described as 'British', but whether they described themselves as such or it was a formula used by a clerk is not clear. By the time of the Civil Survey in the 1650s the term had disappeared, the inhabitants of Ulster being referred to as 'English' or 'Scottish', thereby reflecting the separation of the two entities during the 1640s. Despite the short lifespan of the term 'British' it nonetheless located the project within a much wider discourse about the formation of a new political and social entity after the union of the crowns in 1603: Great Britain. In an Ulster context it represented an experiment in reconciling ethnic differences under a wider 'British' umbrella. The divisions between Scots and English in the plantation should not be minimised since those settlers came from worlds with very different economic and legal systems as well as cultural values. Even Chichester realised that it would be necessary to find points of contact between Scottish and English settlers to create some common bond. As one note on the distribution of lands in the plantation scheme put it 'some course would be taken that English and Scottish may be placed both near and woven one within another'.[30] Thus in 1610 Chichester set out the 'places of intercourse and meeting of the English and Scotch'. These included markets, quarter sessions and assizes 'where they shall be joined in juries and other public services'.[31] This sort of interaction, he presumably thought, would bind English and Scots settlers together around trade and the common law thus preventing friction between them. It proved a remarkably successful strategy as there is little evidence of tension between the two groups in Ulster before the 1640s, the most notable incident being the alleged assassination attempt on the English settler Lord Audley by some of the Scottish tenants of the earl of Abercorn because he had uttered words 'against the Scottish nation'.[32]

The use of the division between 'British' and 'Irish' to create new sets of ethnic identities reveals yet more of the priorities of the planners of the Ulster settlement. While the aim of the scheme had been to equate 'British' with 'Protestant' by requiring grantees and tenants to take the oath of Supremacy, that equation quickly broke down. The Scottish earl of Abercorn, for whom Chichester had much respect, was a Catholic at the time that he received his lands, as were his extended family who obtained lands in the barony of Strabane.[33] Also in Tyrone Lord Audley had become a Catholic by the early 1620s.[34] That such men could be part of a settlement that appeared to exclude the Catholic Irish seems unusual. Viewed, however, in a wider Irish, or more importantly a British, context such men could compare themselves with other Catholics occupying positions of honour, such as the earl of Clanricard.

King James had found Catholics useful in Scotland, most significantly Archibald Campbell, the seventh earl of Argyll and George Gordon, fifth earl of Huntly so long as they subscribed to a code of 'Britishness', which included acceptance of the monarchy, law and honour. According to Chichester, 'it is a matter more of honour and example than for any hope of gain for which this plantation must be undertaken'.[35] Wider contexts could absorb religious problems in the interests of establishing social order.

An awareness of how contemporaries understood the language in which the aims of the plantation were expressed helps to clarify some of the unspoken assumptions about the process. A second strategy in trying to reveal the assumptions of the plantation planners is to consider a wider range of texts than hitherto examined by historians. While the 'Project' and the 'Orders and Conditions' set out the broad lines of the plantation infrastructure in terms of the distribution of land and the introduction of settlers, the documents that gave effect to that broad scheme need to be read more carefully. A case in point is the creation of corporate towns, a central part of the plantation scheme. These corporate charters have been largely ignored and viewed as merely political instruments, creating boroughs to form a Protestant majority in the 1613 parliament but they had a much wider function than this. The charters were also intended to make local urban societies work by establishing a range of local institutions in the same way that land grants attempted to organise rural societies.[36] In the case of Dungannon, for instance, the original charter not only created markets and fairs but also stipulated structures for the government of the town as well as the numbers of men to be settled there, the allocation of land to these men and the organisation and construction of the market place, school and parish church.[37]

While urban charters played a key role in shaping how towns would be governed, the terms of the patents for land granted to the undertakers proved even more crucial. Unfortunately the destruction of the patent rolls in the 1922 Four Courts fire means that many of the details cannot be reconstructed. The nineteenth-century calendars recorded only what appeared to be most important, notably the names of the grantees and their lands, with some comments on tenures.[38] Few of these patents survive in estate collections but these examples are remarkably similar and demonstrate the importance of the details in these grants in fleshing out the ideas of the plantation.[39] Most recited the lands to be granted, the tenure by which they were to be held and confirmed a series of provisions set out in the 'Orders and Conditions'. Thus the patent created the estate as a manor since servitors and undertakers, though not the native Irish, had 'power to erect manors, to hold courts baron twice a year, to create tenures to hold of themselves upon alienation of any part of the said portion'.[40] The manor court, available on settler estates, was to act as a court to enforce customary matters but, in the words of one patent, could also hold pleas 'for all singular actions and transgressions of agreement, accounts, contracts, unpaid debts and for all kinds of demands which in debt or damage do not exceed forty shillings' within the manor.[41] Two seemingly arcane provisions in the undertakers' patents lay at the heart of the plantation scheme. The first of these, allowing for the creation of manors, seems strange in a context of the decline of the significance of the manor in contemporary England with the rise of leasehold tenures. Since manors were not land, however, but bundles of legal rights, and in particular the right to hold manorial courts, they could be made to work in Ulster in a way that they might not elsewhere. For example, they might function in the informal resolution of disputes in locations where the common law was poorly established. In this way tenants learnt the language of the law and some of the rudiments of legal procedure necessary for that world. The evolution of local manorial customs on settler estates shaped by the language of the common law facilitated the formation of local collectives that could transcend other ethnic or social divisions, creating a sense of local identity as part of the plantation process.

Less clear in its import was the fact that the patents set aside the 'qui emptores terrae' provision of the Statute of Westminster (1290). While prohibiting subinfeudation, this statute gave tenants the right to alienate all or part of their holdings without the lord's consent. This allowed landlords to make fee farm grants by subinfeudation, which is an unusual feature of Irish land tenure, and specified the sort of relationship a landlord had with his tenants.[42] The original 'Orders and Conditions' of the plantation stipulated that each undertaker 'within two years plant or place a competent number of English or Scottish tenants upon his proportion' and remove the native Irish.[43] The revised conditions of 1610, however, specifically required the larger undertakers to plant '24 able men of the age of 18 years or upwards being English or inland Scots' within three years. Smaller undertakers had to settle ten families on their land. More significantly, the revised conditions stipulated how to divide the property among the settlers in freeholds, leaseholds and tenancies at will, thus attempting to regulate not only landholding but social structures and relationships.[44] Freeholders in particular were essential for the workings of the jury system within the common law process. Just as the manorial courts provided a local legal structure for resolving problems so the land grants highlighted the legal network which defined the social order. Given this emphasis on law in the construction of society it is difficult not to detect the guiding hand of the Irish attorney general Sir John Davies, who had been in London during the planning stages of the scheme. For Davies, law provided the key to social order so that 'there can never be unity and concord in any one kingdom but where there is but one king, one allegiance and one law'.[45] The Ulster plantation can be seen as the working out of that principle.

While the establishment of manors and social hierarchies set about making micro societies within the plantation, the scheme also had the aim of creating a much broader social world. Thus in 1610 the planners argued that a relatively light tenurial obligation would be 'good preparation to link the undertakers and their issue together in marriage and affection, and so strengthen one another against the common enemy'.[46] Again, as Chichester observed, few would be part of the scheme 'except they are associated with such followers, friends and neighbours as can give them comfort and bring them strength and assistance'.[47] The planners focused on institutions, therefore, to create societies rather than isolated estates. The social significance of the markets is a case in point. A tract on Ireland in the early 1640s described the process of establishing markets: 'they have placed markets and fairs throughout the land whereby the diligent may make use of their labours and whereby the commerce and intercourse would advantage and incorporate the people'.[48] Similar comments appeared in more official contexts. A market created at Dundrum, County Down, in 1629 was to be 'for the public good of the inhabitants residing at or near Dundrum and with the intention that they may have free trade and commerce among themselves and with other liege subjects... by which the rude and country people of that region may be led to a more humane and civil mode of life and the more easily procure a provision of all necessaries'.[49] In 1622 William Parsons described markets and fairs as 'commonwealth meetings'.[50] The logic of establishing markets as part of the plantation scheme lay far beyond the economic imperative and was at the core of forming a new type of society, British in outlook and legal in its articulation.

Clearly if the plantation process was as much concerned with social planning and engineering as with land confiscation and redistribution then the question of the treatment of the native Irish remains a central problem. This subject has not engaged the attention of historians, due in large part to difficulties with the surviving evidence. Very few voices from within the native Irish community have been preserved, a result of the essentially oral nature of Irish society. A body of poetry in Irish from early seventeenth-century Ulster relates to other matters such as the flight of the earls or developments outside the plantation area. The main outlines of the plantation scheme's treatment of the native Irish are well known. The 'deserving Irish' were allocated some 18% of the escheated lands, more than the amount assigned to the servitors but less than that given to the undertakers who received 35%. The undertakers' proportion split equally between Scots and English, placing each of these groups on a par with the native Irish in terms of share of plantation lands, although with more estates of a smaller size.[51]

That the Irish should have received more land than the servitors is a point of some interest given the lobbying on behalf of the latter group by the lord deputy, Sir Arthur Chichester. Servitors, the Church of Ireland and native

Irish landholders could take Irish tenants on their estates but the undertakers had to replace them with British tenants. In August 1610 a proclamation required the Irish on the undertakers' estates to remove themselves to the lands of servitors, other natives or the church but explained that they might be permitted to stay given the low numbers of settlers that had arrived. The older inhabitants received permission to stay on their lands for a further year in order to save the harvest, and this was extended yet again until May 1612. By 1618, little had been done to remove illegal native Irish tenants and the following year fines were introduced for Irish settling on plantation estates. These proved ineffectual as undertakers found it more convenient to keep their native tenants, charging higher rents than they might have got from settlers without the inconvenience and difficulty of finding new tenants.[52] By the 1620s the native Irish were well established on the undertakers' lands and this realisation gave way to a renewed debate as part of the 1622 commission on the state of Ireland on their place in the plantation scheme.[53] In reality, the prospect of removing them seemed remote. One tract from the 1620s railed against the failure of the settlers to drive out the native Irish, claiming that Ulster 'can never be a good plantation' until they did. The Irish, the author argued, proved cheaper to find than expensive settler tenants and would pay more rent because they were 'more servile'.[54] As Francis Blundell explained in his discourse on plantations in 1622, the Irish had been left on the Ulster undertakers' lands 'which being children when they [the undertakers] undertook the lands might have been put from them with little or no trouble but being now grown men it is worth the consideration of this state what course shall be taken with them for I doubt it will be no easy work to free those lands of them'.[55] Technically in breach of the conditions of the plantation, the undertakers attempted to regularise their position by reaching an agreement with the government whereby in return for a fine they could retain native Irish on one-quarter of their estates, removing the rest by May 1629. As with previous schemes, however, this proved almost impossible to enforce in parts of Ulster and significant numbers of Irish tenants remained on settler estates.

What shaped attitudes to the position of the native Irish in the plantation is a difficult question to answer. Even Sir Arthur Chichester conceded that they were necessary for the social and economic order to work but for those intending to settle in Ulster their presence created an immediate law and order problem, especially given the small number of settlers. While the attorney general, Sir John Davies, hoped that in the long term the two communities 'might grow up together in one nation' there was clearly an immediate security problem.[56] Thomas Blennerhasset, for instance, in his pamphlet of 1610 complained of woodkerne, or bandits, although he subsequently became a successful settler in Fermanagh.[57] The fears created by such publications and oral stories could be played on by the government and other undertakers.[58] This paranoia, however, needs to be balanced against the widespread assumption that the native Irish would wish to be part of the scheme. Even the virulently anti-Irish polemicist Barnaby Rich recorded his confidence that a settlement could be created despite all the scare stories about woodkerne.[59] Perhaps the clearest statements about the expectation that the native Irish and the settlers would reach local accommodations came from the Ironmongers Company in the Londonderry plantation.[60] Their agent received instructions that he was 'upon arrival to make publication for all such Irish as will live quietly and manure the ground to entertain, to defend them from the enemy and take no coyne and livery nor cess but what bargain is made to perform the same... there is no doubt but a great number of husbandmen, which the country calls the churls, will come and offer to live under them [the company] and farm the grounds both such as are of the country birth and others, both of the wild Irish and the English pale'.[61] Certainly by 1616 more than a quarter of the rents of the company came from those with Irish names in flagrant breach of the rules of the plantation.[62] In this case at least, and in many others, there were grounds for settler optimism.

In the early stages of the plantation it appears that some form of accommodation or limited assimilation emerged as a common response to the arrival of a relatively small number of settlers. The distribution of the settler population in 1613, for instance, remained concentrated around the main ports of Derry and Coleraine, with little movement into the hinterland, and along the Lagan valley in north Armagh.[63] Large areas of Ulster simply had no settlers with whom the Irish could interact, yet signs begin to appear of the native population becoming involved with the plantation scheme. A case in point is the estate of Sir Claude Hamilton of Schawfield, which lay around what is now the County Tyrone village of Gortin. By 1619 there were allegedly fifty British males on the estate, although in 1622 government commissioners surveying the state of Ireland could find no settlers among the 120 Irish residents there.[64] Before 1613 Hamilton had leased all but five townlands on the estate to a number of native Irish men, including one Patrick Groome O'Dufferne, who presumably had lived there before the plantation. Patrick clearly knew his landlord well and when Hamilton died in 1615 he forgave O'Dufferne his outstanding debts. His executors leased the entire estate to O'Dufferne who became responsible for the rent of the lands including presumably those from other Irish tenants below him.[65] Other examples exist of native Irishmen making settler estates work. The agent for the Ironmongers' estate in County Londonderry, for instance, recorded payments in his accounts to Daniel O'Quig 'for helping to gather the May rent before I came over'. In addition, settlers relied on the assistance of the local population to recover the names of the townlands and to locate the best building stone.[66] This must have been a widespread process of information gathering, which allowed the newcomers to understand the geography of their estates.

While the rental of the Hamilton estate in Tyrone points to considerable continuities across the plantation, with the native Irish being kept on the traditional units of land management (the ballyboe), it also reveals significant innovations. The most important of these lay in the legal basis for the habitation of the land. Tenants now held their property by contract rather than by custom, with social relations spelt out in writing rather than in terms of mutual rights and obligations within a framework of kinship. This drew the native Irish into legal and administrative structures that provided important points of contact between them and the settlers in the early stages of the plantation, a process that occurred across Ireland. The Irish language, for instance, contains a significant number of early seventeenth-century borrowings of legal terminology.[67] In the case of the plantation counties it is possible to trace the interaction of native and settler through legal institutions because of the survival, albeit fragmentary, of the records of some courts. Summonister rolls, for instance, list fines levied on large numbers of individuals, both native and settlers, as a result of actions at the quarter sessions. These include fines for non-attendance at courts, ploughing by tail and failing to maintain roads and bridges.[68] The few surviving gaol delivery rolls of the assizes from the second decade of the seventeenth century provide ample evidence of the engagement of the Irish with the law, both as accused and accuser.[69] They reveal the Irish as involved in theft, mainly of livestock, affray and, occasionally, rape, as well as resolving grievances through the common law process. The ability of woodkerne to manipulate the process of obtaining legal pardons for themselves, sometimes using landlords as intermediaries, also shows a clear understanding of the workings of the law.[70]

The law was not the only place where natives and newcomers came into contact during the early years of the plantation. While much of the evidence for the workings of local government in the plantation scheme has not survived, a few fragments are suggestive, such as the proceedings of the manorial court, which enforced customary law, on the archbishop of Armagh's lands between 1625 and 1627. An estate owned by the church could take Irish tenants and the overwhelming majority of the litigants in the archbishop's manorial court had Irish names, as did about a third of jury members.[71] The sort of cases involved, mainly assaults and slander, are illustrative of local societies grappling with the difficulties of minor rule breaking and local anti-social behaviour through a local common law framework.

The Church of Ireland parish emerged as a particularly important institution in local government and here the attitudes of some of the native Irish proved ambivalent rather than overtly hostile, at least initially, despite their apparent Catholicism. In the aftermath of an abortive rising in 1615 one

Cnougher McGilpatrick O'Mullan, a leaseholder from the Haberdasher's company in County Londonderry, gave a deposition about his very marginal involvement in the conspiracy. In the course of the deposition he recounted another episode that appears to have had nothing to do with the plot. A dispute between Art McTomlen O'Mullen and Brian McShane O'Mullen became violent and 'the said Art uttered these speeches to the said Brian saying "Thou art a churchwarden and yet dost not attend thy office according to thy instructions. Thou had sixteen Masses said in thy house by Gillecome McTeige, abbot, to whom thou gavest a white cow for his service and then relievest the said Gillecome and harbourest him in thy house as well as abroad"'.[72] A good deal of the detail here is vague. The identity of Gillecome McTeige is unclear but he may have been the abbot of the Cistercian house at nearby Coleraine, which had recently been dissolved. The deposition suggests that McTeige had been invited to fulfil some specific task, such as a request in a will, for a number of post-mortem masses. Whatever the problems with the details of this evidence, its main thrust is clear: in the very early stages of the plantation process churchwardens of the established church did not necessarily subscribe to the confession of that church. This may not be as strange as it first appears given that before the canons of 1634 parochial officials were not required to subscribe to any doctrinal statements. Given the close linkages between parish and society before the plantation it is hardly surprising that the native population would wish to maintain links with their own parish and to maintain burial and other rights in the parish graveyard. This implies an ability among at least some native Irish to differentiate between the idea of the parish as a building block of local society and the parish as a religious community. Studies of parish life in better documented areas of Ireland reveal as much. Indeed, Catholic parochial officials remained commonplace at lesser levels in eighteenth-century Church of Ireland parishes. In the very early stages of the plantation the idea of the parish potentially provided an area of common ground, in which local identities might have been formed.

As the plantation proceeded, the balance between accommodation and assimilation may well have shifted towards the former. In addition to the law and other administrative contexts, cultural interchanges also occurred in the areas of language and dress. As early as 1615 two English servants of Sir Toby Caulfield at Charlemont acted as translators for those interrogated about the plot of 1615 and by the 1640s a number of settlers spoke enough Irish to understand their rebel captors.[73] In the area of dress an increasing number of native Irish men and women adopted English fashions so that in the late 1640s an image of the rebel leader Sir Phelim O'Neill published in an English pamphlet portrayed him in the apparel of an English gentleman.[74] While the image of Sir Phelim as a Royalist gentleman carried its own damning message in the late 1640s the publishers clearly thought it was as credible as an image of an Irish lord with gilb and mantle. It is impossible, however, to gauge the full extent of these processes, although one particularly striking piece of evidence does suggest some measure of assimilation. Based on the lists of names of those mustered around 1630, a number of native Irish were sufficiently integrated into settler society to have formed part of the defensive measures of the plantation. Linking names with ethnicities is an inexact science, but in Tyrone at least six people with apparently Irish names, O'Kelly, McMullan, McGowen, McArt, McGill and McCann, appeared at the muster with swords or other arms. In Donegal another dozen can be identified including McClearys and McConogheys.[75]

Against this evidence for some form of working agreement or assimilation there are indications that in the later stages of the plantation some Irish did not integrate well into the emerging social order. Such individuals appeared in the sources in generalised ways early in the process as 'woodkerne', living an ambivalent life on the edge of the plantation. By the 1630s they become more identifiable not simply as bandits but also linked to the plantation scheme itself. In Donegal, Turlogh Roe O'Boyle, who had received plantation lands, was said to have been involved with the Spanish in a plot for a rising in the county and he took part in the 1641 rebellion.[76] More shadowy figures, such as Donn Carrough Maguire and his brother, Edmund Carrough, were also active in Fermanagh in 1641. Donn Carrough lived on the edge of the social order holding land but doing so on annual leases rather than for longer periods as the government wanted in order to promote stability.[77] In Donegal, for instance, government tried to make landlords responsible for their native Irish tenants but this proved unworkable, landlords being unwilling to cooperate.[78]

The activities of such fairly well defined marginal figures suggests a number of forces working to push groups apart in the 1630s, as well as continuing bonds holding them together. Religion was clearly one factor and the more relaxed attitude to Catholicism in Ulster shown by the Dublin administration from the 1620s allowed the Catholic Church to rebuild itself in the province and in so doing so prevented a slide among the native Irish towards the Church of Ireland.[79] In addition to religion, economic trends served to shape the nature of the native Irish community in Ulster at a series of levels. At least some of those granted land in the plantation scheme as 'deserving Irish' had to sell their lands. One of the problems, according to the 1622 commissioners at least, was that 'they do not make certain estates to their tenants but do take Irish exactions as heretofore'.[80] As a result they failed to adapt to the more commercialised world of market rents, fell into debt and had to sell their grants. In Armagh, for instance, the percentage of land held by the native Irish fell from about 25% at the outset of the plantation to about 19%

in 1641, while in Cavan the fall was from 20% to 16% over the same period.[81] At a lower social level the dramatic economic growth that characterised Ulster after the plantation meant that the number of settlers and the demand for high-quality land increased over the first 40 years of the seventeenth century. This inevitably reshaped the patterns of settlement on many estates. On the Balfour estate in Fermanagh, for example, distinct patterns of Irish settlement can be detected by the 1630s. The evidence of the rentals, however, suggests that these were never exclusively Irish and included some intermixed settlers.[82] In some places where there had been substantial early settlement this sorting out process made little difference.[83] In other areas it pushed those Irish who remained on the estates of settlers in the early years of the century on to more marginal land so that by 1660 some 20% of townlands in the core of Ulster, in north Armagh and east Londonderry, had no Irish, while at the edge of the province only between 5% and 10% can be so described.[84] Some caution needs to be exercised in interpreting such figures since they clearly include the effects of war and dislocation in the 1640s and 1650s but the general patterns appear to be clear. Segregation combined with Catholic revitalisation certainly suggests that the levels of assimilation that probably existed in the first two decades of the plantation may have shaded into accommodation by the 1630s.

Despite these changing patterns of response by the native Irish, the plantation scheme in Ulster survived for 33 years without any significant challenge. Nonetheless, at least some contemporaries feared for the fate of plantations. In 1622 Francis Blundell observed how 'the Irish being many, strong and malicious and the undertakers so weak, poor and unprovided of houses, arms and means may easily be surprised by them'.[85] Blundell was correct. The native population greatly outnumbered the settlers, although the distribution of settlers across Ulster varied considerably. Throughout the early seventeenth century, however, the native population showed little inclination to undermine the plantation scheme. Rumours of plots surfaced in 1615 and again in the 1620s when Anglo-Spanish relations reached a low ebb, but there seems to have been little substance to these rumblings of discontent.[86] Other evidence of unrest is hard to find with little in either the records of the Ulster assizes or of the Dublin court of castle chamber, despite the fact that the authorities remained constantly alert. Only a handful of examples can be found in the early stages of the plantation, such as the man in Tyrone brought before the assizes for allegedly stating that 'the king of England was a very poor fellow... and that he did wonder that he should be king of England, for if it should be tried by the histories or chronicles, himself had as much right to be king as he'.[87] Equally, in the early stages of the 1641 rising refugees who deposed to the authorities in Dublin rarely mentioned the plantation as a grievance of those in arms. Indeed it was not until February 1642 that the more radical tract the 'Demands of the Irish' insisted that 'the Scots be removed out of the north of Ireland and the right owners which now beg about Ireland in great want and misery, though of most high blood and birth amongst the nobles of that country' recover their lands.[88] Such sentiments, however, were in short supply a few months earlier. When in October 1641 a number of the O'Reillys of Cavan wrote to the government to explain the reasons for the trouble in their region, the plantation did not feature in their analysis.[89] Instead they proclaimed their loyalty to the king, as did most of those Irish involved in the rising of 1641, and placed the difficulties squarely in the 1630s rather than two decades earlier. They also blamed the 'common sort of people' for the recent violence. The rising of 1641 was a conservative affair, seeking to return Ulster society to the way it had been before the appointment of Thomas Wentworth as lord deputy in 1632 rather than overturn the plantation. In fact Chichester's argument for giving small portions to the native Irish as part of the plantation scheme, so 'that the contentment of the greater number [of Irish would] outweigh the displeasure and dissatisfaction of the smaller number of better blood', would appear to have been correct.[90]

This essay has attempted to demonstrate that questions of success and failure in the Ulster plantation are by no means straightforward. From the perspective of central government the progress of the plantation could not be judged by the simple criteria of whether the undertakers obeyed the strict terms of the 'Orders and Conditions' of the scheme since clearly many did not. The government's aims in Ulster were more complex than the rather literal readings of the key documents by historians have suggested. The plantation was as much about the creation of a new social order as it was about introducing settlers and modifying the landscape, although clearly these elements in the agenda remained closely interlinked. In many ways the fundamental questions addressed by the planners of the Ulster plantation were not new. The end of the Nine Years War in 1603 had posed a series of problems that had to be addressed by the plantation planners. How, for example, was power to be distributed within that society and who would exercise that power? In 1603 the government had answered those questions in part by creating a series of checks on the main magnates, primarily by linking all the inhabitants of Ireland directly to the king.[91] The 'Flight' in 1607 clearly demonstrated that key figures, such as the earls of Tyrone and Tyrconnell, were not prepared to act in the same way as other great Catholic magnates in Ireland, such as Richard Burke, the earl of Clanricard and Sir Randall MacDonnell, later earl of Antrim. Their actions brought questions about the distribution of power and the structuring of society to the fore of the political agenda. The plantation process sought to address these issues rather than being a simple land redistribution exercise. Thus when in 1609 the king declared that the plantation was 'to make those parts not only secure, but a pattern to the other provinces of that kingdom' he did more than declare the royal will towards land. He revealed the crown's vision for the evolution of modern Ulster society.[92]

The full significance of the Ulster scheme is best understood by comparing it with the near contemporary settlement at Jamestown in North America. While there are clear differences in scale and in the migration patterns involved, important lessons can be learnt about the problems associated with establishing a plantation. It is clear that Ulster did not have the initial disastrous problems of the Jamestown settlement. Ulster, for instance, never needed to resort to the type of martial law enforced in Jamestown to control the settler population. Moreover, the settlers in Ulster knew from the outset what it took the Jamestown population almost ten years to discover: colonies were not simply about introducing large numbers of colonists but also about motivating, supporting and controlling those newcomers.[93] The real enduring success of the Ulster plantation lay in these areas. By considering the settlement of Ulster as an exercise in social creation rather than simply one of colonisation, and by ensuring that from the earliest stage each of the participants had a role to play in wider society, the planners hit on the ingredients for success. Those who argued that promoting economic growth would give the native Irish a stake in the new order and reduce the possibility of rebellion may have been closer to the truth than they realised.[94]

NOTES

1 For a critical assessment of this model see Raymond Gillespie, 'The problems of plantations: material culture and social change in early modern Ireland', in James Lyttleton and Colin Rynne (eds), *Plantation Ireland: Settlement and Material Culture, c. 1550–1700* (Dublin, 2009), pp. 43–60.

2 Mark Netzloff, 'Forgetting the Ulster plantation: John Speed's *The Theatre of the Empire of Great Britain* (1611) and the colonial archive', *Journal of Medieval and Early Modern Studies*, 31 (2001), 313–47, and for a refutation of the argument see John Andrews, 'Statements and silence in John Speed's map of Ulster', *Journal of the Royal Society of Antiquaries of Ireland*, 138 (2008), 71–80.

3 Aidan Clarke, 'The genesis of the Ulster rising of 1641' in Peter Roebuck (ed.), *Plantation to Partition: Essays in Ulster History in Honour of J. L. McCracken* (Belfast, 1981), pp. 29–45.

4 For examples of the long-term discontented see Raymond Gillespie, *Seventeenth Century Ireland* (Dublin, 2006), pp. 144–5.

5 George Hill, *A Historical Account of the Plantation of Ulster* (Belfast, 1877); Philip Robinson, *The Plantation of Ulster* (Dublin, 1985). Two studies are of particular significance, T. W. Moody, *The Londonderry Plantation* (Belfast, 1939) and R. J. Hunter, 'The Ulster plantation in the counties of Armagh and Cavan, 1608–41' (M.Litt. thesis, Trinity College Dublin, 1969).

6 For instance see Gillespie, *Seventeenth Century Ireland*, pp. 44–55; Pádraig Lenihan, *Consolidating Conquest: Ireland 1603–1727* (Harlow, 2008), pp. 41–7.
7 *Calendar of State Papers Relating to Ireland, 1603–25* (5 vols, London, 1872–80), 1611–14, p. 146.
8 British Library, London (BL), Harley MS 3292, fols 40–5.
9 Victor Treadwell (ed.), *The Irish Commission of 1622* (Dublin, 2006), pp. 755–63.
10 BL, Additional MS 4777, fol. 64.
11 J. T. Gilbert (ed.), *History of the Irish Confederation and the War in Ireland* (7 vols, Dublin, 1882–91), i, 2.
12 BL, Royal MS 18 A LIII, fols 8v–19. Bell had previously offered unsolicited advice to the government on Ireland, see The National Archives, London (TNA) SP 63/201/156; Raymond Gillespie, 'Plantation and profit: Richard Spert's tract on Ireland, 1608', *Irish Economic and Social History*, 20 (1993), 62–71.
13 Rolf Loeber (ed.), '"Certyn notes": biblical and foreign signposts to the Ulster plantation', in James Lyttleton and Colin Rynne (eds), *Plantation Ireland*, pp. 40–1.
14 For example see the proposal of Sir Francis Shaen for Donegal in *Cal. S. P. Ire.*, 1606–8, pp. 339–42 and BL Cotton MS Titus B XII, ff 112–17; TNA, SP 63/102 pt 4/75.
15 *Cal. S. P. Ire.*, 1606–8, pp. 325–6.
16 Ibid., pp. 276–7. Rumours of expulsion of all the Irish were certainly circulating see p. 286.
17 Ibid., p. 290.
18 T. W. Moody (ed.), 'Ulster plantation papers', *Analecta Hibernica*, 8 (1938), 281–6.
19 Moody (ed.), 'Plantation papers', 297.
20 P. J. Duffy, 'Patterns of landownership in Gaelic Monaghan in the late sixteenth century', *Clogher Record*, 10 (1979–81), 304–22; P. J. Duffy, 'The territorial reorganisation of Gaelic landownership and its transformation in County Monaghan, 1591–1640', *Irish Geography*, 14 (1981), 1–26.
21 Harold O'Sullivan, 'The Magennis lordship of Iveagh in the early modern period', in Lindsay Proudfoot (ed.), *Down: History and Society* (Dublin, 1997), pp. 168–72.
22 *Cal. S. P. Ire.*, 1608–10, pp. 67, 85.
23 Ibid., p. 139. The king attended at least one meeting.
24 *Cal. S. P. Ire.*, 1606–8, pp. 157–61, 501–3.
25 James VI, *Basilicon Doron* (Edinburgh, 1599), p. 43.
26 Moody (ed.), 'Plantation papers', 286–96.
27 Joan Thirsk, *Economic Policy and Projects* (Oxford, 1978), pp. 1, 18, 75–7.
28 *Cal. S. P. Ire.*, 1608–10, p. 259.
29 *Calendar of the Carew Manuscripts* (6 vols, London, 1867–73), 1603–25, pp. 43, 45; *Cal. S. P. Ire.*, 1608–10, pp. 339, 411.
30 Ibid., p. 116.
31 Ibid., p. 406.
32 *Cal. S. P. Ire.*, 1611–14, pp. 538–40.
33 Brian Mac Cuarta, *Catholic Revival in the North of Ireland, 1603–41* (Dublin, 2007), pp. 102–7. For Chichester's regard of Abercorn see R. D. Edwards (ed.), 'Letter book of Sir Arthur Chichester', *Analecta Hibernica*, 8 (1938), 20.

34 H. C. G. Matthew and Brian Harrison, eds., *Oxford Dictionary of National Biography* (60 vols, Oxford, 2004), sub Touchet, Mervin.
35 *Cal. S. P. Ire., 1608–10*, p. 355.
36 The details of the urban planning process are discussed in R. J. Hunter, 'Towns in the Ulster plantation', *Studia Hibernica*, 11 (1971), 40–79.
37 Public Record Office of Northern Ireland, MS D 4183/1/2.
38 The plantation grants were preserved on a number of rolls outside the main series see *Irish Patent Rolls of James I*, ed. M. C. Griffith (Dublin, 1966), pp. 163–7.
39 For example Public Record Office of Northern Ireland MSS D 623/B/1/4, D1928/T/2/1 (undertakers), D3007/A/5/1 (servitor), D1854/11/1 (native grantee).
40 Hill, *Historical Account*, pp. 84, 86.
41 Lord Belmore, *History of Two Ulster Manors* (Dublin, 1903), p. 74. For a wider consideration of the context see W. H. Crawford, 'The significance of landed estates in Ulster', *Irish Economic and Social History*, 17 (1990), 46–53.
42 J. C. W. Wylie, *Irish Land Law* (3rd edn, Dublin, 2005), pp. 74–8.
43 Hill, *Historical Account*, p. 83.
44 T. W. Moody (ed.), 'The revised articles of the Ulster plantation, 1610', *Bulletin of the Institute of Historical Research*, 12 (1935), 181.
45 John Davies, *A Discovery of the True Causes why Ireland was Never Entirely Subdued*…(London, 1612), p. 217.
46 *Cal. S. P. Ire., 1608–10*, p. 357.
47 Ibid., p. 355.
48 BL, Additional MS 4777, fol. 66v.
49 James Morrin (ed.), *Calendar of the Patent and Close Rolls of Chancery in Ireland, Charles I* (Dublin, 1864), p. 452.
50 Treadwell (ed.), *Irish Commission*, p. 760.
51 Robinson, *Plantation of Ulster*, p. 86.
52 T. W. Moody, 'The treatment of the native population under the scheme for the plantation of Ulster', *Irish Historical Studies*, 1 (1938–39), 59–63.
53 Treadwell (ed.), *Irish Commission*, pp. 157–8, 162, 179–80, 609.
54 Bodleian Library, Oxford, Carte MS 30, pp. 52–3.
55 BL, Harley MS 3292, fol. 41v.
56 Davies, *A Discovery of the True Causes*, pp. 280–2.
57 Thomas Blennerhasset, *A Direction for the Plantation of Ulster* (London, 1610); John B. Cunningham, 'The Blennerhassets of Kesh', *Clogher Record*, 15 (1999), 121–6.
58 Raymond Gillespie, *Conspiracy: Ulster Plots and Plotters in 1615* (Belfast, 1987), pp. 9–12, 35–7.
59 Barnaby Rich, *A New Description of Ireland* (Dublin, 1610), sig. B4.
60 Robinson, *Plantation of Ulster*, p. 189.
61 BL, Add. MS 4780, fols 69–69v.
62 BL, Add. MS 4780, fols 48v–50.
63 Robinson, *Plantation of Ulster*, p. 93.
64 Michael Perceval-Maxwell, *The Scottish Migration to Ulster in the Reign of James I* (London, 1973), pp. 344–5.
65 Edinburgh University Library, Laing MS Div II no. 5.

66 BL, Add. MS 4780, fols 32, 38v, 51v.
67 For this wider context of legal borrowings see Liam Mac Mathúna, *Béarla sa Ghaeilge* (Dublin, 2007), pp. 89–128.
68 Public Record Office of Northern Ireland, T808/15090, 15120, 15126, 15130–5, 15139.
69 J. F. Ferguson (ed.), 'The Ulster roll of gaol delivery', *Ulster Journal of Archaeology*, 1st service, 1 (1853), 260–70; 2 (1854), 25–9; R. M. Young (ed.), *Historical Notices of Old Belfast* (Belfast, 1896), pp. 30–9
70 BL, Add MS 3827, fol. 41; Raymond Gillespie, *Conspiracy*, pp. 36–7, 41.
71 T. G. F. Paterson, 'The Armagh manor court rolls, 1625–7', *Seanchas Ardmhacha*, 2 (1956–57), 301–9.
72 *Cal. S. P. Ire.*, *1615–25*, pp. 54–5.
73 Gillespie, *Conspiracy*, p. 47; Nicholas Canny, *Making Ireland British, 1580–1650* (Oxford, 2001), pp. 453–4 for Ulster examples.
74 Gillespie, *Conspiracy*, pp. 47–8; Canny, *Making Ireland British*, p. 487.
75 BL, Add. MS 4770, fols 77, 85, 90, 92, 94v, 108, 187v, 190v, 200, 203.
76 *Cal. S. P., Ire.*, *1625–32*, pp. 382, 383, 393, 396.
77 Raymond Gillespie, 'The murder of Arthur Champion', *Clogher Record*, 14 (1991–3), 56.
78 BL, Add. MS 3827, fol. 62.
79 For this process see Mac Cuarta, *Catholic Revival*, passim.
80 Treadwell (ed.), *Irish Commission*, p. 731.
81 Raymond Gillespie, 'The end of an era', in Ciaran Brady and Raymond Gillespie (eds), *Natives and Newcomers: Essays on the Making of Irish Colonial Society, 1534–1641* (Dublin, 1986), p. 195; for the case of the O'Hanlons see Joseph Canning, 'The O'Hanlons of Orier, 1558–1691', *Seanchas Ard Mhacha*, 18 (2001), 74–5.
82 John Johnston, 'Settlement of a plantation estate: the Balfour rentals of 1632 and 1636', *Clogher Record*, 12 (1985–87), 98, 99, 102, 105, 109.
83 Robinson, *Plantation of Ulster*, 102–3.
84 W. J. Smyth, *Map-Making, Landscapes and Memory: A Geography of Colonial and Early Modern Ireland* (Cork, 2006), p. 350.
85 BL, Harley MS 3292, fol. 44v.
86 For the 1615 plot see Gillespie, *Conspiracy*, passim.
87 Young (ed.), *Historical Notices of Old Belfast*, p. 389. The incidence of these complaints appear lower than in contemporary England, see David Cressey, *Dangerous Talk: Scandalous, Seditious and Treasonable Speech in Early Modern England* (Oxford, 2010), pp. 90–109.
88 J. T. Gilbert (ed.), *A Contemporary History of Affairs in Ireland from 1641 to 1652* (3 vols, Dublin, 1879), i, 382–3.
89 Gilbert (ed.), *A Contemporary History*, i, 364–5.
90 *Cal. S. P. Ire.*, *1608–10*, p. 335.
91 For this Gillespie, *Seventeenth Century Ireland*, pp. 33–42.
92 *Cal. S. P. Ire.*, *1608–10*, p. 245.
93 Karen Ordahl Kupperman, *The Jamestown Project* (London, 2007), especially pp. 322–7.
94 For example BL Add. MS 12490, fol. 119; Add. MS 39853, fol. 12v.

7

The Catholic Church in Ulster under the plantation, 1609–42

BRIAN MAC CUARTA

In 1606, based on his observations while on the assizes in mid-Ulster, the Irish Attorney-General Sir John Davies offered a pessimistic assessment of Ulster Catholicism's chances of survival. The adherence of the lower orders to organised religion was so weak that they were 'apt to receive any faith' if George Montgomery, Church of Ireland bishop of Clogher, Derry and Raphoe 'would come and be a new St Patrick among them'.[1] Davies believed there was little to stop the majority in Gaelic Ulster from embracing the reformed faith – an outcome that prevailed in parts of Gaelic Scotland – provided that preachers were sent to convert them. Yet by the outbreak of the rising in 1641–42, the Roman Catholic Church was so strong that its prelates and clergy soon replaced the Church of Ireland personnel as the de facto ecclesiastical establishment in those large areas of mid- and west Ulster under Irish military control.

This transformation demands exploration, for Catholic life has been a neglected corner in the field of recent plantation scholarship.[2] It is well known that the lack of any serious Protestant engagement with native culture, with the notable exception of Bishop William Bedell in Kilmore in the 1630s, contributed to Catholic survival in Ulster. By contrast, the process of internal reform within northern Catholicism in this era remains less well studied. Down to the eve of the plantation, the existence of a clerical body with forms of maintenance in which the coarb families played an integral role made Ulster distinctive. Military conflict took an increasing toll in the second half of the sixteenth century, culminating in the ravages of the Nine Years War, especially in its later stages. Thus by the onset of the plantation, traditional church structures were doubly traumatised both by the legacy of war, and the state-sponsored imposition of the new ecclesiastical establishment. Starting with this bleak scenario, from about 1609 a Catholic reconstruction emerged.

In this programme, both the friars and the diocesan leadership proved crucial. While the contribution of the Franciscan college of St Anthony's in Louvain (established 1607) to the broader Irish Catholic renaissance has received sustained scholarly attention, its impact in plantation Ulster has not been examined in detail. The role of the new continentally trained diocesan leadership in the northern province has been the subject of even less scrutiny.

In this essay it is proposed to trace how a vibrant Catholic structure emerged in the first three decades of the Ulster plantation, and its role in the collapse of colonial society in the winter of 1641–42. Starting with a survey of church life in the early years of the seventeenth century, the Catholic response to the diffusion of the established church which accompanied the introduction of the plantation will be considered. After the convulsions of the 1610s, the more stable tone of Catholic life in the succeeding decades will be explored, a phase characterised by the re-emergence of limited episcopal leadership, growing clerical assertiveness, and the presence of some Scottish recusant exiles, with the lordly Hamilton family playing a prominent role. Finally the sectarian dimension to Irish attacks on British settlers in the winter of 1641–42 will be outlined, and the emergence of a Roman Catholic Church establishment in the localities will be noted.

By the early years of the seventeenth century, the Catholic Church in general was in an extremely fragile condition. The Nine Years War, together with its uneasy aftermath, contributed to the erosion, if not destruction, of much of the ecclesiastical landscape of Ulster, a patrimony with deep roots in the middle ages. This alteration was most readily observable in the fate of church buildings. In the mid-sixteenth century, this network comprised simple parish churches, together with larger religious houses (generally friaries). Increasing warfare in the later sixteenth century placed this infrastruture under strain, a process greatly intensified by the revolt of the Ulster lords in the 1590s. Both Irish and crown forces saw the convents, the largest edifices in their localities, as strategic targets. Starting in east and south Ulster, with the northern extension of crown authority from the 1570s the friaries were being sequestrated and converted to secular use, if not destroyed. A similar process also occurred with the smaller, and more numerous, parish churches. With the onset of peace in 1603, roofless and ruined churches littered the countryside, although gradually some of these were fitted with thatch roofs, indicative of efforts to revivify communal worship after the turmoil of war. In line with the greater destruction and depopulation east of the Bann, parish structures and religious communities there were more severely attenuated.[3]

By 1609, diocesan life was in almost total disarray. A near full complement of resident bishops had been in place in the northern dioceses at the start of the war. Just prior to the plantation, however, only two bishops appointed by Rome remained. Both were ineffectual. Neil O'Boyle had been bishop of

Raphoe (largely covering Co. Donegal) since 1592, while Cornelius O'Devanny OFM, bishop of Down and Connor, had taken refuge with Cormac McBaron O'Neill (the earl of Tyrone's brother) in Co. Tyrone. The declining role of these prelates paralleled that of their protectors, the Gaelic lords. At a time of crisis for the old ecclesiastical dispensation, episcopal leadership was lacking. In its absence, the vicars-general of the dioceses exercised a limited oversight, though uncertainties over issues of jurisdiction vitiated their authority.[4] Parish boundaries were unclear, contributing to clerical friction. From the early 1590s, if not before, warfare limited the ability of prelates to supervise clergy and parishes, increasing the potential for clerical indiscipline and dispute.

Pastoral provision in Ulster dioceses was rooted in the system of coarbs and erenaghs. These ecclesiastical families fulfilled certain functions in the parish – they saw to the upkeep of the church, they had a care for hospitality, and they supplied priests from among their number. In return, they held the hereditary lands of their sept. By the early seventeenth century, however, this system of maintenance for parish clergy was in terminal decline. Already from the 1580s, the encroaching of crown authority along the fringes of south Ulster resulted in these landholdings being secularised. The Irish lords, under the pressure of maintaining fighting forces during the Nine Years War, also eroded the landholding of the coarbships.[5]

Despite the depopulation arising from war, evidence from south Ulster and the Derry diocese suggests that about three out of four parishes had an incumbent priest. This level of clerical provision may have prevailed across the province, with the exception of the area east of the Bann. Devotion to local saints featured highly, as the series of church dedications for Derry diocese indicates.[6] The vast majority of clergy received their training locally, apprenticed to an established priest. Their literacy, and their ability to speak Latin, equipped them for the role of mediators between Gaelic society and the crown, especially as the plantation got underway. A small minority from clerical families pursued studies outside the region. From the 1570s (if not earlier), those from south Ulster gravitated towards Catholic schoolmasters in Drogheda. The well-established trend for students from Donegal and Derry to travel to Glasgow had come to an end by the start of the plantation, a response to the growing Presbyterian ethos there. As Scotland faded from the educational horizon, continental centres beckoned, often via schools in Connacht. This external influence proved strongest in Raphoe and Derry dioceses, which contained a number of men schooled in Glasgow. These comprised a reform party within Raphoe diocese, and by 1600 tensions had erupted in an open altercation between these reformers and the traditionalist bishop, Neil O'Boyle, exacerbated by the stresses of war. The newer-style clerics wanted to depose O'Boyle but the prelate remained in situ until his death in 1611. Such friction between older and newer style clergy proved characteristic of diocesan life in Ulster, as elsewhere in Ireland, in the early seventeenth century. This undoubtedly contributed to the weakness of the church in Ulster as the plantation got underway.[7]

At the outbreak of the Nine Years War, the Franciscans ran a network of 13 convents across Ulster and north Leinster (the ecclesiastical province of Armagh). Together with the Franciscan Third Order, they represented a bulwark against denominational change, and a resource for Catholic reform.[8] War led to the destruction of the friaries, and the dispersal of community members. The political collapse of the traditional Gaelic polity following the Flight of the Earls contributed to the uncertainty regarding the survival of the old ecclesiastical order. By the onset of the plantation in 1609, the institutional basis of Franciscan life in Ulster was a pale shadow of its former self. Small groups of friars continued to lurk in the vicinity of some former convents. The future vitality of the friars, however, compared to the diocesan clergy, may be traced to their links with the continent, where Ulster friars among others established the Irish Franciscan college of St Anthony, in Louvain, in the Spanish Netherlands, the major Catholic educational centre in northern Europe. On their return to Ulster, men educated in Louvain made a vital contribution to the revival of Franciscan houses in the 1620s and to the renewal of Catholicism more generally through their preaching. The pastoral writings of Louvain-based members provided contemporary resources for preachers, while the historical project launched from St Anthony's helped create a new confidence in Irish Catholic identity. Moreover, Franciscan devotional poetry continued the tradition of popular Gaelic piety.[9]

The emergence from 1609 of the Anglican Church of Ireland in localities across Ulster influenced the evolving profile of the Roman Catholic Church. This denominational divergence left the adherents of the old church bereft of its medieval patrimony. Initially change was purely jurisdictional. Starting in areas of Ulster more strongly under the influence of the Dublin administration, the established church, in line with developments elsewhere in Ireland, absorbed a significant proportion of the existing diocesan clergy.[10] As long as indigenous clergy predominated, few changes occurred in pastoral or liturgical practice. Denominational ambiguity prevailed as a result, particularly in the absence of tridentine clergy, who would insist on a clear demarcation with the state church, thereby rejecting the compromises represented by the church papistry of the traditionalist majority.

The advent, however, of British ministers with the plantation led to a speedy, if geographically uneven erosion of the Gaelic strand in the ministry. This change in personnel sharpened confessional differences. Under the new order, the hereditary church families declined as landowners, while retaining their status as keepers of relics. The appropriation of the existing parish churches by the new regime, and the resulting displacement and destruction of religious images, compounded the sense of loss. Churches were associated with devotion to particular saints, many with purely local cults. It is estimated that about 70% of parish churches in use by the Church of Ireland in Ulster in the seventeenth century were located on a pre-Reformation site. Thus the plantation entailed the privation of a sacred landscape dear to the indigenous population. Evidence exists, however, of some continuity in Catholic worship down to the 1630s in churches in areas of purely Irish settlement.[11]

Tensions existed between the native community and representatives of the established church, including some Irish, over a range of ecclesiastical exactions. Anger at the vigorous and innovative imposition of tithes exploded in a wave of violence against church personnel in mid-Ulster in the winter of 1614–15. Competition between two sets of clergy for maintenance from the local population added to the social friction. As well as tithes, church-related mulcts included recusancy fines (actively imposed in the 1610s), fees associated with the rites of passage (christenings, marriages, funerals), and fines in the church courts for rites performed by recusant clergy. These charges fostered resentment against the established church and this deepened as a result of the collapse of ecclesiastical discipline during the Nine Years War. The poverty of the lower clergy of the Church of Ireland led them to be exigent in demanding fees from the native population. The degree of actual implementation of the legal and financial framework of the state church differed in each locality. This unevenness derived from the varying influence and power exercised by landlords on their estates. Strong resident landlords sought to limit the demands of state and church institutions, thereby making their lands more attractive to tenants. Elsewhere, tenants were exposed to the full rigour of church impositions and migration out of these areas often resulted. Ecclesiastical demands, and chicanery used in the church courts, reinforced native antipathy towards the Church of Ireland. As Bishop Bedell of Kilmore remarked in 1630: 'let us preach never so painfully and piously... let us live never so blamelessly ourselves, so long as the officers in our courts prey upon them [Irish], they esteem us no better than publicans and worldlings'.[12] This animosity contributed to the consolidation of the Catholic Church in Ulster in these decades.

Initially at least, native society responded in a confused, uncertain and ambiguous manner to the rapid appropriation of benefices, cathedrals, parish churches and ecclesiastical lands by the established church. The absence of tridentine priests highlighted the weak state of the existing clerical body, and eased the incorporation of some indigenous clergy into the new establishment. By contrast, the execution of clergy in the wake of O'Doherty's revolt (1608), and of Bishop Cornelius O'Devany and the priest Patrick Loughran in 1612, gave unambiguous and dramatic witness to the cause of resistance to confessional change.[13] By the early 1620s, a clear denominational divide existed, and the drift of Irish clergy into the Church of Ireland ceased. A system of parishes and dioceses, parallel to those of the state church, emerged, staffed by Irish clergy with a clear allegiance to Rome. The two major strands of Catholic renewal during the 1610s consisted of diocesan reorganisation under the direction of David Rothe, the vice-primate, and Franciscan missionaries returning from the recently established Irish Franciscan college in Louvain.

The task of animating and supervising the reform of the dioceses, in the face of the threat posed by a well-endowed and expansionist Church of Ireland, fell to David Rothe. A native of Kilkenny, Rothe served as rector of the Irish College in Douai, until called to Rome in 1602 as secretary to Peter Lombard, archbishop of Armagh. In 1609, Lombard appointed him vice-primate, and he returned to Ireland.[14] His arrival coincided with the beginning of plantation in Ulster, and the panic among diocesan leaders caused by the systematic imposition of conformity on the parish clergy. Those who refused lost their livings. In the opening years of the 1610s, Rothe devoted much time and attention to strengthening the old religion in the northern dioceses. The provincial synod, held over several days in Drogheda, Co. Louth, in February 1614, proved the high point of his endeavours.[15] The diocesan vicars attended, together with representatives of the religious orders active in the Armagh province – the Franciscans, alongside a handful of Cistercians and Jesuits. The programme adopted by the synod focused on strengthening territorially based parish and diocesan life, which inevitably entailed a diminution of the pastoral role of the friars. These had enjoyed wide faculties in the recent past, in response to weakened church structures.

The implementation of renewal devolved largely on the diocesan prelates, hence the need to select suitable men for this role. In the wake of the Reformation, Rome did not normally nominate bishops to head dioceses in the realm of a non-Catholic monarch, appointing vicars-apostolic instead.[16] Vicars-general, elected by the diocesan chapter, also took charge of dioceses: some prelates were the products of their local traditional clerical milieu, without a tridentine formation. Where possible, however, Rothe sought to appoint men who had received training on the continent. Indeed much of the impetus for renewal in Ulster dioceses came from that quarter. Thus in the early 1610s, a trio came to prominence under Rothe's patronage: Patrick Hanratty (vicar-apostolic of Down and Connor 1614–24, and of Dromore, 1625–30); Patrick Duffy (vicar-general of Clogher from 1612); Patrick Matthews (vicar-general of Dromore, Derry, and from 1626 until his death in 1637 of Armagh). Natives of Monaghan, each had studied abroad, indicative of the commitment of some south Ulster families to continental education in the years after 1603. While the work of reform in the northern dioceses called for men with a tridentine training, an ability to function in the Gaelic world of Ulster remained a prerequisite. In 1622 Luke Rochford, a native of the Pale, was appointed as vicar-apostolic in the Derry diocese, but he resigned after three years owing to his inability to preach in Irish.[17]

By visitations, admonitions and statutes they sought to reform the local clergy, and challenge abuses among the laity. Visitations were probably conducted at deanery level. We have an account of one such gathering in Loughinsholin barony, in southeast Derry in October 1613, where a Franciscan missioner, armed with the necessary authority, followed the procedures typical of the role of diocesan vicars at this time.[18] The surviving Catholic gentry of the vicinity attended this public event, suggestive of their close cooperation with the visiting ecclesiastic in the work of reform. Indeed the occasion took place under their patronage, as they provided the venue and encouraged the participation of the clergy and laity of the locality. The friar examined all fourteen priests of the area on a range of possible abuses. He deprived pluralists and deposed those clerics who refused to give up their concubines. Clergy deemed to have been irregularly ordained suffered a similar fate and in an effort to impress on the laity the advent of new standards of clerical life, couples married by them were made to undergo a new ceremony. Clerical maintenance had to be organised afresh in the wake of pressure on traditional forms of church landholding arising from war and plantation. Annual dues to parish clergy, together with fees for clerical services were decreed at the visitation. Significant occasions of donations to the clergy included funerals, where a chorus of women called down blessings on those giving, with more extensive blessings for those making a larger contribution, as at a gentry funeral in Downpatrick, Co. Down, in December 1617.[19]

Drogheda, a town in the Pale second in prominence to Dublin, served as a hub for northern clergy travelling to and from the continent, and as a Catholic educational centre for south Ulster. This prominent role of a Pale town points to the interconnectedness of the entire northern mission, embracing both Gaelic and Old English society. Nevertheless, cultural and political resentments underlay Catholic life in the north in the early seventeenth century. In the recent war, because of their loyalty to the crown the Old English on the northern fringes of the Pale had suffered the depredations of Hugh O'Neill's forces. Their civilizing mission towards the Gaelic-speaking Irish remained part of the Old English identity.[20] Ulster resentment at perceived Old English dominance in church affairs erupted in the debate over the appointment of a successor to Peter Lombard (d.1625) as archbishop of Armagh. One faction cited as a grievance the lack of familiarity of Old English prelates, such as Rothe and Rochford, with the Gaelic world of Ulster. Rothe, however, went to great lengths to inform himself of conditions in Gaelic areas, and prudence dictated that it was wise to supervise northern dioceses from the safety of the Pale. Until the early 1620s, pressure from sheriffs and provost marshals in Ulster made travel there difficult for the prelates.[21] Furthermore, the poverty and relative marginalisation of the surviving Catholic gentry in Ulster left them able to offer only limited support and security to ecclesiastics as they went about their visitations. Thus the northern diocesan vicars began to base themselves in the more secure and comfortable environs of the Pale, under the protection of the Catholic landed elite. Patrick Duffy, vicar-general of Clogher, lived on the estate of the Flemings, barons of Slane, in the early 1610s and by 1622 had established a house in Navan, Co. Meath. These premises served as a base facilitating links between congregations in the north with the Catholic heartlands of the Pale. Other forms of maintenance, unavailable in Ulster, made the Pale a supportive environment for the northern prelates, who acquired papal grants of benefices there. Given that the local gentry controlled church livings in the Pale, the prelates probably derived some benefit from these parishes. Diocesan vicars tended to reside in Meath or Louth over the winter, and only travelled north into their dioceses in the summer. This distance from their charges weakened their authority, leaving problems to fester, and led to a diminished ability to challenge insubordination among clergy.[22]

In various ways the new reformed outlook was diffused across the northern dioceses. Continentally educated prelates transferred from one diocese to another from the later 1610s and into the 1620s.[23] In this way tridentine clerics permeated the leadership cadre, even if the clerical body remained traditionalist. Similarly, Patrick Duffy, the Clogher vicar-general, habitually travelled across the north to assist diocesan leaders in the task of reform. In their perilous, arduous and delicate work of reforming local clergy while on visitation, the diocesan vicars received help and support from individual friars. This was one especially fruitful means whereby Franciscans returning from studies on the Continent assisted the local church. Jesuits offered more limited involvement with visitations, for Ulster remained largely peripheral to their life and ministry. In contrast to Leinster and Munster, they lacked kinship ties in the north, essential to ministry in these turbulent decades for the old faith. As the plantation got underway, diocesan prelates, including David Rothe, sought Jesuit assistance. Shaped by their Old English identity, the vice-primate and the members of the Society of Jesus shared a common cultural and ecclesial outlook. Rothe enjoyed warm relations with the Jesuit leadership during his years in Rome. Thus in 1616 the vice-primate asked that a father of the Society accompany him on his visitations in the north. During the 1610s, Jesuit ministry in Ulster took the form of limited yearly missions by individual men, launched from various bases in the Pale. On their travels, the missioners drew on the support and protection of landed Catholics, such as Randal MacDonnell, subsequently earl of Antrim.[24]

Annual missions to Ulster by northern Franciscans from their new college in Louvain represented a further strand of reform in the plantation decades. This pattern began about 1610, and continued until about 1620. By this time the Franciscan apostolate became more firmly based on the increasing number of informal residences re-emerging in locations of earlier convents across the north. In the early 1610s, these missions originated in representations from the Ulster émigré community in the Spanish Netherlands to the pope's representative, the Brussels inter-nuncio. Missions were held under the patronage of the surviving Irish gentry, and took place on their secluded estates, with people and parish clergy walking to the venue from a wide area.[25] Large crowds came to listen to these preachers from the continent, who could address them in the local Irish dialect. The missioners sought to offer a religious interpretation of the disasters which the Ulster Irish had experienced in recent times, namely the famine and destruction attendant on war, the exile of the Gaelic lords, the desecration of churches, and the arrival of English and Scottish landowners and settlers. The preachers attributed these calamities to divine retribution for the sins of the people, such as drunkenness, whoredom and a lack of devotion and zeal. The over-arching concern, however, of the friars, and presumably of the other Catholic preachers, was to reinforce Ulster Irish loyalty to Rome. To this end they emphasised the pope's interest in their plight, and his intention to send them missionaries (namely the friars) every year. They sought to replace church papistry – a passive acquiescence in the state-sponsored religious changes – with an unambiguously Roman confessional identity. Preachers in the north reiterated the widespread counter-reformation motif linking Protestantism with the devil. The friars denounced attendance at the services of the established church, and warned of the fate of those who would thereby listen to the devil's words. This message made a strong impression on their hearers and hastened the withering of the native Irish tradition within the Church of Ireland. More particularly, attitudes towards Protestantism fostered by preaching contributed to outbreaks of violence against personnel of the established church in mid-Ulster during the winter of 1614–15 and, in more radical fashion, in the breakdown of society in the early months of the 1641 rebellion.[26]

More overtly political concerns also surfaced, as the friars advocated support for the pan-Catholic alliance at the time of the 1613 parliament, and promoted the nationwide collection for the delegation sent to London to present the Irish recusant case at court.[27] The organisation of that collection in the north varied according to the situation of the Catholic community in different localities. In some areas local clergy acted as the main agents, illustrative of a functioning network of parishes with pastors. In other places, where Irish landowners survived, as in the Maguires' territory of Fermanagh, the sheriffs' servants collected the funds. This arrangement pointed to continued native control of aspects of the local administration in the early years of the plantation, which also prevailed in parts of the unplanted counties of Antrim and Down. Thus northern Catholicism became integrated into a broad campaign across the kingdom to achieve limited political ends, embracing both the Old English and Gaelic Irish communities, and relying on close cooperation between clergy and gentry. These aspirations co-existed with a more narrowly focused Ulster Irish agenda, based on the imminent return of Hugh O'Neill with a force from the Continent, a prospect rendered plausible by the brief break in negotiations for a royal match between London and Madrid in 1613.[28] This scenario, couched in apocalyptic terms, drew on prophecy and expressed the revanchist desires of the Ulster Irish.[29] The clerical interest in O'Neill derived from his close association with the Ulster Franciscans, and was at variance with the loyalist sensibilities of the vice-primate (David Rothe), a stance shared by some of the prelates he appointed to Ulster dioceses. Northern support for the nascent Irish colleges abroad provided a further sign of the growing integration of Ulster into the wider context of Irish Catholicism. Fund-raising for the colleges had been part of preaching in the Pale probably since the 1590s. From the early 1610s, merchants accompanied the friars on their missions in Ulster, where they supervised the collection of cattle for the port towns of the Pale. These cattle were slaughtered, salted and shipped abroad to feed the growing number of novices in the Franciscan college at Louvain.[30]

Although limited, the achievement of the Catholic reformers – diocesan vicars, complemented by Franciscan preachers – proved significant. The bulk of the local clergy remained traditionalist, evincing minimal engagement with newer, more exigent standards of clerical life. In place of previous confusion and ambiguity, however, the reformers established a clear allegiance to Rome among clergy and laity, and a corresponding antipathy towards the Church of Ireland. This stance contributed to the demise of indigenous clergy in the established church, hitherto a widespread presence. By the end of the decade, a parallel parochial and diocesan structure, characterised by adherence to Rome, had emerged across the province.

In the early 1620s, Catholicism in Ulster, as elsewhere in Ireland, entered a new phase, shaped by a marked easing of coercion, although a limited crackdown on northern bishops in 1627, and nationwide in 1637, punctuated this era of toleration. A more general anti-recusant campaign by the Dublin administration in 1630–32 involved the temporary closure of religious houses and mass-houses, and renewed efforts to impose the recusancy fine. The most dramatic manifestation of this policy in Ulster occurred with the destruction by Bishop Spottiswood of shrines and the recently erected buildings on the pilgrimage island of Lough Derg, in south Co. Donegal.[31] The bishop's exertions proved atypical. Expansion and consolidation characterised the Catholic

Church in these decades. Clergy increased in number and assertiveness. The Franciscan network of convents, fully developed in Ulster on the eve of the Nine Years War, had re-emerged by the mid-1620s.[32] These new smaller communities were re-established in the vicinity of former foundations across south and east Ulster, albeit in the ordinary dwelling houses typical of the area. The friars in Ulster maintained links with their confreres in St Anthony's College, Louvain, via the port town of Drogheda, and onward through London. Young Ulstermen travelled to Louvain for training, and returned as preachers. The friars organised collections in Ulster to pay for these studies, while preaching, questing and Gaelic devotional verse were regular features of the Franciscan contribution to Catholicism in Ulster during these decades. The convents also participated in the emergence of informal Catholic grammar schools in the years before 1641.[33] The Franciscan mission to the islands and highlands of Scotland from 1619, based in the friary of Bonamargy, in north Antrim, which enjoyed the patronage of the earl of Antrim, provided further signs of vitality.[34]

In line with the reinvigoration of Catholic structures across Ireland from the 1610s, diocesan life in Ulster came under the direction of a cadre of continentally trained clerics. From the mid-1620s resident bishops were active in the dioceses of Armagh, Down and Connor, Kilmore and Raphoe.[35] These men, together with the prelates in charge of the remaining dioceses, sought to re-energise Catholic life. Reform of the pastors was a central aim, and regular synods of clergy became the norm. Marriage in Gaelic Ireland remained in disarray. The turmoil associated with war and plantation vitiated efforts to introduce the reforms arising from the Council of Trent (1545–63). Uncertainty reigned, both as regards the local promulgation of the decrees of Trent and the issue of jurisdiction within the dioceses, at least until the restoration of bishops in the 1620s. Clerical greed contributed to the widespread prevalence of divorce, as both Church of Ireland ministers and Catholic clergy sought the associated fees. Catholic prelates challenged the Church of Ireland's monopoly on marriage jurisdiction, and by 1641 some reports outlined limited progress in introducing tridentine marriage procedures.[36]

Public worship was another area of concern for the bishops. From the early 1620s, a change in the habitual venue for the celebration of mass may be detected in areas where a Catholic gentry network survived. The custom-built mass-house first appeared in the port towns in the 1590s, and had spread throughout the Pale by the 1620s. They also appeared in Co. Cavan during that decade, reflective of the strength of Old English landownership in this south Ulster county. In addition, an improvement occurred in altar plate, and in decorum at services, though these changes remained dependent on the wealth and standing of the Catholic community in the various localities.[37] The most magnificent setting for the Catholic liturgy in Ulster was in the chapel of Dunluce Castle, Co. Antrim, the home of Randall MacDonnell, second earl of Antrim. Under the direction of the countess (Katharine Manners, widow of the duke of Buckingham) the chapel was restored at substantial cost in the late 1630s, and supplied with rich vestments and altar plate.[38]

A further facet of this transformation was the growing assertiveness of Catholic clerics, as they increasingly challenged the monopoly on dispensing sacraments claimed by the established church. In contrast to the fluid denominational borders which prevailed in previous decades, priests now claimed an exclusively Roman identity for their parishioners. Lay people who had made a formal adherence to the established church were reintegrated into the Roman fold, a process associated with the activity of tridentine missioners. Those who had abjured the traditional faith were treated as objects of fear in their communities, arising from the homiletic motif of the alleged link between Protestantism and the devil.[39] Reintegration involved a period of penance, symbolised by the wearing of a white linen gown, with echoes of the baptismal liturgy, and a public acknowledgment of their new status by the parish priest.[40] In growing defiance of the established church, Catholic diocesan authorities were making exclusive claims to ecclesiastical jurisdiction, particularly in the area of the Londonderry plantation. By the early 1630s, clergy in the diocese of Derry exercised jurisdiction very publicly, divorcing a number of couples because a minister of the established church, not a priest, had performed the marriage. Similarly, on the southern borders of Ulster, Catholic diocesan officials excommunicated a woman who brought a marriage case before the ecclesiastical court of the Church of Ireland.[41]

Clerical demands within the Catholic community also increased. In some localities, by the late 1620s priests resorted to the common law to insist on dues from their parishioners. This use of the legal system may have been linked to a favoured relationship with figures in the local administration, as in Co. Derry, or where native Irish landowners survived, such as parts of Co. Down. The Catholic prelates in other parts of Ulster viewed the diocese of Derry with envy, as an oasis of ecclesiastical calm.[42] The Londoners promoted Catholic clergy as tenants on their proportions and they facilitated the erection of mass-houses. They also favoured the priests in their lawsuits. Indeed priests in the Londonderry plantation had laymen whipped who dared to bring lawsuits against clergy, and were able to exact a cow or mare or other large beast for rites of passage such as christenings, marriages, extreme unction and burials.[43]

A similarly protective regime operated in the barony of Strabane, in northwest Tyrone, under a family of Scottish Catholic undertakers, the Hamiltons, with the earl of Abercorn as their head.[44] Among the leading recusant families in Scotland, the Hamiltons exercised a particularly muscular form of seigneurial Catholicism both in their home estates around Paisley (in

Renfrewshire), and on their plantation lands in Co. Tyrone. Strabane became a refuge for recusants fleeing religious coercion in Scotland. Within the colony, a militant strand of Scottish recusancy thrived, with a strong emphasis on gaining conversions to Catholicism from among Scots settlers in the area. Religious pressure on tenants came from a variety of sources. It was alleged that the landlord's agent discriminated against Protestants in the allocation of tenancies. The leading merchant was a recusant exile, as was Dr Barclay, the physician, who railed against the reformed faith. Venues for mass included the homes of these men, and also the more spacious residences of the Hamiltons. A resident Jesuit, who lived in the castle down to the firing of Strabane by Sir Phelim O'Neill's troops in the winter of 1641, further animated the conversion effort. Thanks to Hamilton protection, Strabane became the de facto headquarters of Derry diocese, with the vicar-general living there by the late 1620s, and regular assemblies of the diocesan clergy taking place in the town. The government's anti-recusant crackdown in 1629–30 further highlighted this seigneurial protection of Catholicism. The royal bishop of Derry, following government orders, sought to convict prominent residents on charges of recusancy, but to no avail, as under Abercorn leadership, local Catholics ignored the bishop's summons with impunity. The court connections of this distinguished Scottish family provided a bulwark against religious coercion emanating from the Dublin administration and their agents in the Ulster localities.[45]

Another successful magnate family, the MacDonnell earls of Antrim, did much to further Catholic interests on their vast estates in north Antrim, a fact acknowledged by a group of prelates in 1632. The MacDonnells provided refuge for their co-religionists fleeing coercion in Scotland, sometimes temporarily, as in the case of the Jesuit, Patrick Anderson, in 1621. From the early 1610s, Sir Randal MacDonnell facilitated the mission of individual Jesuits to the islands and highlands of west Scotland, and to communities of Scots in north Ulster. As noted above, the earl's engagement with the Catholic revival in the west of Scotland intensified with his support of the Irish Franciscan mission there. The peer's kinsman, Bishop Bonaventure Magennis OFM, stayed at Dunluce Castle on pastoral visits to the northern part of his diocese, and in the summer of 1639 he confirmed about seven hundred Scots who had sailed across to Antrim to avail of his ministry.[46] A similarly protective stance characterised the other lesser Ulster Catholic peers, the Magennises (viscounts Iveagh) and the Maguires (barons Enniskillen) though Maguire in particular rapidly sank into debt during the course of the 1630s.

Wide local variations in the social standing of the old faith developed across Ulster. Strong resident landlords protected their tenants from the exactions of outside agencies, including officials and collectors associated with the established church. In this way tenants on the Bagnall estates in south Down, which incorporated the territory of the medieval Cistercian monastery of Newry, remained largely exempt from the ecclesiastical charges facing tenants on less closely managed lands. Thus the Catholic community there avoided the full burden of recusancy fines, and the various mulcts associated with church courts.[47] By contrast, tenants on the absentee earl of Essex's lands in Farney barony, in Co. Monaghan (bordering Co. Louth) were left exposed to the depredations of local parsons and prelates.[48] A significant number of recusant landowners survived in Cavan, to a degree exceptional in the planted counties. Old English families acquired lands in the territory following the dissolution of the monasteries in 1540. By 1641, the Plunketts, earls of Fingall, and the Nugents, earls of Westmeath were among the largest landowners in the county. The diffusion of masshouses in Cavan by the late 1620s reflected patterns of Catholic life which began in the port towns in the 1590s, and spread outwards into the Pale countryside thereafter. Tensions existed, however, between the Catholic bishop of Kilmore and these magnates. The latter retained the church livings acquired at the Dissolution, to the impoverishment of the parish clergy, while the Catholic diocesan authorities hoped that these parish benefices would be restored to the diocese. The arrival of some British gentry families in the county by about 1639, recusant refugees from political instability in Scotland and England, underlined the stability and continuity of Catholic landowning. Yet the lack of tridentine clergy (apart from a few friars and the diocesan leadership) pointed to the limits of elite support for a renewed Catholicism. Continentally trained clergy remained clustered in the Pale, rather than the outlying Cavan estates of seigneurial families whose principal residence lay elsewhere.[49]

By 1641, Catholic renewal had made significant progress in Ulster to the extent that the province became a place of refuge for Scots recusants fleeing religious coercion in their homeland. This trend points to the links between the Catholic communities in the Jacobean realms.[50] The migration arose from geographical proximity, the vastly increased inflow of Scots in the years after 1603, and the webs of patronage spanning Scotland and Ulster maintained by several lordly families such as the Hamiltons. It included the movement of islanders across the Straits of Moyle to Antrim, a traditional response to crown expansion within the context of the Gaelic lordship of the isles, and in the 1610s and 1620s embraced recusants in Dumfries and Paisley.[51] English Catholics also arrived in Ulster with the plantation. In Scotland, the 'personal rule' of King Charles (1629–40) witnessed renewed coercion against recusants, and several Catholics of high social standing found a haven in the north of Ireland, facilitated by the Ulster Franciscan missioners active in the west of Scotland. It would appear that the troubles of the late 1630s also gave rise to the migration of a small number of recusant gentry families (English and Scottish) to Ulster. Thus, in the early seventeenth century, Ulster served as a permissive frontier, where religious adherence, unacceptable to the authorities in the metropolitan area, could be more freely practised.[52]

The endemic poverty of the Ulster Irish, however, intensified by social decline under the plantation, put clear limits on the progress of revival within institutional Catholicism.[53] Deprivation reinforced conservative tendencies, and accounted for the lack of a variegated tridentine culture. As in other parts of Ireland, poorly trained vagrant clergy remained as a clerical substratum, beyond formal ecclesiastical control. Erenagh families continued to provide parish clergy, and survived as revered custodians of artefacts venerated for their association with local saints. The prelates' lack of means left them dependent on often impoverished Catholic landed families. As a result, Bishop John O'Cullenan left his diocese of Raphoe to live in the Pale, and even contemplated emigration to France.[54] Ecclesiastics without powerful relations within the diocese lacked the authority to challenge traditionalist behaviour among clergy and laity. To a greater degree than in the Catholic heartlands of the Pale, northern prelates were thus exposed to charges of exercising foreign jurisdiction brought by the recalcitrant clergy, leaving bishops subject to harassment and occasional imprisonment by the authorities.

Ulster's experience of the Catholic revival in the early seventeenth century contrasted sharply with that of communities in Leinster and Munster where a relatively wealthy landed Catholic elite survived. On the eve of the 1641 revolt, the material poverty of the indigenous northerners, together with the persistence there of conservative forms of church life, reinforced existing provincial differences within Irish society. These broader cultural divergences underlay the political and military tensions which emerged during the 1640s.[55] As early as January 1642, this fractiousness was already surfacing in the liturgical sphere, with the arrival in south Ulster of Old English refugees from the Pale. The pastor in Cavan would not permit clergy among the followers of Christopher Plunkett, earl of Fingall, to celebrate mass in the local church, while the newcomers refused to attend mass with the incumbent of the parish. Pastimes after mass also reinforced the cultural divide; while the Old English played tennis, the Cavan people amused themselves by drinking, piping and dancing.[56]

In addition to the tensions within the Catholic community, the role of Catholicism in the collapse of Ulster society in the winter of 1641–42 needs to be considered. By the outbreak of the Irish rising in 1641, significant concentrations of English and Scottish settlers (overwhelmingly Protestant) were to be found alongside the indigenous population in parts of Ulster. This extensive co-existence gave a particular intensity to the Catholic revanche which was a major feature of the crisis in the province. A degree of continuity may be detected between the growth in Catholic assertiveness in the face of the established church across Ireland during the course of the 1630s, and the ecclesiastical restoration underway in the north after October 1641, as expressed in the commandeering of the sizeable homes of some prominent settlers for the use of Catholic clergy, and the re-appropriation of churches.[57] Some priests proved reluctant, initially at least, to celebrate mass in these premises, presumably because of fears arising from their polluted status. In March 1642 the Ulster synod decreed that priests could celebrate there without scruple, using the portable altars customary hitherto at masses on hills, or in woods, or in private homes.[58] Thus in the localities, with the collapse of royal rule, the Irish moved towards replacing the Anglican Church with a Roman ecclesiastical establishment.

The breakdown of restraints allowed latent anti-Protestant feeling, gestating in Irish society since the mid-sixteenth century, to flourish. These atavistic animosities provided a purgative focus and rhetorical justification, couched in religious terms, for violent conduct by the Irish towards their neighbours. This process soon spread across Ireland but was intensified in the north by the dense, and relatively recent influx of Scots and English settlers, and the consequent prominence of Protestant communities in many localities.[59] Sectarian hostility was based on the premise of intrinsic links between Protestantism and the devil; the prevalence of this view in the reported words of the insurgents across south Ulster in 1641–42 indicates that it had been a staple subject of Catholic preaching at least since the early seventeenth century.[60] With the collapse of social constraints characteristic of those months, the practical consequences of this theological categorisation proved brutal, and encompassed a range of actions. The Irish perceived the Bible as the central symbol of Protestantism and targeted copies of the book for destruction and ritual abuse. This could lead, at times, to a generalised attack on all books found in the possession of settlers, linked perhaps to a desire to destroy ledgers which recorded debts owing by the Irish to local settlers. The ministers' role as money-lenders, together with a significant increase in clerical income in the 1630s, contributed to the animosity of native attacks on the pastors and prelates of the established church.[61] Thus as in 1614–15, attacks on church personnel had roots in social and economic grievances, which aggravated sectarian tensions. The rebels also destroyed or desecrated the interiors of some church buildings, including benches for seating, and royal armorial plaques. A desire to purge the locality of the taint of Protestantism intensified a popular concern with the conversion of Protestant neighbours, and contributed to the pressure on English and Scots settlers. Disinterment of Protestant graves and a refusal to allow the burial of Protestants were further expressions of a desire to purify the locality of all vestiges of the alien religion.[62]

Catholic preaching in the plantation era served to give a sectarian veneer to animosities which by 1641 were as much political, social and economic, as religious, reflective of the marginalisation and indebtedness of the Ulster Irish within the new colonial society.[63] The increasing political instability within the three kingdoms from the late 1630s heightened Catholic fears of religious coercion, anxieties arising from the growing political prominence of more extreme forms of Protestantism in Scotland and England. The numerous and ever-mobile friars, with their ready links to London as part of their route to and from the continent, diffused these anxieties, which included a focus on the fate of the Catholic Queen Henrietta Maria, throughout Ulster.[64] As the Irish insurgency got underway, the Louvain-educated Franciscans active as preachers in Ulster portrayed the conflict as a struggle of national and religious purification. To this end they drew on various parts of the Hebrew Bible recounting the struggle against the dilution of Jewish identity in the Hellenistic era. Pride of place was given to the Jewish epic of the Maccabbees, as recorded in the eponymous books of the Old Testament. This family led a successful revolt – national, cultural and religious in inspiration – against the hellenizing programme of the Jewish kings in the second century BC.[65] Thus the friars provided a scriptural frame of reference, replete with stirring stories, to support the insurgents' cause.

The initial response of the lower clergy tended to reflect the frustrations and actions of their parishioners and kinsmen. In their preaching, they supported the popular desire to rid the localities of Protestants, either through forced conversions or expulsion and they claimed ecclesiastical premises for the Catholic Church. Protestant refugees named a few clergy (Hugh Mac an Deaganaigh Maguire in Fermanagh, and James O'Hallagan in north Armagh) as leaders in violence against the settlers, suggesting that the majority of priests did not assume a prominent role in the attacks. Others sought to exercise a restraining influence, in defence of settlers under threat.[66] Faced with the collapse in social restraint at the end of October 1641, senior Catholic clergy proved anxious that law and order be reimposed. To that end, a synod of northern diocesan representatives took place at Kells, Co. Meath, in March 1642; a series of edicts sought to combat social disorder in the recent revolt, decreeing the excommunication of those involved in killings, thefts and robberies, and condemning those who occupied the lands of others. The synod ordered a three-day fast in the parishes, to conclude with confession and communion. The ecclesiastics also sought to foster a united front with the local Catholic gentry in establishing a system of local government, an initiative which culminated in the foundation of the Catholic Confederation in Kilkenny in October 1642.[67] This new-found public role was yet one more indication that the Roman Catholic Church in parts of Ulster and throughout much of the kingdom had assumed the de facto position of the established church, an outcome which would have seemed highly unlikely only a few decades earlier.

NOTES

1 'Observations of Sir John Davys', 4 May 1606, *Calendar of the State Papers Relating to Ireland* (24 vols, London, 1860–1911) (CSPI), *1603–6*, p. 468.
2 However, see the sections on the early seventeenth century in surveys by M. Elliott, *The Catholics of Ulster: A History* (London, 2000), and O. Rafferty, *Catholicism in Ulster 1603–1983: An Interpretative History* (London, 1994); Catholic ecclesiastical developments in early seventeenth-century Ireland are outlined in P. Corish, *The Catholic Community in the Seventeenth and Eighteenth Centuries* (Dublin, 1981), pp. 18–42.
3 On the ecclesiastical situation of mid-sixteenth-century Ulster, see Henry Jefferies, *Priests and Prelates of Armagh in the Age of Reformations, 1518–1558* (Dublin, 1997); the expansion of crown influence in Ulster in the late sixteenth century is treated in H. Morgan, *Tyrone's Rebellion: The Outbreak of the Nine Years War in Tudor Ireland* (Woodbridge, 1993); on ecclesiastical decline by the early seventeenth century, see B. Mac Cuarta, *Catholic Revival in the North of Ireland 1603–41* (Dublin, 2007), pp. 17–36.
4 On bishops in 1600, see J. Hagan (ed.), 'Some papers relating to the Nine Years War', *Archivium Hibernicum*, 2 (1913), 301; for the emergence of vicars apostolic, see F. M. Jones, 'Canonical faculties on the Irish mission in the reign of Queen Elizabeth, 1558–1603', *Irish Theological Quarterly*, 30 (1953), 152–71; for the career of one Ulster bishop, see I. Fennessy, 'Richard Brady OFM, bishop of Kilmore, 1580–1607', *Breifne*, 36 (2000), 225–42.
5 On the coarbs and erenaghs, see K. Simms, 'Frontiers in the Irish church – regional and cultural', in T. Barry, R. Frame and K. Simms (eds), *Colony and Frontier in Medieval Ireland* (London, 1995), pp. 177–200; grant of coarbship of Clones, Co. Monaghan, to Sir Henry Duke in 1588, *Irish Patent Rolls of James I* (Dublin, 1966), p. 11; for an example of O'Neill's appropriation of a coarbship in the 1590s, see T. Moody and J. Simms (eds), *The Bishopric of Derry and the Irish Society of London, 1602–1705* (Dublin, 1968), i, 182.
6 On church life in northwest Ulster just prior to the plantation, see 'Bishop Montgomery's survey of the bishoprics of Derry, Raphoe and Clogher', in A. Alexander (ed.), 'O'Kane papers', *Analecta Hibernica*, 12 (1943), 69–127; H. Jefferies, 'Derry diocese on the eve of the plantation', in G. O'Brien and W. Nolan (eds), *Derry and Londonderry: History and Society* (Dublin, 1999), pp. 175–204.
7 Mac Cuarta, *Catholic Revival*, pp. 27–9; developments in Raphoe are outlined in E. Maguire, *A History of the Diocese of Raphoe* (Dublin, 1920), i, pp. 124–32.
8 Colm Lennon suggests that the diffusion of Franciscan Third Order convents, especially in Connacht and Ulster, and the survival of those in northwest Ulster down to the early seventeenth century, contributed to an openness towards Catholic reform, C. Lennon, 'The dissolution to the foundation of St Anthony's College, 1534–1607', in E. Bhreathnach, J. McMahon and J. McCafferty (eds), *The Irish Franciscans 1534–1990* (Dublin, 2009), pp. 3–36, at 15–16.
9 For an overview of Franciscan Ireland from the mid-sixteenth century, see C. Lennon, 'The dissolution to the foundation of St Anthony's College, Louvain,

1534–1607'; on the historical, hagiographical and pastoral dimensions of the Louvain project, see the essays by Bernadette Cunningham, Pádraig Ó Riain and Salvador Ryan in the above collection; for examples of Franciscan devotional poetry prior to 1600, see C. Mhag Craith (eag.), *Dán na mBráthar Mionúr* (2 vols, Dublin, 1967–1980).

10 On the absorbtion of native clergy into the nascent Church of Ireland in Ulster from the 1590s until the early 1610s, see B. Mac Cuarta, 'Catholicism in the province of Armagh, 1603–1641' (Ph.D. thesis, Trinity College Dublin, 2004), pp. 33–9; this trend is also reflected in 'A presentment of clergy', Co. Monaghan, July 1606, *Report on the Manuscripts of the Late Reginald Rawdon Hastings* (HMC, 4 vols, London, 1928–47), iv, pp. 154–6, and in scattered references to the presence of Irish clergy in individual parishes, e.g. 'Grievances of tenants…in Farny' [1622], National Library of Ireland, Dublin (NLI), MS 8014 (x).

11 On the emergence of the Church of Ireland in Ulster, see A. Ford, *The Protestant Reformation in Ireland, 1590–1641* (Dublin, 1997), pp. 127–54; on the location of parish churches, see William Roulston, 'The provision, building and architecture of Anglican churches in the north of Ireland, 1600–1740' (Ph.D. thesis, Queen's University Belfast, 2003), pp. 122–3.

12 W. Bedell to Archbishop Ussher, 15 Feb. 1629–30, James Ussher, *The Whole Works of…James Ussher*, ed. C. Elringrton and J. Todd (17 vols, Dublin, 1847–64), xv, 467; for a study of these developments in one Ulster diocese, see A. Ford, 'The reformation in Kilmore to 1641', in R. Gillespie (ed.), *Cavan: Essays on the History of an Irish County* (Dublin, 1995), pp. 73–98; on the poverty of church livings in sixteenth-century Ireland, relative to Wales, see S. Ellis, 'Economic problems of the church: why the Reformation failed in Ireland', *Journal of Ecclesiastical History*, 41: 4 (1990), 239–65, at 248–52.

13 On the growth of Catholic structures in one northern diocese, see B. Mac Cuarta, 'Catholic revival in Kilmore diocese, 1603–41', in Brendan Scott (ed.), *Culture and Society in Early Modern Breifne/Cavan* (Dublin, 2009), pp. 147–72; K. Devlin, 'Conor O'Devany OFM and Patrick O'Loughran', in P. Corish and B. Millet (eds), *The Irish Martyrs* (Dublin, 2005), pp. 107–37; on the martyrdom of Donogh McRedie, dean of Derry, see *Congregation for the causes of saints…Cause for the beatification and canonization of the servants of God, Richard Creagh et al.* (1998), ii, 733–51.

14 Thomas O'Connor, 'Rothe, David (1573–1650)', in James McGuire and James Quinn (eds), *Dictionary of Irish Biography* (9 vols, Cambridge, 2009) (DIB), viii, pp. 621–3.

15 On early seventeenth-century synods in Ireland, see Alison Forrestal, *Catholic Synods in Ireland, 1600–1690* (Dublin, 1998).

16 F. Jones, 'Canonical faculties on the Irish mission', passim.

17 Mac Cuarta, *Catholic Revival*, 'Appendix I: vicars-apostolic and vicars-general, province of Armagh, 1603–41', pp. 247–52; on Rochford's linguistic limitation, see Thomas O'Connor, *Irish Jansenists 1600–70: Religion and Politics in Flanders, France, Ireland and Rome* (Dublin, 2008), pp. 113–14.

18 The National Archives, London (TNA), SP 63/232/21, Deposition of Teag Modder M'Glone, 21 Oct. 1613 (*CSPI, 1611–14*, pp. 429–31).

19 B. Mac Cuarta (ed.), 'A Catholic funeral in Co. Down, 1617', *Archivium Hibernicum*, 60 (2006-7), 320-5.
20 On the ecclesiastical dimension to Pale identity in the sixteenth century, see James Murray, *Enforcing the English Reformation in Ireland: Clerical Resistance and Political Conflict in the Diocese of Dublin, 1534-1590* (Cambridge, 2009).
21 Sir Moses Hill, provost marshal in Ulster from 1603 until 1629, was a noted opponent of Catholicism, Mac Cuarta, *Catholic Revival*, p. 20.
22 See row over replacement of a rural dean, Derry diocese, in the early 1620s, Moody and Simms (eds), *The Bishopric of Derry*, i, pp. 143-4.
23 Mac Cuarta, *Catholic Revival*, appendix 1, 'Vicars-apostolic and vicars-general, province of Armagh, 1603-41'.
24 On Irish Jesuit missionary activity in the west of Scotland in the early seventeenth century, see Fiona Macdonald, *Missions to the Gaels: Reformation and Counter-Reformation in Ulster and the Highlands and Islands of Scotland, 1560-1760* (Edinburgh, 2006), pp. 44-53.
25 Franciscan missions in Ulster are noted in two contemporary examinations, TNA SP 63/232/21-2, 'Deposition of Teag Modder M'Glone...21 Oct. 1613', and British Library, London, Cotton MSS, Tit. B.X., 'The examination of Shane McPhelomy O Donnelly...22 Oct. 1613', fol. 236 r-v (*CSPI, 1611-14*, pp. 429-31).
26 On the attacks on church personnel in 1614-15, see Mac Cuarta, *Catholic Revival*, pp. 57-9; on the confessional dimension in the breakdown of 1641-42, see Mac Cuarta, 'Religious violence against settlers in south Ulster, 1641-2', in David Edwards, P. Lenihan and C. Tait (eds), *Age of Atrocity: Violent Death and Political Conflict in Ireland, 1547-1650* (Dublin, 2007), pp. 154-75.
27 For the broader political context, see Aidan Clarke, with R. D. Edwards, 'Pacification, plantation and the Catholic question, 1603-33', in T. W. Moody, F. X. Martin and F. J. Byrne (eds), *The New History of Ireland*, iii: *Early Modern Ireland, 1534-1691* (Oxford, 1976), pp. 187-231, at 214-17.
28 On this rupture in the royal match, see Glyn Redworth, *The Prince and the Infanta: The Cultural Politics of the Spanish Match* (New Haven and London, 2003), pp. 10-11.
29 On the use of prophecy in popular political discourse, see R. Gillespie, *Devoted People* (Manchester, 1997), pp. 137-42.
30 C. Giblin (ed.), *Liber Lovaniensis* (Dublin, 1956), pp. 58-9; for customs exemption (Sept. 1617) on salted beef, fish, butter and domestic fabrics imported from Ireland for the use of St Anthony's, Louvain, see B. Jennings (ed.), *Louvain Papers* (Dublin, 1968), pp. 58-9.
31 Henry Jones, *St Patrick's Purgatory* (London, 1647); on the evolution of devotion on this pilgrimage site, see B. Cunningham and R. Gillespie, 'The Lough Derg pilgrimage in the age of the counter-reformation', *Éire-Ireland*, 34: 3-4 (2004), pp. 169-79; for an articulation of official thinking which led to this campaign, see B. Mac Cuarta (ed.), 'A document concerning the Irish Privy Council's policy towards Catholicism (1629)', *Archivium Hibernicum*, 62 (2009), 150-3.
32 On the growth of Franciscan convents in the early seventeenth century, see R. Gillespie, 'The Irish Franciscans, 1600-1700', in Edel Bhreathnach *et al.* (eds), *The Irish Franciscans*, pp. 45-76, at 45-57.

33 *The humble petition of the Protestant inhabitants of the counties of Antrim, Downe, Tyrone, etc* (London, 1641), p. 6; Connor Maguire, Baron Enniskillen, was educated by the friars at Lisgoole, Co. Fermanagh, see Brian Mac Cuarta, 'Maguire, Connor, second baron of Enniskillen (c.1612–1645)', in *Oxford Dictionary of National Biography* (Oxford, 2008); online edition (*ODNB*), www.oxforddnb.com/view/article/17790 (accessed 3 October 2011).

34 On this enterprise of the Irish Franciscans, see C. Giblin (ed.), *The Irish Franciscan Mission to Scotland, 1619–46: Documents from Roman Archives* (Dublin, 1964), and F. Macdonald, *Missions to the Gaels*, pp. 55–96.

35 On Catholic episcopal activity in Ireland in the 1630s, see Tadhg Ó hAnnracháin, *Catholic Reformation in Ireland: the Mission of Rinuccini, 1645–1649* (Oxford, 2002), pp. 41–68; on their social and educational profiles, see Dónal Cregan, 'The social and cultural background of a counter-reformation episcopate, 1618–60', in A. Cosgrove and D. McCartney (eds), *Studies in Irish History Presented to R. Dudley Edwards* (Dublin, 1979), pp. 85–117.

36 Progress in implementing tridentine norms was a focus of episcopal reports to Rome; see series of diocesan reports ('Relationes status'), including Down and Connor, and Kilmore, 1629–39, in J. Hagan (ed.), 'Miscellanea Vaticano-Hibernica, 1580–1631', *Archivium Hibernicum*, 3 (1914), 227–365; for material illustrative of the episcopate of Bonaventure Magennis OFM, of Down and Connor (1630–40), see James O'Laverty, *The Bishops of Down and Connor* (Dublin, 1895), pp. 402–39.

37 On developments in the setting for worship, see Mac Cuarta, *Catholic Revival*, pp. 114–17; for the situation in the Pale heartland, see Colm Lennon, 'Mass in the manor-house: the counter-reformation in Dublin, 1560–1630', in J. Kelly and D. Keogh (eds), *History of the Catholic Diocese of Dublin* (Dublin, 2000), pp. 112–26.

38 On the accoutrements of this chapel, see Hector MacDonnell (ed.), 'A seventeenth-century inventory from Dunluce Castle, county Antrim', *Journal of the Royal Society of Antiquaries of Ireland*, 122 (1992), 109–27; the restoration of this place of worship, undertaken in 1637–40, was probably by the English builder William Parrott, who was paid £1,800 at a bond of £3,500, reflecting the exceptional scale of the enterprise, R. Loeber, *A Biographical Dictionary of Architects in Ireland 1600–1720* (London, 1981), pp. 83–4; see also James Stevens Curl, *The Honourable the Irish Society and the Plantation of Ulster 1608–2000* (Chichester, 2000), pp. 143, 147.

39 On this motif, see N. Canny, *Making Ireland British* (Oxford, 2003), pp. 424–5; for the alleged remark of an Irish minister in Cavan in the 1630s, that he would be damned for his involvement with the established church, see 'A sum of the matters objected against Murtagh King', *CSPI, 1633–47*, p. 206.

40 On this process, see Mac Cuarta, *Catholic Revival*, pp. 199–200.

41 'Notes of speech of crown counsel in case against Londoners' [12 May 1635], *CSPI, 1647–60*, p. 199; Bishop Bedell to Archbishop Ussher, 15 Feb. 1629–30, in *The whole works of...James Ussher*, xv, 467–8.

42 Both the bishops of Raphoe and of Kilmore were seeking a transfer to Derry diocese in the 1630s, R. Nugent to Jesuit general, 15 Sept. 1636, Archivum Romanum Societatis Iesu, Anglia 6a, Missio Hibernica, fols 1–2; B. Jennings (ed.), 'Acta... Propaganda...1622–50', *Archivium Hibernicum*, 22 (1959), 93, 102.

43 W. Burke (ed.), 'The diocese of Derry in 1631', *Archivium Hibernicum*, 5 (1916), 1–6; 'Notes of speech of crown counsel'.
44 M. Perceval-Maxwell, *The Scots Migration to Ulster in the Reign of James I* (London, 1973), pp. 193–5, 325–7, 344–6; see map on p. 369.
45 On Scots Catholics in plantation-era Ulster, see B. Mac Cuarta, 'Scots Catholics in Ulster, 1603–41', in David Edwards (ed.), *The Scots in Stuart and Cromwellian Ireland 1600-1660* (Manchester University Press, forthcoming); for examples of Catholic migration to Ulster in the early 1630s, see Macdonald, *Missions to the Gaels*, pp. 65, 79; on the Catholicism of another magnate family in the early Stuart era, see Michael Questier's study of the Brownes viscounts Montague, *Catholicism and Community in Early Modern England: Politics, Aristocratic Patronage and Religion, c.1550–1640* (Cambridge, 2006).
46 On the Catholicism of the MacDonnells earls of Antrim in the 1620s, see J. Ohlmeyer, *Civil War and Restoration in the Three Stuart Kingdoms: The Career of Randal MacDonnell, Marquis of Antrim* (2nd edn, Dublin, 2001), p. 27.
47 Raymond Gillespie, *Colonial Ulster: The Settlement of East Ulster 1600–1641* (Cork, 1985), pp. 84–112; Trinity College Dublin (TCD), MS 550, pp. 241–4.
48 NLI MS 8014 (x), 'Grievances of the inhabitants of Donamaine [1622]'.
49 Developments in Cavan are discussed in Mac Cuarta, 'Catholic revival in Kilmore diocese, 1603–41'.
50 On some issues within Catholicism in England in the early Stuart era, see John Bossy, 'The English Catholic community 1603–1625', in A. G. R. Smith (ed.), *The Reign of James VI and I* (London, 1973), pp. 91–105; for a historiographical survey, see Caroline Hibbard, 'Early Stuart Catholicism: revisions and re-revisions', *Journal of Modern History*, 52 (1980), 1–34; on English Catholics in Ireland before 1641, see David Edwards, 'A haven of popery: English Catholic migration to Ireland in the age of plantations', in A. Ford and J. McCafferty (eds), *The Origins of Sectarianism in Early Modern Ireland* (Cambridge, 2005), pp. 95–126.
51 For the relationship between Ulster and the Scottish Isles in the late medieval era, see Simon Kingston, *Ulster and the Isles in the Fifteenth Century: The Lordship of the Clann Domhnaill of Antrim* (Dublin, 2004).
52 On these Catholic exiles, see Macdonald, *Missions to the Gaels*, pp. 65, 79; on the permissive character of colonisation in early modern Ireland, see Nicholas Canny, 'The permissive frontier: social control in English settlements in Ireland and Virginia 1550–1650', in K. Andrews, N. P. Canny and P. E. H. Hair (eds), *The Westward Enterprise: English Activities in Ireland, the Atlantic and America, 1480–1650* (Liverpool, 1978), pp. 17–44.
53 For the decline of Ulster Gaelic families in the plantation era, see the studies by Joseph Canning, *The O'Hanlons of Orior* (Armagh, 2006), and Jerrold Casway, 'The decline and fate of Donal Ballagh O'Cahan and his family', in Micheál Ó Siochrú (ed.), *Kingdoms in Crisis: Ireland in the 1640s* (Dublin, 2001), pp. 44–62.
54 B. Millett (ed.), 'Catalogue of Irish material...Scritture originali...Propaganda', *Collectanea Hibernica*, 12 (1969), 42–3.
55 On tensions within the Catholic Confederation in the late 1640s, see Micheál Ó Siochrú, *Confederate Ireland, 1642-9: A Constitutional and Political Analysis* (2nd edn, Dublin, 2008); on the link between the notable poverty of the Catholic

Church in Ulster, and political intransigence in the Confederation, see Ó hAnnracháin, *Catholic Reformation in Ireland*, pp. 49–50; on an earlier church dispute involving provincial sensibilities, see the discussion of the Drogheda row (1619–24) in Mac Cuarta, *Catholic Revival*, pp. 218–25.

56 For an exploration of fragmentation among insurgents in south Ulster in 1641-2, see Joseph Cope, 'The experience of survival during the 1641 Irish rebellion', *Historical Journal*, 46:2 (2003), 295–316, at 308–10.

57 In December 1641, Bishop Sweeney re-consecrated the cathedral at Kilmore, Co. Cavan, see TCD MS 832, Deposition of Thomas Crant, undated, fol. 77v; he was keen to take over the residence of Bishop Bedell, see E. Shuckburgh (ed.), *Two Biographies of William Bedell, Bishop of Kilmore* (Cambridge, 1902), pp. 65–6, 188–90, 204; Philip McHugh McShane O'Reilly, a leader of the revolt in Cavan, seized the home of a wealthy settler, and installed the local Franciscan community, see TCD MS 832, Addendum to Deposition of Simon Greame, 13 Feb. [1643], fol. 107r.

58 'Decrees of the provincial synod of the province of Armagh, held at Kells [Co. Meath], the 22nd March 1642', in P. Moran (ed.), *Spicilegium Ossoriense* (2nd series, Dublin, 1878), pp. 2–8.

59 On sectarian attacks in Leinster, see N. Canny, 'Religion, politics and the Irish rising of 1641', in J. Devlin and R. Fanning (eds), *Historical Studies XX: Religion and Rebellion* (Dublin, 1997), pp. 40–70; on the Catholic revanche in one predominantly Old English county, see Jason McHugh, '"For our owne defence": Catholic insurrection in Wexford, 1641-2', in B. Mac Cuarta (ed.), *Reshaping Ireland 1550–1700: Colonization and its Consequences: Essays Presented to Nicholas Canny* (Dublin, 2011), pp. 214–40.

60 On the Catholic view of Protestantism in Gaelic society see M. Mac Craith, 'The Gaelic reaction to the Reformation', in S. Ellis and S. Barber (eds), *Conquest and Union: Fashioning a British State, 1485–1725* (London and New York, 1995), pp. 139–61.

61 Ministers' incomes are treated in R. Gillespie, 'The Church of Ireland clergy, c.1640: representation and reality', in T. Barnard and W. Neely (eds), *The Clergy of the Church of Ireland, 1000–2000: Messengers, Watchmen and Stewards* (Dublin, 2006), pp. 59–77, at 61–2, 72–5; clerical incomes in south Ulster are considered in Mac Cuarta, 'Religious violence against settlers', pp. 159–61; for a study of one Ulster minister's possessions, see R. Hunter, 'The bible and the bawn: an Ulster planter inventorised', in C. Brady and J. Ohlmeyer (eds), *British Interventions in Early Modern Ireland* (Cambridge, 2005), pp. 116–34.

62 For details illustrating these forms of behaviour, see Mac Cuarta, 'Religious violence against settlers in south Ulster, 1641-2'.

63 On social and economic tensions in Ulster society by 1641, see R. Gillespie, 'The end of an era: Ulster and the outbreak of the 1641 rising', in C. Brady and R. Gillespie (eds), *Natives and Newcomers: The Making of Irish Colonial Society, 1534–1641* (Dublin, 1986), pp. 191–213.

64 The broader political context is explored in M. Perceval-Maxwell, *The Outbreak of the 1641 Rebellion in Ireland* (Dublin, 1994); on the Catholic dimension in English politics, see Caroline Hibbard, *Charles I and the Popish plot* (Chapel Hill,

1983); Caroline M. Hibbard, 'Henrietta Maria (1609–1669)', in *ODNB*, www.oxforddnb.com/view/article/12947 (accessed 3 October 2011).

65 On this use of scripture in the service of the native Irish cause in the first half of the seventeenth century, see J. Casway, 'Gaelic Maccabeanism: the politics of reconciliation', in J. Ohlmeyer (ed.), *Political Thought in Seventeenth-Century Ireland: Kingdom or Colony* (Cambridge, 2000), pp. 176–88; in similar vein, at a preaching event in the courthouse of Dungannon, Co. Tyrone, in Lent 1642 the friar recounted the story of the wily Jewish princess, from the Book of Esther, TCD MS 839, Deposition of George Burne, 12 Jan. 1643–4, fol. 7r–v.

66 Mac Cuarta, 'Religious violence', pp. 167–8.

67 'Decrees of the provincial synod of the province of Armagh'.

8

Randal MacDonnell and early seventeenth-century settlement in northeast Ulster, 1603–30

COLIN BREEN

While the historical context of the Ulster plantation has been well documented archaeological research on the opening decades of the seventeenth century has, until recently, been piecemeal and lacking in coherence. Yet the potential of this subject to offer new insights into the landscape, buildings and material culture of the time is enormous. This chapter will provide a review of the work undertaken to date on the official Londonderry plantation. Despite the lack of a broad coherent research strategy in addressing the archaeology of this period a number of bawn sites and village settlements have been investigated as part of individual or conservation-led projects, in order to discern their architecture, settlement forms and material culture. In addition, a recent research project on a number of sites associated with the 'unofficial' MacDonnell plantations demonstrates that these undertakings were as structured and formulaic as the official schemes west of the River Bann. Towns were laid out in a grid-like fashion, with a high level of planning, and the houses conformed to contemporary architectural norms. This research is still in its early stages but it is apparent that the settlements associated with the projects in Antrim compare favourably to the London companies' ventures. Scottish merchants provide the central focus of these schemes, although evidence exists of a considerable level of native Irish integration and involvement. Furthermore, the settlements constituted an integral part of the east Ulster landscape and bear testimony to the earl of Antrim's political standing and the continual renegotiation of his identity in the opening decades of the seventeenth century.

The historical contexts of the plantations are well known but nonetheless it is probably worth repeating the pertinent elements relating to this study. The ending of the Nine Years War and the subsequent submission of Hugh O'Neill, earl of Tyrone, in early 1603 initiated a period of tense peace in

Ulster. A fragmentation of the Gaelic polities ensued as Hugh O'Neill moved to consolidate his power base in Tyrone and internal rivalries beset the O'Donnell lordship in west Ulster. Significant complications arose across the lordships when Donal O'Cahan, occupying territory west of the Bann, objected to the payment of rents to O'Neill and sought legal redress for his grievances. From 1604, a restless Rory O'Donnell, earl of Tyrconnell, became involved in a conspiracy to aid insurrection, ultimately leading to the flight of the earls three years later. The departure of O'Neill and O'Donnell from Rathmullan in September 1607 left a large portion of Ulster lands open to formal plantation.[1]

The native people of Donegal and Tyrone remained on the land as King James VI and I moved to confiscate the earls' patrimonies by legal means. Sir Cahir O'Doherty of Inishowen rose in revolt in 1608, capturing and burning the nearby citadel of Derry.[2] This insurrection facilitated the further confiscation of large tracts of Gaelic Ulster and the initiation of a formal plantation scheme, which envisaged three classes of beneficiaries: English and Scottish undertakers with their tenants (English and Scots); servitors with settler or Irish tenants; and finally native freeholders, the so-called 'deserving Irish'. In the decades prior to the plantation, the MacDonnells had already wrested control of much of the territory of the Route and Dunluce Castle, incorporating the north-western portion of Antrim, from the McQuillians, becoming in the process the dominant political family in the region. The MacDonnell chief, Randal, emerged from the Nine Years War relatively unscathed having succeeded his brother James as head of the family in 1601. Although the MacDonnells sided with O'Neill during the latter stages of that conflict, Randal received a pardon and resumed control of his lands. The plantation provided him with further opportunities to consolidate and indeed develop his family's possessions.

Although the plantation process has been subject to extensive historical analysis, this involved very limited systematic archaeological investigation of contemporary sites and landscapes. Until recently, there was a general acceptance that the surviving documentary and cartographic record provided adequate evidence for a detailed reconstruction of settlement and industry associated with the movement and placement of peoples across the Ulster landscape in the seventeenth century. Indeed, a number of archaeologists queried the validity of excavating sites from this time given the nature and range of information associated with them.[3] Recent investigative work, however, has added significantly to our understanding of this complex period, yielding extensive and exciting information on contemporary architecture and material culture, thereby facilitating the exploration of the social relationships between planter and Gael. Significantly, work across the Antrim estates, an area with only limited cartographic and associated historical references in comparison to the areas of formal plantation, provides a suitable template for future research west of the Bann.

Much of this early work was pioneered by Nick Brannon of the Department of the Environment in Northern Ireland, who conducted a series of excavations, prompted in most cases by construction or refurbishment projects. No defined research agenda existed as few academics saw any value in investigating this period. The archaeologists involved tended to come from the government sector and operated largely in a reactive environment. In Coleraine, for example, during the course of a building project the remains of a 1611 timber framed building on stone foundations were recovered in New Row, underlying a house erected in c.1674.[4] This house constituted one of a number of buildings constructed on the street in the earliest phase of the plantation scheme.

As part of a programme of architectural conservation carried out at Dungiven priory over two seasons in 1982–83, Brannon discovered the foundations of the manor house established by Sir Edward Doddington in 1611[5] An English garrison had been stationed there in 1602 and the excavations demonstrated the reuse by English authorities of the pre-existing O'Cahan tower house, reconstructed at the west end of the original Augustinian priory. Bannon uncovered the main long range of the manor house running south from the tower house, as well the bases of two substantial fireplaces and a stone-flagged scullery. The adjacent portion, centred on what became the town of Bellaghy, had been granted to the Vintners' Company as part of the City of London undertakings. Sir Baptist Jones subsequently rented the land and undertook the construction of a bawn, the Vintners' Hall, which was completed in 1619. Jones died in 1624 and Henry Conway leased the property, before surrendering it to the rebellious Irish in 1641. Brannon also conducted excavations at this site from 1989 to 1990, as part of an architectural conservation project. He uncovered a gun platform and an associated earthen platform reveted with stone on the west bawn wall, as well as a two-roomed house located along the west wall, separated internally by an H-plan fireplace.[6] Excavations at the north end of the house uncovered sections of the bawn's northwest circular flanking tower, complete with stone footings and an angled doorway. He also recorded evidence for an early medieval rath but suggested that Jones may not have known of its existence as it had been abandoned in the centuries prior to plantation.

Other archaeologists conducted similar rescue excavations, often in very difficult circumstances and politically charged times. From 1975 to 1984 Brian Lacey, working with Derry City Council and the University of Ulster, coordinated excavations across the Plantation Citadel, uncovering a range of structural remains, including wells, two seventeenth-century houses (in Linenhall Street) and many assorted finds, which shed light on the material culture of the plantation-period town.[7] The IRA bombing campaign throughout the 1970s and 1980s caused significant damage across the city centre, necessitating much of this excavation and conservation work.

Further afield, on the shores of Lough Neagh, the American archaeologist Orloff Miller conducted two short seasons of research excavation at Salterstown in 1988 and 1989.[8] Miller had become fascinated by the nature of the surviving remains during a visit to Ulster and subsequently undertook a Ph.D. on the site, examining English colonial settlement from an archaeological perspective. The original village as depicted on Raven's 1622 map is now a greenfield site with only partial structural elements of the original bawn still surviving. The excavations focused on a lot associated with either Walter Walton or Rowland Warbanks and exposed the stone foundations of a timber house. The original occupant may have been engaged in glass production given that a glasshouse occupied the site from 1608. Miller's research provides evidence of trading links with the local Irish, thereby initiating an important discussion on the cultural interplay between settler and Gael. Only a small percentage of the site was examined but the results highlight the extraordinary potential that plantation sites have in terms of the preservation of archaeological material.

More recently the Leicester based archaeologist Audrey Horning brought a rigorous American tradition of theoretically informed historical archaeology to the investigation of a number of early seventeenth-century sites across the Ulster landscape. In particular, her research develops important lines of enquiry including emerging identities, meaning and contemporary political constructs arising from the plantation. While engaged in fieldwork in the Roe Valley, she initially focused her research at Movanagher, the site of the Mercer's company village on the western banks of the River Bann.[9] Excavating within the area of the plantation village, adjacent to the still partially standing bawn, she located the remains of a sub-rectangular house site built in a vernacular Irish style with external cobbling and a fenced entry, which led her to conclude that English settlers had occupied the original house. The location of such a house and associated material culture questions traditional assumptions about the role and involvement of the existing Irish population during the early years of plantation.

Horning also undertook extensive landscape analysis and limited excavation at an upland settlement at Goodlands in northeast Antrim.[10] This site consists of numerous house platforms dating in age from the Neolithic through to the seventeenth century. Sidebottom and Case, the original excavators of the site, suggested a post-medieval provenance, although both late medieval and prehistoric evidence was also found.[11] Horning believes that this complex comprised the scattered homesteads of Scottish highlanders who settled in this part of Antrim at the beginning of the seventeenth century.[12] Alexander Magee from Islay had been granted 280 acres of land in this area in 1620 by Randal MacDonnell. Alexander and his brother Donal subsequently acquired leaseholds on the townlands of Goodland, Torglass,

Bighouse, Knockbrack and Tornaroon in 1630. Although the houses on the site have a clear late medieval architectural provenance, a style replicated across Scotland and Ireland, no clear artefactual evidence exists that the cluster represents a Scottish settlement. Certainly its haphazard arrangement of scattered homesteads does not correlate to the structured formality of the settlements established elsewhere by Randal MacDonnell across the Antrim estates now being excavated. There is certainly evidence of both late medieval and early seventeenth-century occupation but its form and function awaits further elucidation.[13]

It is also important, in the context of this paper, to include Tom McNeill's excavations at Dunineny, a coastal cliff top fort immediately west of Ballycastle in north Antrim in pursuit of his research into late medieval Gaelic lordship and in part a rescue dig accompanying conservation work undertaken on the castle's masonry. The site has an extensive history. Initially fortified in prehistory and possibly periodically occupied at various times during the medieval period, a small English garrison had been placed there in 1585.[14] In December 1603, Hugh O'Neill received the constableship of the 'fort of Dunaneny', as part of the Antrim estate grant, a lease subsequently reissued to MacDonnell in 1612 to include the market town of Ballycastle. The excavations showed that the surviving standing masonry and the visible internal features had been constructed in the early seventeenth century, with a castle-like gatehouse and a series of timber-framed structures erected in the fort's interior. Trenches uncovered two such structures measuring 6 × 3m and at least 5 × 4m respectively. Both originally consisted of timber frame superstructures built on low stone footings with windows holding paned glass. A fine external, cobbled pathway with a central drain circumvented the house platforms. Limited evidence of material culture included four shards of everted-rim ware and two shards of slip-ware. McNeill plausibly suggests that the site functioned as the administrative centre of the barony of Cary, a hypothesis that sits very comfortably with the emerging picture of the structure and management of the Antrim estates following recent discoveries at other MacDonnell centres including Dunluce, Ballycastle and Glenarm.[15]

The intense programme of plantation west of the Bann from 1608 was replicated east of the river in the MacDonnell territories. Scholars can access a wealth of cartographic and documentary evidence for the official plantations. The numerous visits, surveys and map evidence comprise a substantive paper-based record of the west Ulster plantation in early years of the seventeenth century. Settlement east of the Bann is far less visible historically, with few known documentary references and no cartographic material to rival Raven's maps. Archaeological investigation in this area, therefore, is proving invaluable and has begun to clarify the morphology of these unofficial plantations across northeast Antrim. Following the ascension of James to the throne of England in 1603, Randal MacDonnell received letters patents for the Route and Glynns.[16] The original grant in May 1603 covered 16 toughs and included the 4 baronies of Dunluce, Kilconway, Cary and Glenarm totalling 333,907 acres (see Table 1).[17] As part of this grant MacDonnell would provide

Table 1 Details of Randal MacDonnell's 1603 grant consisting of sixteen toughs equating to the modern baronies of Dunluce, Kilconway, Carey and Glenarm, originally comprising 333,907 acres

Territory	Tough	Description
Route	Between the Bann and Bush	Included the parishes of Coleraine, Ballyaghran, Ballywillan, Ballyrashane, Dunluce and Kildollagh
Route	Dunseverick and Ballintoy	Part of Billy parish in the barony of Cary and part of the parish of Ballintoy containing Dunseverick Castle
Route	Ballylough	Part of Billy parish in lower Dunluce and the parish of Derrykeighan
Route	Loughguile	Parish of Loughguile
Route	Ballymoney and Drumard [Dromart]	Parish of Ballymoney, parish of Kilraghts
Route	Kilconway	Part of barony of Kilconway
Route	Killyquin	13 townlands in western parish of Rasharkin parish
Route	Killymurris	Concentrated around Dunloy in Finvoy parish
Route	Dunaghy	Concentrated around parish of Dunaghy including village of Clough and Oldstone castle
Glynnes	Munerie (Mowbray, Mowberry)	Parish of Ramoan and Grange of Drumtullagh, included town of Ballycastle
	Cynamond of Armoy and Rathlin	Parish of Armoy and Rathlin
	Carey (Cary)	Originally smaller than Cary barony but included Fairhead and Murlough Bay
	Glinmiconogh	The Middle Glens including Cushendall
	Largie (Largy)	Including part of Ardclinis parish
	Park	Included Tickmacrevan, Templeoughter and Solar. Tuogh name refers to the demesne of Glenarm castle
	Larne	Included parishes of Carncastle, Killyglen, Kilwaughter and Larne.

Source: (PRONI D2977/5/1/1/1). The identification of the toughs is taken from W. Reeves, *Ecclesiastical Antiquities of Down, Connor and Dromore, Consisting of a Taxation of Those Dioceses* (Dublin, 1847).

'120 foot soldiers, 60 to be good shot, and the rest swordmen and pikemen, and 24 horsemen', an important illustration of his military capacity.[18] This effectively recognised the original grant (June 1586) of part of the Route to Sorley Boy MacDonnell, a territory extending from the River Bush to the Bann and three tuoghs of Dunseverick, Loughguile and tough of Ballymoney which included all of the lands of the McQuillans and the constableship of Dunluce castle.[19] Following decades of internecine conflict, the MacDonnells finally displaced the McQuillans in the 1580s. Through astute political manoeuvring the MacDonnells ensured their survival and maintained their territories into the opening years of the seventeenth century in no small part due to their relationship with James. Problems with the 1603 grant necessitated a reissue in 1604, with the added proviso that he divided the territory into 2,000-acre precincts and built a 'castle' or manor house on each smaller 500 acre unit.[20] This structured approach was replicated in the later official plantations across Londonderry.

In granting these lands to a traditional Gaelic lord James at once rewarded loyalty, displayed an acute understanding of the political situation in Ulster and demonstrated that he had clearly learned a number of valuable lessons from the ill-fated Lewis plantations in the 1590s.[21] The Scottish islands continually posed a political problem for Edinburgh, with a number of clans essentially operating semi-autonomous territories, and hindering trade and economic activity with lowland Scotland. His plantation schemes attempted to break the power of the island families and bring civility, modernity and trade to these marginal lands. In November 1598, the Fife Adventurers set out with 500 men to establish a burgh on the island of Lewis, initially founding a settlement at Stornoway.[22] Almost immediately the undertaking met with stiff resistance from the islanders and in 1601 the MacLeods captured the settler's fort. The majority of the planters left the island although settlement continued in various guises until 1609 when the Mackenzies finally bought out the rights to the scheme.[23] It is unclear whether MacDonnell had any direct experience of the Lewis enterprise but he would certainly have been familiar with its history and outcomes. Kintyre had been earmarked for settlement in 1598 but the scheme did not commence until 1607. By 1609 the earl of Argyll had established a burgh at Lochead, later renamed Campbeltown, which drew inhabitants from Ayrshire and Renfrewshire.[24]

The Lewis settlement floundered and the Lochead scheme only succeeded due to Argyll's patronage. Consequently James appreciated the importance of gaining local support and participation in any future plantation in Ireland. Furthermore, the imposition of central bureaucratic control and governance, whether from Edinburgh or London, could not succeed without local support. By granting lands to MacDonnell James at once groomed an ally and ensured the continuation of political and economic stability in an area of key strategic interest. For his part, MacDonnell's primary interests centred on the maintenance of his personal standing with the crown and the development of his estates. The MacDonnells had already demonstrated a particular talent for survival in the maelstrom of Irish politics. It is significant that the refurbished architecture of their principal residence at Dunluce Castle reflects this continual renegotiation of their political and social status, reflecting in turn Gaelic, Scottish and finally English forms. James recognised that MacDonnell's territories would create a buffer zone between the still volatile Gaelic territories of mid and west Ulster. He also appreciated the fact that MacDonnell would both facilitate and encourage the future expansion of trade activity throughout the north Irish Sea region. This scion of the Clan Iain Mhòir would dilute Argyll's influence across the region and check over-zealous English bureaucrats who governed the new colonies. Consequently the MacDonnell grants generated significant opposition. Lord Deputy Chichester who had particular designs on north Antrim himself, complained bitterly to the king.[25] MacDonnell prevailed and in July 1610 he received a new patent confirming the original grant, which followed his surrender of nine townlands in Coleraine to the formal plantation.

MacDonnell's estates constituted a very varied landscape. The Route, concentrated immediately east of the River Bann, ran southwards towards Lough Neagh and still contains good agricultural land positioned for the most part in a river valley. Barnaby Rich, an Elizabethan solider described the territory as good corn country, while Marshall Bagenal's 1586 *Description of Ulster* depicted the Route as a 'pleasant and fertile country'.[26] In contrast, lying further east, the Glynnes consisted of a series of upland glens sweeping down to the coast. It would have been heavily wooded in places, a valuable resource in the early seventeenth century. An account written in 1598 recorded the Glens as being boggy and heavily wooded while in 1586 Bagenal described them as being 'full of rocky and woody dales' and a place where Scottish galleys regularly landed.[27] Historically, coastal and salmon fisheries had been of crucial importance and MacDonnell obtained the fishing rights to the Bann in 1603. He later lost his grant of the tidal portion of the river in 1610 as part of his surrender of the Coleraine townlands.[28]

Other resources including coal deposits near Ballycastle are referenced in a 1639 estate lease which referred to the coal mines and salt pans at Bonamargy.[29] The importance of these two resources tends to be overlooked despite the declining availability of woodland across Britain and Ireland and an upsurge in demand for salt as a preservative. Again, MacDonnell demonstrated considerable nuance in moving to exploit both; the adoption of particular salt extraction technologies in the form of pans at Ballycastle indicated his willingness to engage with new technologies and adapt to the developing patterns of seventeenth-century proto-industrialisation. It would be a mistake, therefore, to regard Randal as a static landowner rooted in late medieval agrarian practice. Rather, he showed himself willing to continually change, learn and adapt to new circumstances, operating within the official and unofficial sphere in order to continue generating capital and trade. In 1631, the authorities fined Robert Millar and John Lochlarne, both servants of the earl and presumably operating under his orders, for bringing a boat from Ireland and selling a cargo of oats to traders who were not burgesses of the town of Dumbarton.[30] Both were required to sell all of their salt and timber to cover the cost of the fine. Agriculture remained the main activity conducted throughout the estates and MacDonnell actively supported new practices of enclosure. The county Cavan poet Lochlainn Ó Dálaigh's poem decried the change in the Ulster landscape at this time with 'the mountain all in fenced fields; fairs are held in places of the chase; the green is crossed by girdles of twisted fences'.[31] Ohlmeyer has demonstrated the anxiety of the MacDonnells to be viewed as improving landlords through their agricultural practices and use of long-term leases.[32]

Plantation emerged as a key element in the strategy of improvement and development. Randal MacDonnell established a network of settlements across his estates, centres of trade and small-scale industrial activity, which would be populated by Scottish settlers and the indigenous Irish inhabitants. In May 1608 Lord Deputy Chichester authorised him to make denizens of Ireland of all the 'Scotchmen' he would require.[33] In developing this network MacDonnell directly replicated developments in areas of the government-sponsored plantation. Indeed, he tried to enact an innovative and ambitious plantation to rival these official schemes, establishing a novel settlement pattern across other parts of east Ulster and in parts of Munster. MacDonnell evidently had a clear idea of the type of infrastructure required for his settlements following discussions with both the king's bureaucrats and the architects of the official plantations. By 1606 he had received grants to hold a fair at Cloghmagherdunaghy [Clough] on the feast of John the Baptist; a fair at Dunkerd in the Route on St Michael's Day; a Saturday market at Dunluce; a Tuesday market at Dunineny and a Thursday market at Glenarm in the Glynnes.[34] These grants demonstrated MacDonnell's interest in implementing economic frameworks and the fairs themselves served as new centres of settlement and administration. According to Gillespie, non-corporate east Ulster towns used markets, and to a lesser extent fairs, to legitimise the power of the landowner and develop forms of governance and bureaucratic control outside of the official plantation areas.[35] By May 1608 Randal had agreed to divide up his territory in accordance with the terms of the earlier grant and gave an undertaking to build 'castles' and 'capital messuages' in each of the listed manors.[36] Each of these sites contained a preexisting castle or fortification and probably small-scale associated settlement, thereby ensuring a degree of cultural continuity. Nevertheless, the first two decades of the seventeenth century represent an important break with the past, not least because of the number of medieval fortifications (Dunluce, Glenarm and Dunineny) had been extensively refurbished or rebuilt. More importantly from a plantation perspective, MacDonnell established a number of new towns on these late medieval sites. Despite limited documentary evidence, archaeological research is now beginning to elucidate important information about their morphology and settlement patterns.

The 1611 report on the work undertaken by servitors in the counties of Down, Antrim and Monaghan reported that the town of Dunluce consisted 'of many tenements, after the fashion of the Pale, peopled for the most part with Scotsmen', and that Randal had built 'a good house of stone with many lodgings and other rooms'.[37] Recent survey work significantly enhances our understanding of this town. An integrated archaeological project conducted by the University of Ulster, Queen's University Belfast and the Northern Ireland Environment Agency, extensively surveyed and excavated the site, thereby elucidating the evolution and cultural chronology of a town well established by 1611. MacDonnell's new town took on a significantly more formalised structure to the unplanned late medieval settlements, with buildings laid out to a regular sub-rectangular plan and a new cobbled street surface running from the castle entrance and outer enclosure. This street led southwards to an east–west running junction, where a large courthouse at its highest topographic point served as a new central focus for administrative activity in the locale. A second street, orientated on an east–west axis, connected to a western street which ran parallel to the castle entrance street. A cobbled, concave profile allowed surface water to run off into two drains on either side of the road. Rows of stone and timber built houses lined the 10 m wide street. Pathways and clearly delimitated lines of paving marked the plots and entrances of individual houses and each house also had its own specific drainage needs incorporated into the ground design. One large two-storey masonry house, excavated in 2009 and possibly the residence of a Scots merchant who certainly occupied the house by 1614, contained an internal gable fireplace, opposing central entrances, glazed windows and an internal privy. Finds included Scottish coinage, early seventeenth-century ceramics and a mid sixteenth-century Polish coin, possibly kept and worn as a souvenir from earlier Scottish migrations to Europe. A large private garden plot, located at the rear of the house was protected by a low earthen bank. Geophysical investigations and a topographic survey identify this building as one of a number of similarly constructed houses in this area. A range of other building types including timber-framed structures completed the site.

An ambitious undertaking, MacDonnell's new town required significant planning and investment. As a mercantile scheme, it aimed to further his economic interests throughout the region by attracting both Scottish and Irish settlers to the site. Its establishment mirrored the planning and architecture of the new plantation town at Coleraine with a number of key differences. In contrast to Coleraine and indeed Derry, there appears to be no defensive enclosure in Dunluce town, possibly indicative of MacDonnell's more secure mindset following the end of the Nine Years War. Dunluce, however, ultimately became a folly, reflecting MacDonnell's inability to break with the past and dispense with some of the traditional aspects of Gaelic society. Throughout the early decades of the seventeenth century Randal sought to maintain his pivotal place within the Gaelic maritime world of north Ulster and western Scotland. Nonetheless, he inexplicably failed to take due cognisance of changing economic practice as Dunluce lacked a sufficient harbor to develop trade. Furthermore, by establishing his new town in close proximity to Coleraine MacDonnell also invited direct competition with a settlement situated on an important fishery and navigable waterway (the Bann) to the sea. The rich vested interests based in the Coleraine settlement placed it at a distinct advantage to the largely self-financed town at Dunluce. The town of Dunluce represents a Gaelic lord's attempt to bridge the gap between a traditional past and a changing future. Archaeological evidence suggests that mercantile activity faded on the site by the 1630s and the residents reverted to subsistence agriculture. These developments mirrored the earl's own decaying fortunes at that time. Native Irish rebels subsequently burnt portions of the town to the ground during the 1641 revolt and the settlement never really recovered. Although probably totally abandoned by the 1690s, Dunluce fair continued to be held in the ruins of the town into the eighteenth century.

Elsewhere, in November 1611 McDonnell leased the constableship of Dunineny and Ballycastle to one Hugh McNeill.[38] Excavations at the former site point to significant reconstruction of the fortifications, not least the construction of a gatehouse and a number of internal buildings. This became one of MacDonnell's chosen locations for development and it is likely that a small settlement subsequently emerged on the immediate outskirts of the fort. Local oral testimony refers to the existence of a cobbled street with earthworks, possibly representing house sites located on either side of the modern road prior to the development of a caravan park. Hopefully, the buried archaeology remains intact and future research can quantify the nature and extent of sub-terrestrial structures. As in the case of Dunluce, however, MacDonnell chose a site of historic note but with little potential to become anything other than a localised administrative and market centre. Dunineny's position on a cliff edge, the absence of a suitable sheltered harbour and its proximity to Ballycastle probably militated against expansion. At Ballycastle, a new street appears to have been laid out running southwards from the late medieval castle site, originally located in the current Diamond. Although Hugh Boyd realigned the town's main street in the eighteenth century and the castle was levelled in the nineteenth century a number of important architectural clues point to the existence of an earlier seventeenth-century plan. These include surviving, later medieval windows in the rear annexes to existing main street buildings and a number of seventeenth-century architectural fragments that either remain *in situ* or repositioned away from their original context. At Glenarm a street was apparently established running seawards from the late medieval towerhouse or castle, the line of which appears in the contemporary town plan. Later in the 1630s the MacDonnells moved their primary residence across the river to a new site but the street pattern survived.

Randal MacDonnell's activities in the early decades of the seventeenth century earned particular favour with James VI and I, who commented to Chichester in 1613 on McDonnell's 'dutiful behaviour to the state and the example of his civil and orderly life endeavours very much of the reformation and civilizing of those rude parts...where he dwells'.[39] In 1617, the Council Book of Ayr records his admission as a free burgess of the town, indicative of his continued connections to Scotland, increasingly centred on economy and mercantilism.[40] In May 1618 James granted Randal the title of viscount of Dunluce, and two years later he became first earl of Antrim.[41] Over the following years MacDonnell consolidated his position and showed increasing confidence. A letter to the Lord Deputy in December 1625, seeking further funds to improve his military capacity, loudly proclaimed his loyalty.[42] Two years later he took the decision to put the crown bailiff in stocks and hold court sessions partly through the medium of Irish. He complained that his tenants had to travel to Carrickfergus for quarter sessions and proposed instead holding two sessions at Oldstone [Clogh], where his people would build a session house.[43] Critics claimed that Antrim's son was being educated in France as a papist and that he should be returned to England.[44] Despite these controversies his economic activities continued to expand. In July 1628 he obtained a licence to hold a fair at Ballycastle and in 1629 received confirmation of his estates in the Route, the Glynnes and Rathlin, except for three parts of the Bann fishery, the castle of Olderfleet [Larne] and land in the possession of the Bishop of Down and Connor.[45] These lands were created into the manors of Dunluce, Ballycastle, Glenarm and Oldstone, alias Cloghinaghene Donaghie [Clogh]. As a measure of his success the earl of Clanricard noted that he 'hath good tenants and is very well paid his rents'.[46]

The family's political and military fortunes, along with their newly established towns, declined dramatically in the 1630s but that story is outside the scope of this essay. Instead, the contemporary archaeological record bears testimony to a thriving estate by the closing years of the third decade of the seventeenth century. This strongly suggests that Antrim's 'unofficial' plantation had been largely successful, its vibrant new settlements representing diverse communities of Irish, Scots and English. This portion of Antrim once again assumed its prominent position within the broader maritime province of northeast Ulster and western Scotland. The sea acted not as a barrier but rather as a facilitator of trade and communications between areas culturally linked for many thousands of years. The exact mechanisms, however, by which the earl of Antrim established his estates requires further scrutiny. How, for example, did he finance these developments? A combination of personal resources, the mortgaging and leasing of land, loans, speculative investment, official support and a variety of other sources must all have contributed to some degree but further research is necessary to fully understand his methods. The MacDonnell plantations, however, were not unique. Other major landowners actively sought to develop their estates. Immediately to the south of the MacDonnell lands, both the Hamilton and Montgomery estates in south Antrim and north Down underwent similar landscape transformations, while the same process can be observed in Munster. The MacDonnell estates provide a suitable, if neglected, case-study for plantation and its attendant socio-economic and cultural change in the opening decades of the seventeenth century. Archaeological research can assist in constructing a more detailed and inclusive picture of life and landscape in this part of Ulster during a formative and dynamic period.

NOTES

1 J. S. Curl, *The Londonderry Plantation, 1609–1914: The History, Architecture and Planning of the Estates of the City of London and Its Livery Companies in Ulster* (Chichester, 1986).
2 H. A. Jefferies, 'Prelude to plantation: Sir Cahir O'Doherty's rebellion in 1608', *History Ireland*, 17:6 (2009), 16–19.
3 J. P. Mallory and T. E. McNeill, *The Archaeology of Ulster* (Belfast, 1991).
4 N. F. Brannon, 'Where history and archaeology unite', in A. Hamlin and C. J. Lynn (eds), *Pieces of the Past* (Belfast, 1988), pp. 78–9.
5 N. F. Brannon, 'Archaeological excavations at Dungiven priory and bawn', *Benbradagh*, 15 (1985), 15–18; N. F. Brannon and B. S. Blades, 'Dungiven bawn re-edified', *Ulster Journal of Archaeology*, 43 (1980), 91–6.
6 N. F. Brannon, 'Bellaghy bawn', *Excavations 1989* (Bray, 1990), pp. 16–17, 91.
7 Brian Lacey, 'Two seventeenth-century houses at Linenhall Street, Londonderry', *Ulster Folklife*, 27 (1981), 52–62; Brian Lacey, 'Archaeology and war in an Irish town', *History Ireland*, 17:6 (2009), 60–1.
8 C. J. Donnelly, *Living Places; Archaeology, Continuity and Change at Historic Monuments in Northern Ireland* (Belfast, 1997), p. 121.
9 A. Horning, '"Dwelling houses in the old Irish barbarous manner": archaeological evidence for Gaelic architecture in an Ulster plantation village", in P. Duffy,

D. Edwards and E. Fitzpatrick (eds), *Gaelic Ireland 1250–1650: Land, Lordship, and Settlement*, (Dublin, 2001), pp. 199–215.
10 A. Horning, 'Archaeological explorations of cultural identity and rural economy in the north of Ireland: Goodland, County Antrim', *International Journal of Historical Archaeology*, 8:3 (2004), 199–215.
11 J. M. Sidebottom, 'A settlement in Goodland townland, Co. Antrim'. *Ulster Journal of Archaeology*, 13 (1950), 44–53; H. J. Case, 'Settlement patterns in the north Irish neolithic', *Ulster Journal of Archaeology*, 32 (1969), 3–27.
12 Horning, 'Archaeological explorations', p. 209.
13 B. Williams and P. Robinson, 'Bronze Age cists and a medieval booley hut at Glenmakeeran, County Antrim, and a discussion of booleying in north Antrim', *Ulster Journal of Archaeology*, 46 (1983), 29–40.
14 T. E. McNeill, 'Excavations at Dunineny castle, Co. Antrim', *Medieval Archaeology*, 48 (2004), 167–204.
15 Ibid., p. 193.
16 P. Robinson, *The Plantation of Ulster* (Belfast, 1984), p. 52.
17 Public Record Office of Northern Ireland (PRO NI) D2977/5/1/1/1; G. Hill, *An Historical Account of the MacDonnells of Antrim* (Belfast, 1873), p. 196.
18 *A repertory of the Inrolments on the Patent Rolls of Chancery in Ireland, commencing with the Reign of James I*, ed. J. C. Erck (1 vol, 2 parts, Dublin, 1846–52), ii, 1603, p. 8.
19 J. S. Brewer and W. Bullen (eds), *Calendar of the Carew Manuscripts* (2 vols, London, 1867–68), ii, 427–8.
20 *Cal Patent Rolls James I*, 1604, pp. 52, 137.
21 A. Mac Coinnich, 'Siol Torcail and their lordship in the sixteenth century', in *Crossing the Minch: Exploring the Links between Skye and the Outer Hebrides* (Lewis, 2007), pp. 7–32.
22 D. Dobson, *Scottish Emigration to Colonial America, 1607–1785* (Athens, GA, 2004).
23 MacCoinnich, 'Síol Torcail and their lordship in the sixteenth century', p. 16.
24 Dobson, *Scottish Emigration*, p. 12.
25 N. Canny, *Making Ireland British, 1580–1650* (Oxford, 2001), p. 188.
26 Gillespie, *Colonial Ulster*, p. 11.
27 E. Hogan, *A Description of Ireland c.1598* (Dublin, 1878), p. 15; Hill, *MacDonnells of Antrim*, p. 15.
28 Gillespie, *Colonial Ulster*, p. 16.
29 PRO NI D2977/3A/2/36/1.
30 *Burgh Council Minutes* (ref. 1/1/1).
31 C. Maxwell, *Irish History from Contemporary Sources 1509–1610* (London, 1923), p. 291.
32 J. H. Ohlmeyer, *Civil War and Restoration in the Three Stuart Kingdoms* (Cambridge, 1993), pp. 25, 36–7.
33 *Calendar of State Papers Relating to Ireland*, 1606–8 (London, 1874), p. 578.
34 *Cal. Patent Rolls James I*, p. 274.
35 R. Gillespie, *Colonial Ulster, The Settlement of East Ulster 1600–1641* (Cork, 1985), pp. 189–92.

36 *Cal. Patent Rolls James I*, 1608, p. 527.
37 PRO NI T811/3, fol. 13; Ohlmeyer, *Civil War and Restoration*, p. 25.
38 T. E. McNeill, 'The stone castles of northern county Antrim', *Ulster Journal of Archaeology*, 46 (1983), 101–28.
39 British Library, Additional MSS, 4794, fo. 233; J. H. Ohlmeyer, 'A laboratory for empire? Early modern Ireland and English imperialism', in K. Kenny (ed.), *Ireland and the British Empire* (Oxford, 2004), p. 47.
40 B6/11/4, fols. 583v–586v.
41 *Calendar of State Papers Relating to Ireland 1615–25* (London, 1880), p. 199; *Cal. Patent Rolls James I*, p. 373; *Calendar of Patent and Close Rolls Charles I* (London, 1861–63), p. 520; *Irish Patent Rolls of James I: Facsimile of the Irish Record Commission's Calendar* (Dublin, 1966), p. 492.
42 *Calendar of State Papers Relating to Ireland, 1625–32* (London, 1900), p. 64.
43 Ibid., p. 490.
44 Ibid., p. 81.
45 *Cal. Pat Close Rolls Charles I*, 1629, p. 490.
46 Ohlmeyer, *Civil War and Restoration*, p. 25.

9

Educating the colonial mind: Spenser and the plantation

ANDREW HADFIELD

Nicholas Canny makes a powerful case that Edmund Spenser set the agenda for the age of plantation and the transformation of Ireland from its independent identity as Hibernia to a British nation.[1] More than three centuries after Spenser's death W. B. Yeats asked himself 'Did that play of mine send out/Certain men the English shot?', and one wonders whether Spenser would have expressed similar fears had he lived rather longer.[2] Sometimes, perhaps, Anglo-Irish writers on either side of the colonial divide have either given themselves or been given rather too much credit/blame. The principal means of the fundamental change of Irish society and its identity was the establishment of the Ulster plantation by James VI and I in 1609, an act made possible by the union of the crowns of England, Scotland and Ireland in 1603.[3] The plantation witnessed a massive influx of people from Scotland and England into Ireland, and its ethnic mixture has impacted on the nature of the British Isles ever since.[4] The numbers rival those who migrated from Spain to the Americas at the same time, which is truly astonishing given the relatively low population of the British Isles.[5] According to Michael Perceval-Maxwell, about 12,000 adults from England and Scotland moved to Ireland before 1622 – slightly more Scots than English and more men than women (4:3).[6]

Ever since the experiment in Laois and Offaly during the reign of Queen Mary, plantation emerged as one of the principal means for imposing English government on Ireland.[7] Successive Tudor monarchs encouraged plantation, albeit rather half-heartedly and with little real conviction, as a relatively cheap policy, especially when compared to the cost of maintaining an army. Theoretically at least plantations did not require victualling from England, a logistical nightmare, or the direct exploitation of the local population, a policy that caused understandable resentment and risked provoking further rebellion.[8] There were sporadic attempts to colonise Ireland through private enterprise, most notably in the Ards Peninsula in Ulster in the early 1570s, before the establishment of the ill-fated Munster plantation in the mid to late 1580s.[9] With the benefit of hindsight, plantation looked like an ever more significant policy in the minds of English governors and administrators, but the course of plantation was hardly smooth, and, as has often been noted, its purpose altered according to circumstances and there was probably never a coherent long-term plan. In times of confidence undertakers were imagined as figures who would help transform the local population and bring them to proper farming techniques and civility; in more dangerous periods, they appeared to be lone defenders of law and order rather than beacons of civilized values. Planters were always supposed to be wary of employing native labour for fear that they would be culturally swamped by their malign influence, an inevitable process that had eroded the values of the Old English and transformed them into the worst enemies of the crown rather than its supporters, a process commonly known as 'degeneration'. The practicalities of everyday life in Ireland invariably militated against such purism and almost every planter made use of native labour.[10]

Spenser spent at least ten years on the Munster plantation (1588–98), perhaps a little longer, as he may well have taken up residence there before he signed the title deeds for Kilcolman. He named his estate Hap-Hazard, with what was undoubtedly conscious irony, given the tortuous path by which he had acquired the property. The title could also stand as an accurate description of English plantation policy in Ireland up to that point, something else that Spenser probably had in mind as well.[11] Here, he started to think about and perhaps write the notorious *View of the Present State of Ireland*, the dialogue which Nicholas Canny argues determined later events. The process shows how Spenser's work developed out of his own experience as a colonial administrator in Ireland, when he eventually obtained an estate beyond anything he could have hoped to acquire as reward for his labours had he remained in England. The *View*, written as the Nine Years War gathered pace, is a defence of the rights of English planters to remain in Ireland and an account of how they should, in turn, be defended by the English crown. It is a work that provided the rationale for the future development of the policy of plantation, albeit not quite as Spenser could have imagined or planned. Read one way, the process has a clear logic; read another way, it is haphazard.

Evidence would suggest that the dialogue did have an immediate impact. Spenser's support for Robert Devereux, second earl of Essex, surely helped facilitate the earl's involvement in Irish affairs soon after a *View* was written. Irenius, in his concluding speech, argues that a powerful Lord Deputy needed to be sent to Ireland, but that 'over him there were placed a Lord Lieutenaunt of some of the greatest personages in England, such a one I could name, upon whom the eye of all England is fixed, and our last hopes now rest'.[12] In 1598, after the spectacular success of the Cadiz expedition in 1596, Irenius could only have been referring to the earl of Essex, the most visible figure in the Privy Council into whose circle Spenser was clearly moving.[13] The large number of manuscript copies of a *View* that survive – about twenty, an unusually high number, especially for a political work written in the period – indicates that it had a wide readership, including professional politicians eager to know what was happening throughout the dominions of the English crown at an especially fraught time.[14] Accordingly, Spenser's words probably influenced the appointment of Essex as Lord Lieutenant on 12 March 1599, although it should be noted that the earl was actually sent to Ireland to defeat Tyrone, rather than supervise events from England, as Spenser suggested.[15] A further sign of a *View*'s significance for early readers is the attempt made by Matthew Lownes to publish the work in 1598 when it was entered into the Stationers' Register. Arguments continue to rage as to whether a *View* was censored, as it did not appear in print until 1633, by which time the Ulster Plantation was well established. Nonetheless, someone clearly thought the work worth publishing back in 1598, and it made a major impact thereafter despite remaining in manuscript form.[16]

Given Spenser's role as an administrator in Ireland and his close involvement with the men who developed military strategy, the most influential section of a *View* was the second part detailing the amount of military investment required to protect the English presence in Ireland, a section of the text that was later summarised in the desperate 'Brief Note of Ireland' delivered to the authorities – probably by Spenser himself – in December 1598.[17] Spenser noted the expenditure of upwards of twenty thousand pounds a year necessary to keep Ireland quiet, and argued instead for a huge army of ten thousand footmen and a thousand cavalry to sort out any rebellion quickly. Despite the initial large outlay to cover costs, the strategy would eventually save considerable sums of money. Then, in his final recommendations as to how Ireland should be reformed Irenius contended that the country needs to be leveled, flattened and civilised:

> And first I wish, that order were taken for the cutting and opening of all places through woods, so that a wide way of the space of 100 yards might be layde open in every of them for the safety of travellers, which use often in such perilous places to be robbed, and sometimes murdered. Next, that bridges were built upon the rivers, and all the fordes marred and split, so as none might passe any other way but by those bridges, and every bridge to have a gate and a gate house set thereon, whereof this good will come that no night stealths which are commonly driven in by-wayes, and by blinde fordes unused of any but such like, shall not be conveyed out of one country uinto another, as they use, but they must passe by those bridges, which they may either be happly

encountered, or easily tracked, or not suffered to passe at all, by meanes of those gate-houses thereon: Also that in all straights and narrow passges, as between 2 bogges, or through any deepe foord, or under any mountaine side, there should be some little fortilage, or wooden castle set, which should keepe and command that straight, whereby any rebells that should come into the country might be stopped that way, or passe with great peril. Moreover, that all high ways should be fenced and shut up on both sides, leaving onely 40 foote bredth for passge, so as none shall be able to passe but thorough the high ways, whereby theeves and night robbers might be the more easily pursued and encountered, when there shall be no other way to drive their stolne cattle, but therein, as I formerly declared. Further, that there should bee in sundry convenient places, by the high ways, townes appointed to bee built, the which should be free Burgesses, and incorporate under Bayliffes, to be by their inhabitants well and strongly intrenched, or otherwise fenced with gates on each side thereof, to be shut nightly, like as there is in many places in the English Pale, and all the ways about it to be strongly shut up, so as none should passe but through those townes: To some of which it were good that the priviledge of a market were given, the rather to strengthen and inable them to their defence, for the re is nothing doth sooner casuse civility in any countrie then many market townes, by reason that people repairing often thither for their needs, will dayly see and learne civil manners of the better sort (*View*, pp. 156–7).

Spenser clearly thought carefully about what needed to be done in Ireland and his proposals were exactly in line with what happened in the establishment of the Ulster plantation – the removal of the great forests that once covered Ulster; the introduction of small farms and small towns; the cutting/laying of proper roads, crossings and bridges; and, crucially, the establishment of garrisons throughout the north.[18] Spenser's comments were in part a response to the failings of Munster Plantation, a scheme designed to establish an English presence on the confiscated lands forfeited to the crown after the Desmond Rebellion. Unfortunately for Spenser and his fellow settlers, the spread of the law which enforced their rights to the land had proved intensely problematic, with many Irish landowners, like Spenser's immediate neighbour, Lord Maurice Roche, producing old title deeds and defending their prior right to their lands.[19] Accordingly, most English settlers found themselves exposed and surrounded in isolated and undefended homesteads, one reason why they become so vociferous and demanding as Hugh O'Neill's forces approached from the north.[20] Given his own experiences, Spenser proved cagey about the rule of law in a *View* and argued strongly in the dialogue that what looked like a good idea in England would simply prove a disaster in Ireland. In his case against Lord Roche Spenser had focused on the Irish lord's treasonable activities, whereas Roche concentrated on the prior entitlement to the lands in dispute, a sign of the vast gulf that separated natives and newcomers. Conflict over land titles was an issue that Spenser, along with his fellow settlers, hoped to avoid in the future through stern military intervention and the vigilant protection of the colonists.[21]

Spenser's text was surely read by his fellow poet and official in Ireland, Sir John Davies, author of *A discouerie of the true causes why Ireland was neuer entirely subdued, nor brought vnder obedience of the crowne of England, vntill the beginning of his Maiesties happie raigne* (1612), especially as we know that the Lord Deputy for whom Davies worked, Sir Arthur Chichester, possessed a copy of the dialogue.[22] The very title of Davies's work sounds like a response to Spenser: at last, an answer to mere observations, a dig beneath the surface rather than another survey. Davies reversed Spenser's insistence on grand military intervention and argued instead for the spread of law, flattering King James by demonstrating that only now, with Ireland properly subdued, could the rule of law be inaugurated in earnest and the country transformed into another part of Britain. The lessons of the Munster Plantation had finally been learned and in his analysis Davies echoed Spenser:

> if the English had builded their Castles and Towns in those places of fastnesse, and had driuen the Irish into the Plaines and open Countries, where they might haue had an eye and obseruation vpon them, the Irish had beene easily kept in Order, and in short time reclaimed from their wildnesse; there they woulde haue vsed Tillage, dwelt together in Towne-ships, learned Mechanicall Arts & Sciences. The woods had bin wasted with the English Habitations, as they are about the Forts of *Mariborough* and *Phillipston*, which were built in the fastest places in *Leinster*, and the wayes and passages throughout Ireland, would haue boene as cleare and open, as they are in England at this day.[23]

Davies's text is a mirror image of Spenser's, arguing, perhaps with the premises of his predecessor in mind, that what was required had now been done. The woods been cleared and opened, 'discovered' rather than hidden, and as a result the settlers could thrive, free from attack. The issue of 'waste', a legal term that meant letting land lie unused and unproductive, preoccupied both writers. Inevitably, this involved transferring ownership from the Irish to the English.[24] Sir James Ware changed Spenser's text when publishing the work in 1633, convinced, as he argues in the preface, that had the author lived in quieter times, after the victory at Kinsale and the establishment of the Ulster plantation, he would have tempered his excesses, rained in his harsh language and not singled out certain key families for criticism (*View*, pp. 5–7).[25] In a sense, Davies had already done the same twenty years before Ware.

If, then, Spenser set the agenda, as both Davies and Ware suggest – and there were few enough treatises of real insight and substance to rival a *View* – did he do so simply because he observed developments from his vantage point as an official in the Dublin government and then on the Munster plantation? Or is there a history to what he wrote that tells us a great deal about the process of colonial theory and practice? Put another way, can the history of colonialism be explained in practical terms or is there an intellectual history that needs to be uncovered?

Spenser's key influence at Cambridge was his tutor, Gabriel Harvey, probably only a couple of years older than himself. Harvey, in turn, looked up to Sir Thomas Smith (1513–77), a scholar, diplomat, political theorist, and like Harvey, a native of Saffron Walden near Cambridge.[26] Smith developed a particular interest in colonial theory in relation to Ireland and acquired land in the Ards Peninsula in 1571. Smith's illegitimate son, Thomas, went over to lay claim to the land in 1573, but fell victim to angry locals, who boiled his head and fed it to the dogs – at least according to Thomas Churchyard.[27] Spenser's experiences in Munster might well have ended the same way had the poet not fled to Cork and then on to the safety of London, where he died.[28]

Smith, as his pamphlet, *A letter sent by I.B. Gentleman vnto his very frende Maystet [sic] R.C. Esquire vvherin is conteined a large discourse of the peopling & inhabiting the cuntrie called the Ardes, and other adiacent in the north of Ireland, and taken in hand by Sir Thomas Smith one of the Queenes Maiesties priuie Counsel, and Thomas Smith Esquire, his sonne* (1572), demonstrates, was a practical thinker who did a great deal of research before taking action. The pamphlet is very thorough, as one might expect from a work designed to persuade people to undertake a colonial enterprise. The printer, Henry Bynemman, also published works by Spenser and Gabriel Harvey.[29] Smith relied heavily on Roman models of colonisation and thought very carefully about the key issues in colonial theory, particularly the question of whether colonies should be set apart from the people they sought to control. He argued that the Irish churls were very keen to work for the English, which created a potential conflict with colonists anxious to take over their own labourers in order to alleviate demographic problems at home.[30] Smith staged debates at his house, Hill Hall, near Saffron Walden, a purpose-built humanist mansion, with edifying classical narratives designed to inspire virtuous behaviour adorning the walls and windows.[31] Surviving records describe one debate that took place there in 1570 or 1571, clearly a prelude to the attempted colonisation of the Ards, involving Sir Humphrey Gilbert, who had been employed in active service in Ireland for the past five years. Harvey records the debate, as follows:

> Thomas Smith junior and Sir Humphrey Gilbert [debated] for Marcellus, Thomas Smith senior and Doctor Walter Haddon for Fabius Maximus, before an audience at Hill Hall consisting at that very time of myself, John Wood, and several others of gentle birth. At length the son and Sir Humphrey yielded to the distinguished Secretary: perhaps Marcellus yielded to Fabius. Both of them

worthy men, and judicious. Marcellus the more powerful; Fabius the more cunning. Neither was the latter unprepared [weak], nor the former imprudent: each as indispensable as the other in his place. There are times when I would rather be Marcellus, times when Fabius.[32]

The debate follows the style of a formal dispute, the principal method of educating pupils in grammar schools, and the chief means of teaching and examining at a university like Cambridge, each side arguing a case, *in utramque partem*, a habit which dominated every aspect of intellectual life.[33] Indeed, in order to understand works such as a *View* and Davies's *Discovery* it is necessary to look at the education system in English universities and grammar schools. The colonisation of Ireland had begun, if not on the playing fields of Eton, then in the classrooms of England.

The arts curriculum at Cambridge was designed to prepare an educated elite for service in government, national and local, as well as the church. Universities throughout Europe sought to meet the requirements of a new social order, which demanded a wider series of roles for graduates. The complete course in the arts – which included what are now defined as the sciences – started with the trivium, the study of grammar, logic and rhetoric; proceeded to the quadrivium, which covered arithmetic, geometry, astronomy and music; and concluded with the study of the three philosophies, moral, natural and metaphysics.[34] By the later sixteenth century this basic structure remained in place – much less rigidly at Cambridge than at Oxford – but with important variations. Elizabeth produced a series of Statutes for the universities in 1570, which set out what should be taught by lecturers and the forms of examinations, as well as the government and organisation of the institution. The university lecturer in rhetoric was required to offer Cicero, Quintilian, and Hermogenes; the lecturer in Greek, Homer, Demosthenes, Isocrates and Euripides; and the lecturer in philosophy, Aristotle, Pliny and Plato, with every author to be taught in their original language. All examinations were oral disputations, with students expected to argue with their teachers at least twice a year, as well as declaim in class a set number of times. The faculty present decided the results.[35]

The Cambridge degree demanded that the student concentrate on rhetoric, dialectic and philosophy over the four years for the BA, with the Statutes requiring students to focus on rhetoric in the first year, logic in the second and third and philosophy in the final year.[36] Inspired by Quintilian and Erasmus, who placed great emphasis on the student's ability to argue a case, all undergraduates had to participate in a number of public debates, making university a natural progression from grammar or public school.[37] The universities placed a new emphasis on the study of modern languages alongside Latin and Greek, in order to train graduates to suit the needs of society outside college.[38] By the time Spenser graduated he would have read and attended lectures on Aristotle, Plato, Pliny (philosophy); Strabo, Ptolemy, Pomponius Mela (mathematics); Cicero, Aristotle, Quintilian (dialectic/ rhetoric); Virgil, Horace (grammar); Euclid (geometry); Boethius (music); and many other classical authorities.[39]

In addition, tutors such as Gabriel Harvey, who pursued his own course of learning and recommended numerous modern authors to his students, encouraged extra-curricular reading. Harvey noted in his letter-book the enthusiasm of students for recent works of controversial political history:

> You can not stepp into a scholars studye but (ten to on) you shall litely finde open other Bodin de Republica or Le Royes Exposition upon Aristotles Politiques or sum other like Frenche or Italian Politique Discourses.
>
> And I warrant you sum good fellowes amongst us begin nowe to be prettily well acquayntid with a certayne parlous booke called, as I remember me, Il Principe di Niccolo Macchiavelli, and I can peradventure name you an odd crewe or tooe that ar as cuninge in his Discorsi sopra la prima Deca di Livio, in his Historia Fiorentina, and in his Dialogues della Arte della Guerra tooe.[40]

Harvey undoubtedly encouraged students to read these books, as they reflected his taste in modern political theory and he certainly owned copies of Bodin, Guicciardini and Machiavelli.[41] The wills of dead students also reveal their ownership of these sort of books, along with Castiglione's *Book of the Courtier*, Camden's *Britannia* and Buchanan's *History of Scotland*.[42] Given Spenser's relationship to Harvey it can be no accident that *A View of the Present State of Ireland* refers to Aristotle and Machiavelli's *Discourses*, and is also undoubtedly informed by the reading of Bodin and Guicciardini.[43] That Spenser would have been a voracious reader given his education at the Merchant Taylors' School and Cambridge is hardly surprising, but Harvey's particular theories of careful, targeted reading helped form his particular habits of arguing carefully, methodically assembling a host of relevant examples to make a case, driving an opponent into submission with the force of his logic and the breadth of his knowledge, exactly what Harvey attempted in his published writings.[44]

Perhaps the most significant change to the student experience in the late sixteenth century occurred in the teaching of rhetoric and logic. Students no longer had to master scholastic logic but were taught instead using the works of Aristotle, and, controversially, through the work of his modern, Protestant interpreter, Peter Ramus (1515–72), who perished in the Saint Bartholomew's Day Massacre.[45] Ramus's status rose as a result of his martyrdom.[46] *Logic* was translated in 1574, just before Spenser took his MA, and a collection of other books followed in the years immediately afterwards, including works on grammar (1585); geometry (1590); and arithmetic (1592).[47] Ramus had an enormous influence on virtually every aspect of English culture in the next half century, from science to law, from history to the science of memory, and, above all, in enabling Calvinist divines to develop a system of logic that explained how they saw God's plans for the world and the choices that he presented to humankind.[48] Ramus sought to simplify logic, basing it on a method of practical reasoning that could deal with any situation and could be transported easily enough from one to another. Rhetoric and logic were no longer seen as divergent fields, but rather as ways of thinking and arguing that complemented each other, to be combined and deployed as the user saw fit. Roll's translation of Ramus's *Logic*, which argues for the need to produce such works in the vernacular, opens with a characteristically lucid statement outlining divisions for the reader to follow:

> Dialecticke otherwise called Logicke, is an arte which teachethe to dispute well.
> It is diuidyd into two partes: Inuention, and iudgement or disposition.
> Inuention is the first parte of Dialecticke, which teachethe to inuente arguments.
> An argumente is that which is naturally bente to proue or disproue any thing, suche as be single reasons separately and by them selues considered.
> An argumnte is eyther artificial, or without arte.
> Artificiall is that, which of it self declare and is eyther first, or hathe the beginning from the first.
> The first is that which hathe the beginning of it self: is eyther simple or compared.
> The Symple is that, which symplie and absolutelie is considered: and is eyther agreeable or disagreeable.
> Agreeable is that, wich agreethe with the thing that it prouethe: and is agreeable absolutely, or after a certaine fashion.
> Absolutely, as the cause and the effecte.[49]

Ramist logic involved carefully dividing up different definitions and opposites, often leading to the construction of charts that show how any one concept could be divided into two, two into four, and so on, with the hope that the reader would be able to see the apparently complex, but actually straightforward relationship between a host of ideas and forms.

As an actual method Ramist logic probably proved less revolutionary than either its proponents or its detractors claimed, certainly for the majority of non-specialist students taught at Cambridge.[50] Often the forms of argument described and recommended were the same as those of the scholastic logic they supposedly replaced, although stripped of the sophisticated levels of content and forms of proof.[51] The basic building block of logical remained the syllogism, derived from Aristotle's statement, 'discourse in which, certain things being posited, something else necessarily follows', but which by the middle ages had become more obviously formulaic, involving three categorical propositions, with the third (the conclusion), following from the first two (the premises).⁵² The most simple form of syllogistic logic was represented as 'All men are mortal; Socrates is a man; therefore, Socrates is mortal.' Although Ramists attacked logical forms such as tautologies as a simple exercise in exploiting the definitions of words, they, nevertheless, invariably relied on syllogistic reasoning.⁵³ Dudley Fenner's explicitly Ramist *The Artes of Logike and Rhethorike* (1584), for example, an attempt to outline a usable manual for Protestants, establishes a series of Ramist charts to outline the options available for logical investigation. He classifies the syllogism with a series of relevant examples:

> *The Affirmatiue generall:*
> All the iustified shalbe saued:
> Al the iustified shal raigne with Christ: Therefore
> Some that raigne with Christ, shalbe saued.
> *The Negative with the Proposition generall:*
> No hypocritical caller upon God shalbe saued.
> All hypocritical callers vpon God, say, Lorde, Lorde. Therefore
> Some that say, Lord, Lord, shal not be saued.
> *Affirmatiue speciall:*
> Some who fel in the wilderness heard the word.
> Al who fell in the wilderness, tempted God. Therefore
> Some that heard the word, tempted God. Heb. 6. 3.⁵⁴

The Calvinist nature of Fenner's examples, demonstrating to readers that outward signs of God's grace would not necessarily place them among the elect, is clear enough.⁵⁵

A *View* follows a straightforward logical argument, its form as a dialogue owing much to the educational system Spenser experienced as a university student, augmented by Gabriel Harvey. Its argument is simple enough: Ireland needs drastic reform through military intervention and the establishment of English colonies, an argument that the text claims is beyond the understanding of most Englishmen who have never been there. The bold stylistic flourishes; the wealth of examples used; and the relentless development of the implications of the basic argument are products of the concentration on practical rhetoric and logic in schools and the two universities in Spenser's lifetime. Other public debates in Ireland and England took place with a conscious acknowledgement of the forms of logical and argument acquired through a university education.⁵⁶ The point is further made by noting that Abraham Fraunce employed the works of Sidney and Spenser – albeit not a *View* – to illustrate his books on Ramist rhetoric and logic, a case of *post hoc ergo propter hoc* logic.⁵⁷ Irenius constantly reminds Eudoxus that his propositions might seem reasonable in England, but do not work in Ireland because of the strange and original nature of the country, meaning that proper experience was required in order to argue logically about it.[58] The argument between the two invariably takes the form that Eudoxus statements would be true if Ireland was as he believed it to be, but it is different and so the answer has to be different, a form of negative syllogistic reasoning. Hence, Eudoxus contends that the spread of the law should render Ireland loyal to the English crown, but is told by Irenius that his [Irenius's] experience qualifies his logic and renders it inadequate on its own to understand a strange land in which fundamental principles need to be re-established. To take just one example: Eudoxus is staggered at the size and cost of the army that Irenius recommends but is then persuaded that such expense will save money in the long run (*View*, p. 130). Therefore, the dialogue might be read as one large, provocative syllogism:

> Ireland needs to be governed properly.
> Proper government will involve methods that will seem shocking.
> Therefore:
> Ireland requires methods that will seem shocking.

Much more could be said on the subject that relates the theory and practice of colonialism, but the plantation of Ulster required intellectual arguments as well as practical methods, the one reinforcing the other.[59] Davies's *Discovery*, the work of a lawyer who clearly understood contemporary logic, as well as legal theory and practice, responds to Spenser in kind and shows that Ireland, now conquered, is ready to be reformed which places it in a syllogistic relationship to Spenser:

> Spenser argues that Ireland will not be amenable to legal reform until it has been conquered.
> Ireland has been conquered.
> Therefore:
> Ireland is amenable to legal reform.

The task of establishing the plantation had begun, secure in the knowledge that the fate of its southern counterpart would now be avoided. The Treaty of Mellifont signed at the end of the Nine Years War in 1603, and the Flight of the Earls four year later, would establish a far more secure base for the effective colonisation of Ireland than the rather more tenuous confiscation policy practised in the wake of the Desmond Revolt which ended in 1583.[60] In a crucial sense Spenser's logic of confiscation became accepted practice, with the native Irish in general declared traitors and their land forfeit, an over-riding legal principle that saw English law come down on Spenser's side rather than that of Lord Roche.[61]

There is a neat irony in the logic of migration advocated in a *View*. Spenser argued that the Irish were really Scots, having come over as earlier colonisers.

In doing so, he followed Buchanan's *History of Scotland*, which he would have read soon after it appeared in 1582, undoubtedly inspired by Harvey's strong recommendation of Buchanan's political ideas, especially his savage denunciations of Mary Stuart.[62] Spenser added his own particular anthropological speculations:

> And first of their armes and weapons, amongst which their broad swordes are proper Scythian, for such the used commonly, as you may read in Olaus Magnus. And the same also the old Scots used, as you may reed in Buchanan, and in Solinus, where the pictures of them are in the same forme expressed. Also theire short bowes, and little quivers with short bearded arrowes, are very Scythian, as ye may reade in the same Olaus. And the same sort, bothe of bowes, quivers, and arrowes, are at this day to be seene commonly amongst the Northerne Irishe, ... Moreover, their longe broad shieldes, made but with wicker roddes, which are comonly used amongst the said Northerne Irishe, but especially of the Scots, and brought from the Scythians, as you may read in Olaus Magnus, Solinus, and others; likewise their going to battale without armor on their bodies or heads, but trusting to the thicknes of their glibbs, the which (they say) will sometimes beare off a good stroke, is meere Scythian, as you may see in the said images of the old Scythes or Scots, set foorth by Herodianus and others. Besides, their confused kinde of march in heapes, without any order or array, their clashing of swords together, their fierce running upon their enemies, and their manner of fight, resembleth altogether that which is read in all histories to have beene used of the Scythians. By which it may almost infallibly be gathered together with other circumstances, that the Irishe are very Scots or Scythes originally, though sithence intermingled with many other nations repairing and joyning unto them (*View*, pp. 61–2).

The link between Scots, Irish and Scythians is made with repetitive, emphatic logic throughout a *View*.[63] After Spenser's death the Ulster plantation replaced many of these Scots with lowland Scots and English. The defeat of the most recalcitrant and aggressive section of the native Irish population pleased that lowland Scot, James VI and I, who had little time for his Highland compatriots and indigenous Irish Catholics.[64] Initially at least, James had favoured a 'cautious course in all matters concerning Ireland' but after Chichester – who, as already noted, owned a copy of a *View* – 'recommended that Scottish as well as English Protestants should be invited to become undertakers in Ulster', the plantation became a 'pet project' of the king, part of his drive to 'create a greater Britain'.[65] This neat and pleasing transformation, established by educated thinkers informed by a knowledge of British history and, first or second hand, an experience of colonial government, provides a splendid example of colonial theory and practice being used together to drive fundamental change. And, of course, it established the basis for the division of the former kingdom in 1920.[66]

It is, of course, hard to assess the exact impact of a *View* on later policy in Ireland, in particular the establishment of the Ulster plantation. Intellectual traffic is notoriously difficult to chart. Equally, I would suggest, it would be foolish to imagine that Spenser's dialogue played no part in the subsequent course of Irish history. The work circulated widely after Spenser's death in January 1599, and was read by those with power and influence, most notably, the Lord Deputy who oversaw the establishment of the plantation, Sir Arthur Chichester. What is important to acknowledge is that discussions of colonial policy by soldiers, statesmen and academics, developed out of an intellectual culture that helped to put theory into practice. Reading Latin and Greek literature at school and university led to an understanding that the establishment and maintenance of colonies was a central element of a responsible government's concerns, which in turn led to the selection of passages from classical texts that provided useful advice on such matters. The use of Livy in the debate at Hill House illustrates this point nicely.

But it was not just the content of classical texts that was important; it was also their methods and styles of argument and the ways in which they supported and illustrated their claims. A *View* is a calculating text that plays upon the expectations of its audience, constantly reminding its readers that the author knows exactly how they have been trained to think and satisfying and thwarting those expectations by turn in order to persuade them that they will have to accede to the logic of its arguments. Perhaps Spenser simply stated what was generally understood and what he argued about plantation would have happened anyway; perhaps he did make a crucial difference to the course of Irish history. Either way, it is clear that there is a symbiotic relationship between the theory and practice of colonialism, one that indicates the influence of intellectual on military culture. Spenser, of course, had been convinced by his experience in Ireland, then turned back to his intellectual resources in order to make the best possible case. Equally, it is surely true that he experienced Ireland in terms of what he had already learned. It is easy to over- or under-estimate the impact of texts on actions, unless we are such hard-headed empiricists that we believe that any relationship between the two can only be coincidence.

NOTES

1 Nicholas Canny, *Making Ireland British, 1580–1650* (Oxford, 2001), ch. 1. See also Nicholas Canny, 'Spenser and the development of an Anglo-Irish identity', *Yearbook of English Studies*, 13 (1983), 1–19.
2 W. B. Yeats, 'The Man and the Echo', lines 11–12, in *Collected Poems* (London, 1990), p. 393.
3 T. W. Moody, *The Londonderry Plantation, 1609–41: The City of London and the Plantation in Ulster* (Belfast, 1939); Cyril Falls, *The Birth of Ulster* (1936; repr.

London, 1996); David Harris Willson, *King James VI and I* (London, 1956), pp. 322–6.
4 See, for example, J. H. Andrews, 'A geographer's view of Irish history', in T. W. Moody and F. X. Martin (eds), *The Course of Irish History* (Cork, 1967), pp. 17–30, at pp. 24–5.
5 Canny, *Making Ireland British*, p. 211.
6 Michael Perceval-Maxwell, *Scottish Migration to Ulster in the Reign of James I* (London, 1973), pp. 217–28.
7 The best analysis of the early colonial experiments remains D. G. White, 'The Tudor plantations in Ireland before 1571' (Ph.D. thesis, Trinity College Dublin, 1967), a work that should have reached a much wider audience than it has.
8 On the cost of maintaining an army in Ireland, see Cyril Falls, *Elizabeth's Irish Wars* (1950; repr. London, 1996); on the perils of victualling an army in Ireland, see Ciaran Brady (ed.), *A Viceroy's Vindication: Sir Henry Sidney's Memoir of Service in Ireland, 1556–78* (Cork, 2002).
9 On the Ards Penninsula, see Hiram Morgan, 'The colonial venture of Sir Thomas Smith in Ulster, 1571–5', *Historical Journal*, 28 (1985), 261–78; on the Munster plantation, see Michael McCarthy-Morrogh, *The Munster Plantation: English Migration to Southern Ireland, 1583–1641* (Oxford, 1986).
10 See Andrew Hadfield, 'Rethinking early modern colonialism: the anomalous state of Ireland', *Irish Studies Review*, 7 (1999), 13–27; idem, 'Irish colonies and the Americas, 1560–1610', in Robert Appelbaum and John Wood Sweet (eds), *Envisioning an English Empire: Jamestown and the Making of the North Atlantic World* (Philadelphia, 2005), pp. 172–91.
11 Ray Heffner, 'Spenser's acquisition of Kilcolman', *Modern Language Notes*, 46 (1931), 493–8. See also Julia Reinhard Lupton, 'Mapping mutability: or, Spenser's Irish plot', in Brendan Bradshaw, Andrew Hadfield, and Willy Maley (eds), *Representing Ireland: Literature and the Origins of Conflict, 1534–1660* (Cambridge, 1993), pp. 93–115.
12 Edmund Spenser, *A View of the State of Ireland*, ed. Andrew Hadfield and Wily Maley (Oxford, 1997), p. 159. All subsequent references to this edition are in parentheses in the text.
13 Rudolf B. Gottfried, 'Spenser's view and Essex', *PMLA*, 52 (1937), 645–51. More generally, see Paul E. J. Hammer, *The Polarisation of Elizabethan Politics: The Political Career of Robert Devereux, 2nd Earl of Essex, 1585–1597* (Cambridge, 1999).
14 For a list, no longer complete, see Edwin Greenlaw, Charles Grosvenor Osgood, Frederick Morgan Padelford and Ray Heffner (eds), *The Works of Edmund Spenser: A Variorum Edition* (11 vols, Baltimore, 1932–45), x, 507. For the provenance of one particular manuscript, copied out in a volume with other important contemporary material, see James McManaway, 'Elizabeth, Essex, and James', in *Elizabethan and Jacobean Studies Presented to Frank Percy Wilson* (Oxford, 1959), pp. 219–30.
15 T. W. Moody, F. X. Martin and F. J. Byrne (eds), *A New History of Ireland: IX, Maps, Genealogies, Lists: A Companion to Irish History*, pt 2 (Oxford, 1984), p. 487.
16 For discussion, see Andrew Hadfield, *Spenser's Irish Experience: Wilde Fruit and Salvage Soyl* (Oxford, 1997), ch. 2, and the excellent account in Steven K. Galbraith,

'Edmund Spenser and the history of the book, 1569-1679' (Ph.D. dissertation, Ohio State University, 2006), ch. 5.

17 Raymond Jenkins, 'Spenser: the uncertain years, 1584-89', *PMLA*, 53 (1938), 350-62; Vincent P. Carey, 'Atrocity and history: Grey, Spenser and the slaughter at Smerwick (1580)', in David Edwards, Pádraig Lenihan, and Clodagh Tait (eds), *Age of Atrocity: Violence and Political Conflict in Early Modern Ireland* (Dublin, 2007), pp. 79-94; Ciaran Brady, '*A Briefe Note of Ireland*', in A. C. Hamilton (ed.), *The Spenser Encyclopedia* (Toronto, 1990), pp. 111-12.

18 Canny, *Making Ireland British*, ch. 4. On Irish woods, see C. Litton Falkiner, *Illustrations of Irish History and Topography, Mainly of the Seventeenth Century* (London, 1904), pp. 143-59.

19 Patricia Coughlan, 'The local context of mutabilitie's plea', in Anne Fogarty (ed.), *Irish University Review: Spenser in Ireland, 1596-1996* (Autumn/Winter 1996), pp. 320-41.

20 Anthony J. Sheehan, 'The overthrow of the plantation of Munster in October 1583', *Irish Sword*, 15 (1982), 11-22. For one representative text, see Willy Maley (ed.), 'The Supplication of the blood of the English, most lamentably murdred in Ireland, Cryeng out of the yearth for revenge, c.1598', *Analecta Hibernica*, 36 (1994), 3-91.

21 Edmund Spenser, *Selected Letters and Other Papers*, ed. Christopher Burlinson and Andrew Zurcher (Oxford, 2009), pp. 215-20.

22 Ray Heffner, 'Spenser's view of Ireland: some observations', *Modern Language Quarterly*, 3 (1942), 507-15. The copy is now in the British Library, BL Additional MS 22022. See also Canny, *Making Ireland British*, p. 58. On Davies's role in Ireland, see Hans S. Pawlisch, *Sir John Davies and the Conquest of Ireland: A Study in Legal Imperialism* (Cambridge, 1985).

23 Sir John Davies, *A Discouerie of the True Causes Why Ireland was Never Entirely Subdued, nor Brought under Obedience of the Crowne of England, untill the Beginning of his Maiesties Happie Raigne* (London, 1612), p. 161.

24 B. J. Sokol and Mary Sokol, *Shakespeare's Legal Language: A Dictionary* (London, 2000), pp. 408-10.

25 See also Andrew Hadfield, 'Historical writing, 1550-1660', in Andrew Hadfield and Raymond Gillespie (eds), *The History of the Irish Book, Vol. iii: The Irish Book in English, 1550-1800* (Oxford, 2006), pp. 250-63, at pp. 255-7.

26 On Harvey, see Virginia F. Stern, *Gabriel Harvey: A Study of his Life, Marginalia, and Library* (Oxford, 1979); on Smith, see Mary Dewar, *Sir Thomas Smith: A Tudor Intellectual in Office* (Cambridge, 1964).

27 Andrew Hadfield and John McVeagh (eds), *Strangers to That Land: British Perceptions of Ireland from the Reformation to the Famine* (Gerald's Cross, 1994), p. 99.

28 On Smith, see Morgan, 'Colonial venture of Sir Thomas Smith'; Richard Bagwell, *Ireland under the Tudors* (1885-90; repr. 3 vols, London, 1963), ii, 246-7.

29 Henry R. Plomer, 'Henry Bynemann, printer, 1566-83', *The Library*, new service, 9 (1908), 225-44.

30 Sir Thomas Smith, *A Letter Sent by I.B. Gentleman unto his very frende Maystet [sic] R.C. Esquire vvherin is conteined a large discourse of the peopling & inhabiting*

the cuntrie called the Ardes, and other adiacent in the north of Ireland, and taken in hand by Sir Thomas Smith one of the Queenes Maiesties priuie Counsel, and Thomas Smith Esquire, his sonne (London, 1572), Sig. D2r.

31 On the debates, see Lisa Jardine and Anthony Grafton, '"Studied for Action": how Gabriel Harvey read his Livy', *Past & Present*, 129 (1990), 30–78; on Hill Hall, see Paul Drury, with Richard Simpson, *Hill Hall: A Singular House Devised by a Tudor Intellectual* (London, 2009).

32 Cited in Lisa Jardine, 'Encountering Ireland: Gabriel Harvey, Edmund Spenser, and English colonial ventures', in Bradshaw et al. (eds), *Representing Ireland*, pp. 60–75, at p. 63.

33 Peter Mack, *Elizabethan Rhetoric: Theory and Practice* (Cambridge, 2002), ch. 1; David Cressy, *Education in Tudor and Stuart England* (London, 1975), pt 5; Joel B. Altman, *The Tudor Play of Mind: Rhetorical Inquiry and the Development of Elizabethan Drama* (Berkeley, 1978), ch. 2.

34 Mark H. Curtis, *Oxford and Cambridge in Transition, 1558–1642: An Essay on the Changing Relations between the English Universities and English Society* (Oxford, 1959), p. 86; Cressy, *Education in Tudor and Stuart England*, pp. 132–5; J. M. Fletcher, 'The Faculty of Arts', in James McConica (ed.), *The History of the University of Oxford: Volume III, The Collegiate University* (Oxford, 1986), pp. 157–212, at pp. 171–81.

35 See the description of Cambridge examinations in *The Diary of Baron Waldstein: A Traveller in Elizabethan England*, trans. G. W. Groos (London, 1981), pp. 93–101.

36 Fletcher, 'Faculty of Arts', p. 173.

37 Curtis, *Oxford and Cambridge in Transition*, pp. 88–9; Fletcher, 'Faculty of Arts', pp. 167–9; *Collection of the Statutes for the University and the Colleges of Cambridge* (London, 1840), pp. 294–9.

38 Curtis, *Oxford and Cambridge in Transition*, pp. 137–9; Warren Boutcher, '"A French Dexterity, & an Italian Confidence": new documents on John Florio, learned strangers and protestant humanist study of modern languages in Renaissance England from c.1547 to c.1625', *Reformation*, 2 (1997), 39–110.

39 Fletcher, 'Faculty of Arts', p. 172.

40 *Letter-Book of Gabriel Harvey, 1573–1580*, ed. Edward John Long Scott (Camden Society, London, 1884), pp. 79–80.

41 Stern, *Harvey*, pp. 226, 239, 246, 265, 268.

42 Curtis, *Oxford and Cambridge in Transition*, pp. 136–7.

43 Spenser, *Variorum*, x, 121, 229; Edwin A. Greenlaw, 'The influence of Machiavelli on Spenser', *Modern Philology*, 7 (1909), 187–202.

44 See, for example, Gabriel Harvey, *Ciceronianus*, ed. and trans. Harold S. Wilson and Clarence A. Forbes (Lincoln, 1945); idem, *Rhetor* (1577), trans. Mark Reynolds, http://comp.uark.edu/~mreynold/rheteng.html.

45 Norman Jones, *The English Reformation: Religion and Cultural Adaptation* (Oxford, 2002), p. 178.

46 Lisa Jardine, 'Humanistic logic', in Charles B. Schmitt and Quentin Skinner (eds), *The Cambridge History of Renaissance Philosophy* (Cambridge, 1988), pp. 173–98, at p. 185; Fletcher, 'Faculty of Arts', p. 178. Roll's translation of Ramus's *Logic*, hails him as 'Martyr to God' (p. 12).

47 *The logike of the moste excellent philosopher P. Ramus martyr*, trans. M. Roll (London, 1574); *The Latine grammar of P. Ramus: translated into English* (London, 1585); *Elementes of geometrie. Written in Latin by that excellent scholler, P. Ramus*, trans. Thomas Hood (London, 1590); *The art of arithmeticke in whole numbers and fractions in a more readie and easie method then hitherto hath bene published, written in Latin by P. Ramus; and translated into English by William Kempe* (London, 1592).

48 Walter J. Ong, *Ramus, Method, and the Decay of Dialogue: From the Art of Discourse to the Art of Reason* (Cambridge, MA, 1958); Frances A. Yates, *The Art of Memory* (London, 1966), passim. For attempts to apply Ramus's logical methods to Calvinist theology, see Dudley Fenner, *The artes of logike and rethorike plainelie set foorth in the English tounge, easie to be learned and practised: togeather with examples for the practise of the same, for methode in the gouernment of the familie, prescribed in the word of God: and for the whole in the resolution or opening of certaine partes of Scripture, according to the same* (Middleburg, 1584); H. C. G. Matthew and Brian Harrison (eds), *Oxford Dictionary of National Biography* (60 vols, Oxford, 2004), sub 'Fenner, Dudley'.

49 Ramus, *Logike*, sig. B1r-v.

50 Although one should be wary of underestimating the achievements of Renaissance logic: see Jardine, 'Humanistic logic', pp. 173–4.

51 E. J. Ashworth, 'Traditional logic', in Schmitt and Skinner (eds), *Cambridge History of Renaissance Philosophy*, pp. 143–72; D. P. Henry, *Medieval Logic and Metaphysics* (London, 1972).

52 'Syllogism', in Ted Honderich (ed.), *The Oxford Companion to Philosophy* (Oxford, 1995), p. 862.

53 Ramus, *Logike*, pp. 82–4; Lisa Jardine, 'Humanism and the sixteenth century Cambridge Arts Course', *History of Education*, 4 (1975), 16–31, at 20–1; Ashworth, 'Traditional logic', pp. 162–72.

54 Fenner, *Artes of Logike and Rethorike*, sig. C3v.

55 Jean Calvin, *Institutes of the Christian Religion*, ed. and trans. John T. McNeill and Ford Lewis Battles (2 vols, Philadelphia, 1965), ii, 968–70 (III.24, 4); François Wendel, *Calvin: The Origins and Development of His Religious Thought*, trans. Philip Mairet (London, 1965), pp. 234–42.

56 Brian Jackson, 'The construction of argument: Henry Fitzsimon, John Rider and religious controversy in Dublin, 1599–1614,' in Ciaran Brady and Jane Ohlmeyer, (eds), *British Interventions in Early Modern Ireland* (Cambridge, 2005), pp. 97–115.

57 Fraunce, *Arcadian Rhetorike*. Trinity College, Dublin, founded in 1592, was notable for the Ramist character of its curriculum: Alan Ford, '"That Bugbear Armenianism": Archbishop Laud and Trinity College, Dublin', in Brady and Ohlmeyer (eds), *British Interventions in Early Modern Ireland*, pp. 135–60, at p. 154.

58 See Hadfield, *Spenser's Irish Experience*, ch. 2.

59 For discussion, see M. I. Finley, 'Colonies: an attempt at a typology', *Transactions of the Royal Historical Society*, 26 (1976), 167–88.

60 See Steven G. Ellis, *Tudor Ireland: Crown, Community and the Conflict of Cultures, 1470–1603* (Harlow, 1985), ch. 9.

61 Pawlisch, *Davies*, pp. 3–14.
62 Stern, *Harvey*, pp. 204–5, 246, passim. See also Andrew Hadfield, 'Spenser and Buchanan', in Roger Mason and Caroline Erskine (eds), *George Buchanan* (Farnham, 2012).
63 Hadfield, *Spenser's Irish Experience*, ch. 2.
64 Willson, *James VI and I*, pp. 96–115.
65 Canny, *Making Ireland British*, p. 192.
66 R. F. Foster, *Modern Ireland, 1600–1972* (London, 1988), chs 1–3.

10

Responses to transformation: Gaelic poets and the plantation of Ulster

MARC CABALL

Woe to the heart that meditated, woe to the mind that conceived, woe to the council that decided on the project of their setting out on this voyage, without knowing whether they should ever return to their native principalities or patrimonies to the end of the world. (*Annals of the Four Masters*, sub anno 1607)[1]

Although composed at a remove from the flight of the northern earls to the continent in 1607, the *Annals of the Four Masters* (1632–36), a chronicle of the history of Ireland from the biblical deluge to contemporary times, encapsulate a quintessentially apocalyptic element in Gaelic perceptions of political and cultural transformation in early seventeenth-century Ulster in their haunting lamentation of the earls' departure. In effect, the response of the Gaelic intellectual elite to the plantation of Ulster is defined by a powerful and pervasive sense of trauma, alienation and communal dispossession.[2] In an anonymous poem extant in the 'Book of O'Conor Don' (1631), which reflects bitterly on the exile of the northern earls from Ireland, several of the key themes and motifs which inform and articulate a complex interpretative pattern of despondency, dislocation and personal agency find early expression. In this poem beginning 'Mochean don loing si tar lear' ('Good luck to that ship'), the full impact and implications of a fundamental shift in the perception and exercise of authority and ethnic integrity are acutely and clinically delineated. While focusing on the immediate fate of O'Donnell and O'Neill, the unidentified author is unambiguous in his portrayal of a tragic but emblematic development which reverberates ominously across the island of Ireland. The departure of these advocates of Gaelic suzerainty signals the demise of collective autonomy ('na Gaoidhil aniú sa neart, go fiú an aoinfhir ar n-imtheacht').[3] In fact, the literal and actual supremacy of the *Gaoidhil* accompanies the passengers of this vessel across the seas to the continent and collective disgrace has resulted ('Do chuaidh oireachus bhfear bhFáil...tar sál re lucht na luinge, tár as a ucht oruinne'). Ireland lies inanimate and passive such is the debilitating nature of the disaster consequent on the earls' and their kinsmen's exile ('Ní mhaireann déis an eathair...dé don Bhanbha ar beathughadh').[4] Imploring protection from the Holy Trinity for the ship's passengers, the poet also stresses the depth of his own distress. In summary, therefore, the impact of this transformation in Ireland is manifest on a variety of levels: national, provincial, dynastic, family and personal.

These sentiments are echoed in another anonymous poem also extant in the 'Book of O'Conor Don'. In the composition beginning 'Cáit ar ghabhadar Gaoidhil?' ('Where have the *Gaoidhil* gone?'), the author laments an apparently dramatic reversal of fortune for aristocratic Gaelic society.[5] Traditional Gaelic haunts and pursuits are neglected following the dispersal and exile of the warrior elite from all parts of Ireland. Their concealment from the poet is neither the consequence of what he calls a supernatural incantation ('briochd síodhuidhe') nor a deceitful magical mist ('ní ceó doilbhthe draoidheachta'), rather their exile ('ionnarbthoigh dhíobh do dhéineamh') and their present familiarity with foreign noble households is rooted firmly in reality. The place of the Gaelic nobility has been usurped by arrogant *arriviste* interlopers of base origin from Scotland and England ('dírim uaibhreach eisiodhan ...Saxoin ann is Albonaigh').[6] These newcomers now divide the island among them and denominate their holdings in 'acres' ('Roinnid í eatorra féin, an chríoch-sa chloinne saoirNéill'). Ireland resembles a chessboard robbed of its authentic chessmen that have been replaced by upstarts. Change for the worse has occurred. Places of worship occupied by intruders, religious services conducted in the open, clergymen's vestments used as bedding for cattle, fairs have taken the place of hunts, agricultural cultivation has displaced horse racing, and former aristocratic residences have been replaced by new homesteads. Nobody among the Gaelic Irish has cause for joy. Moreover, the cultural rituals of high Gaelic civility have also been submerged. Praise poetry, harp and organ music, dynastic tales and genealogical scholarship no longer appeal to the *Gaoidhil* such is the severity of the oppression which they endure ('Daoire na mbreath bhíos orra, gadaidh asda a n-anmanna').[7] Ultimately, however, the banishment of the Gaelic Irish should not be ascribed to foreign military superiority: divine wrath is the actual cause of their expulsion ('fearg Dé ré ccách dá ccolgadh, is é is fháth dá n-ionnarbadh'). The Gaelic Irish are not alone in having incurred the wrath of God. Other peoples, such as the 'sons of Israel' ('meic Israhel') and the 'great race of Maccabaeus', incurred divine punishment on account of their transgressions.[8] Each of these, however, had secured forgiveness through condign repentance. The poet returns to the case of his own people and laments their failure to atone for sinful misdeeds ('Aithrighe a-nois dá nós sin, truagh nach déanuid meic Mhílidh'). Atonement promises deliverance but, in the meantime, divine retribution has enabled the 'men of Scotland' and the 'youths of London' to supplant the Gaelic Irish in their own country ('Díoghaltas Dé as adhbhar ann-fir Albon, ógbhaidh Lunnand, do anadar 'na n-áit sin').[9] In his edition of this poem, William Gillies suggests that it provides evidence of the viewpoint of dispossessed swordsmen driven into exile or local banditry.[10] In fact, this piece is the product of an elite and learned milieu composed in the bardic syllabic metre known as *deibhidhe* and drawing on the conventions and thematic repertoire of medieval and early modern Gaelic praise poetry. In another poem *Mo thruaighe mar táid Gaoidhil!* ('The fate of the *Gaoidhil* saddens me') composed in the aftermath of plantation in Ulster, Fearflatha Ó Gnímh explicitly links the fate of Gaelic Ireland with its nobility.[11] These works and the poems discussed below operate at several different but intersecting levels. They are concurrently cultural, social, political and ideological in their texture and focus. The cultural historian Lynn Hunt cautions that 'documents describing past symbolic actions are not innocent, transparent texts' and that they were produced by authors 'with various intentions and strategies'. In response, Hunt advises historians to devise their own strategies in reading these documents.[12] In the case of praise poems, it is necessary to remember that they, like other historical texts, are neither disinterested, dispassionate nor without strategic intent.[13]

This essay will examine literary work by the professional cadre of Gaelic praise poets in the aftermath of the plantation of Ulster. In particular, poems by three poets, Eochaidh Ó hEódhusa, Fearghal Óg Mac an Bhaird and Eoghan Ruadh Mac an Bhaird, are discussed with a view to discerning an outline of their responses to unprecedented social, cultural and economic upheaval. During the first decade of the seventeenth century, Ulster, traditionally considered a bastion of Gaelic society and culture, was transformed in a relatively short timeframe by the military defeat and subsequent departure in September 1607 of the northern earls, Rory O'Donnell and Hugh O'Neill, to the continent. Moreover, the unsuccessful revolt in the following year of Cahir O'Doherty of Inishowen resulted in a decision by the government to diminish drastically the authority and influence of the remaining Gaelic lords in the province. The proposed new political and social settlement entailed a wholesale process of confiscation of lands and the introduction of a large cohort of settlers from Britain. Philip Robinson's description of the Ulster plantation as 'one of the most politically significant mass migrations to have taken place in western Europe since medieval times' underlines the transformative nature of this process.[14] Designed to underpin stability in hitherto intractable west Ulster, the plantation also proved highly attractive in other respects to its sponsor James VI and I. In addition to extending English and Scottish commercial reach, the allocation of land to military personnel by way of reward for service entailed minimal cost to the government. Additionally, demographic pressures in Scotland and England might be alleviated through the transfer of population to the north of Ireland.[15]

Philip Robinson has succinctly defined the concept of plantation as a state-endorsed process which 'involved the transfer of an entire package of personnel, laws and materials into a new territory'.[16] Essentially, the plantation aimed in theory to create a cultural and social *tabula rasa* in a region which had been characterised by a long-established and highly-codified indigenous culture.[17] Bardic poets were in many respects key custodians of this ancient cultural patrimony. Composed by a professional cohort broadly between 1250 and 1650, praise poetry reflects the political ideology and outlook of the seigneurial milieu in which it was produced and disseminated. The primary function of the poet centred on the provision of dynastic, cultural and social validation for the seigneurial elite. Drawing on a common and generic stock of literary motifs and conventions, poets established and articulated the validity of a ruling or potential lord before fellow noblemen and the inhabitants of his territory.[18] Praise poetry enshrined pivotal concepts and tropes that underpinned and informed conventional notions of lordship and its exercise. Extant evidence suggests strongly that praise poets were intimately involved in the dynastic affairs of Gaelic lordships. In a poem beginning 'Mór an t-ainm ollamh flatha' ('Great is the title chief's poet'), Eochaidh Ó hEódhusa drew the attention of Hugh Maguire (*d*.1600), lord of Fermanagh, to the semi-diplomatic role of a lord's poet. In this eloquent vindication of the emoluments, professional status and entitlements of a chief's poet, Ó hEódhusa is quite specific in his attribution of the function of political emissary to the latter. In addition to right of access to his patron and his enrichment with appropriate material reward, he asserts the entitlement of the chief poet to accompany his lord in the context of dynastic treaty negotiations.[19]

The provision by a poet of counsel and advice to his patron, and perhaps also his assumption of an instrumental function as emissary or political mediator, is confirmed elsewhere. For example, the Munster nobleman Florence MacCarthy Reagh (*d.c.*1641) in a letter addressed to Sir Robert Cecil in 1602 identified two groups as having particular status in Ireland: priests and poets. Intriguingly, he suggested that some of the poets might be employed in the service of the crown. According to MacCarthy only priests and poets were 'of greatest ability, and authority to persuade that country gentlemen, which of all other sorts, and sexes, doth most distaste and mislike the state, and government of England'. MacCarthy, however, glossed his remarks by claiming that because of the hostility of the Catholic clergy towards the crown authorities only praise poets could be trusted with official business, and even then, the poets' primary loyalty apparently rested with their patrons.[20] When Giolla Riabhach Ó Cléirigh congratulated Cú Chonnacht Maguire (*d*.1589) and the grandees of Fermanagh on their knowledge of poetry ('do-chuaidh ar chléir ccomhadhaigh buain fá t'fhéin a n-ealadhain'), he highlighted both the pivotal role of the poets in the articulation of Gaelic ideology of lordship and their symbiotic relationship with the Gaelic elite.[21] Notwithstanding the growing importance of print culture, the communication of information in early modern Europe remained essentially verbal in nature and in this regard Gaelic Ireland proved no exception.[22] Such orality allowed for intersecting channels of exchange between elite and popular cultural media. The public declamation of praise poetry at times of collective assembly and communal festivity suggests that its influence extended beyond the lordly hierarchies of the Gaelic elite. A fleeting comment written by an anonymous scribe in 1620 in a manuscript into which he was copying poems composed for Pilib O'Reilly (d.1596) provides a fascinating glimpse of their oral transmission. The scribe, apparently an untrained amateur, in seeking to have his lack of scribal experience excused by readers, remarks that the only source he had for the poems was a blind old man who had committed them to memory thirty years previously.[23] In explicitly addressing his remarks to a putative reader, the scribe highlights the complex interplay of verbal and scribal cultures which informed the dissemination and reception of praise poetry.

The first of the three poets discussed in this essay has been recognised both by his contemporaries and modern scholars as a master of his craft. Born circa 1568, Eochaidh Ó hEódhusa was a son of the poet Maoileachlainn Ó hEódhusa. The family had emerged as professional poets in the early fourteenth century and by the end of the sixteenth century they are recorded as resident in the townland of Baile Í Eódhusa in Fermanagh and affiliated to the Maguire lords.[24] References to Eochaidh occur in several official documents of the time, namely in a series of pardons granted to the Maguires and their associates in the period from 1586 to 1592. Crucially, Ó hEódhusa successfully negotiated the legal and administrative challenges deriving from the consolidation of the crown's writ in Ulster in the first decade of the seventeenth century, and he was included in a general pardon to the Maguires and denizens of Fermanagh granted in 1607. Despite the previously noted bardic emphasis on Gaelic dispossession and dislocation, Eochaidh Ó hEódhusa is nonetheless listed among the 'natives names & proportions' on the basis of a plantation grant in 1610–11 of 100 acres in the baronies of Coole and Tirkennedy and again in 1611 he was assigned a grant of 210 acres in the 'precinct of Clinawly' (Clanawley barony) in Fermanagh.[25] Accordingly, local intellectuals such as Ó hEódhusa were clearly not entirely without social significance or administrative capacity in the context of the new order.

Eochaidh probably received his initial bardic formation under the tutelage of his father. In what is apparently his first composition as praise poet to Cú

Chonnacht Maguire (*d*.1589) beginning 'Anois molfam Mág Uidhir' ('Now I will praise Maguire'), he seeks the understanding of his patron for the delayed assumption of his poetic duties on account of his relative youth and consequent obligation to complete his professional training. Eochaidh also remarks that he studied among the court poets of the north of Ireland. His inclusion of a quatrain in honour of Hugh Maguire in this poem, a feature which was to become his professional hallmark, is an early indication of his close relationship with Cú Chonnacht's son and successor as lord of his name.[26] He also composed poems for Hugh's half brother and successor, Cú Chonnacht Óg Maguire. In his inaugural ode for Hugh beginning 'Suirgheach sin, a Éire ógh' ('This is a case of courtship, O virgin Ireland!'), presumably composed in or shortly after 1589, he argues that Hugh's accession signals an end to Ireland's woes. He recounts an exemplum of a young Greek knight who chances on a troubled but hideous young woman. Previously ravishing, she had been transformed into her present unfortunate state while bathing at a waterfall by a bewitching shower. It was subsequently prophesied that she would only be restored to her original beauty when a handsome hero arrived to wash her countenance. The Greek knight having duly bathed her face and witnessed her transformation, wins her affection and marries her: the couple live happily together henceforth.[27] The young woman symbolises Ireland, the bewitching shower is the English while the Greek knight represents Hugh Maguire.

In fact, the themes of Ireland's territorial sovereignty and foreign intrusion are prominent in Ó hEódhusa's work. In the poem beginning 'Fada re urchóid Éire' ('Ireland has long been subject to harm'), he adopts a highly formal tone in rendering homage to Hugh following his death in a military encounter outside Cork city in 1600.[28] Maguire's passing is presented as a blow to Ireland and he is compared to the pelican who gives its life blood to revive its young who have been killed by serpents. Hugh's blood will similarly revive the people of Ireland. The pelican was commonly considered a symbol of Christ and its usage in this context has been described by James Carney as 'apparently the first nationalistic application of the idea of the redemption through blood-sacrifice'.[29] Ó hEódhusa's sense of national consciousness was, however, informed by a strategic awareness of broader political opportunities and challenges. In a poem composed to celebrate the accession of James VI of Scotland to the English throne in 1603 beginning 'Mór theasda dh'obair Óivid', he brilliantly articulates his perception of political renewal and evolution, which he argues would have been worthy of inclusion by Ovid in his verse chronicle of great transformations, *Metamorphoses*.[30] The succession of James to Elizabeth's throne inaugurates a fresh beginning for the people of Ireland who may now put behind them their previous troubles and anticipate a new era of benevolence under a benign Stuart monarchy.

Two poems attributed to Eochaidh Ó hEódhusa reflect salient aspects of his political and personal outlook in the aftermath of the flight of the earls and ensuing plantation. The first of these compositions, 'Beag mhaireas do mhacraidh Ghaoidheal' ('Few remain of the Gaelic youth'), was addressed to Sir Brian MacMahon (*d*.1622) of Oriel apparently after the flight of the earls and possibly in the initial years of the plantation.[31] Son of Aodh Óg MacMahon, Brian married Mary, a daughter of Hugh O'Neill, earl of Tyrone. The latter had endorsed Brian's efforts to secure the MacMahon lordship and he had supported Tyrone during the Nine Years War.[32] Pardoned in 1603, Brian secured a substantial grant of land on foot of the 1606 settlement of Monaghan.[33] Along with Antrim and Down, Monaghan was one of three Ulster counties excluded from the provisions of the plantation settlement. As the Irish freeholders of Monaghan were not implicated in the flight of the earls, their landholdings were not subject to confiscation or surrender.[34] The poem is indelibly marked by heightened political pessimism. The Gaelic Irish are in a state of pronounced despondency, their everyday existence resembles a living death ('geall re mbás a mbeatha ghnáth') and everything within the author's purview provides scant reason for joy. Indeed, today all have their appointed time for mourning.[35] The supporting bulwarks of the *Gaoidhil* though extant are inert and without vigour ('Tuir chothaighthe Chlainne Míleadh, gé mhairid ní mhairid siad'). Overcome by a destructive wave, the Gaelic warrior band is acquiescent.[36] Hearts burdened by enveloping weariness, the men of the Greek-like Gaelic Irish are deeply dispirited ('léir nach fuil anam i n-aoinfhear, dá marann d'fhuil Ghaoidheal nGréag'). The comparison of the Irish to the ancient Greeks appears to infer a historical analogy for the present demise of a once vibrant culture. Indeed, the distressed state of the *Gaoidhil* is instructive in its sermon-like nature ('Seanmóir linn dáríribh é').[37] The juxtaposition of references to the pagan ancient world and the quintessentially Christian medium of the sermon are disconcerting and suggestive of contemporary turmoil. The period of prosperity allocated to the nobles of Ireland has been spent and once again the notion of a catastrophic wave is invoked to emphasise transformation and profound dislocation. The Gaelic noblemen are in a state of passive acquiescence. In two quatrains beginning powerfully with the verbal construction 'Do dhearmaid said…' ('They have forgotten…'), Ó hEódhusa describes how the warrior elite had abandoned its martial discipline and routine. Indeed, their previous belligerence currently strikes him as having been possibly imaginary ('dob aisling uair éigin é').[38] The passivity of the *Gaoidhil* is not merely evidenced by their lack of martial vigour, it is also apparent in their neglect of traditional pursuits such as hunting, athletics and horsemanship. Then, rather startlingly given the bleak despondence and apparent fatalism of the preceding quatrains, Ó hEódhusa declares that there yet remains an unextinguished spark among the Gaelic Irish which will revive their spirits ('sgaoilfeas ceas do ghlainfhréimh Ghaoidhil').[39] In some respects, this is a rather enigmatic poem requiring further textual and thematic elucidation before it is possible to offer an authoritative interpretation of its narrative.[40] Nonetheless, Ó hEódhusa's insistent emphasis on the passivity of the Gaelic nobility is remarkable. Their indifference to martial endeavour appears to imply a process of demasculinisation. Moreover, he also seems to suggest that the emollient quality of the majority will be redeemed by the regenerative intervention of a resurgent minority.

Yet both personal disappointment and political despair broadly characterise Ó hEódhusa's mindset at this time. Indeed, a sense of personal tragedy resonates throughout the poem beginning 'Dá ghrádh tréigfead Máol Mórdha' ('I will abandon Máol Mórdha out of love'), in an extended reflection on patronage, frustrated loyalty and institutional stasis.[41] This poem, while not definitively ascribed to Ó hEódhusa in the late seventeenth-century manuscript now known as British Library Additional 40766, has been attributed to him on stylistic grounds.[42] Ostensibly addressed to Máol Mórdha O'Reilly (*d*.1617), the poem is arguably more concerned with its author's acute sense of personal and intellectual humiliation than with its somewhat episodic focus on an obscure east *Bréifne* nobleman.[43] In fact, such is the poem's overwhelming impression of futility and personal despair, one can only speculate as to how it was received by its erstwhile addressee. In any case, Ó hEódhusa begins the composition by stating that he will abandon O'Reilly because of his affection for him and not as a result of dislike or enmity. It is the peculiar fate of the poet that those to whom he extended his professional affections did not subsequently live for long ('an té is carthanach fám cheann, nach marthanach é acht aithghearr'). Curiously, the poet deploys two specific terms, *cinneamhuin* or 'fate' and the arcane and somewhat elliptic *fácbála* or saints' ordinances and injunctions, to describe his predicament.[44] Both nouns are strongly suggestive of an inescapable and preordained outcome. Ó hEódhusa asks rhetorically but effectively which of his patrons has not proved a source of torment for him ('Cía an compán do char misi...nach bhfúair mé comhchrádh fá [a] cheann?'). He states that God has not permitted his patrons to become his companions. Among these individuals were members of the O'Neill, O'Donnell and Maguire families. It is difficult for the poet because of his affection for Máol Mórdha to associate with him: for those beloved of him encounter misfortune ('an té charuim 's é a olc é, ní thaghuim é re hannsocht').[45] Instead, he asks O'Reilly to disregard his poetry and not to offer him his friendship. Building on the references already noted, the poet announces bluntly that the design of fate is to be feared ('eaguil cor na cinneamhna') or more precisely the death of O'Reilly. In fact, Ó hEódhusa declares that henceforth he will not offer his professional services to any man of the *Gaoidhil* ('Ní thiubhra mh'aire d'aoínfhear, d'úaislibh fhola fhionnGhaoídheal').[46]

Ó hEódhusa then proceeds to recount an exemplum by way of elucidation of his situation. The exemplum was a literary device traditionally employed by praise poets to elaborate on an argument. Described by Miri Rubin as 'the performative tokens of medieval culture, the persuasive enactments of cultural tropes', exempla also functioned as a means of adducing a historical precedent for a contemporary scenario or course of action. Ó hEódhusa's use of an exemplum at this point in his narrative emotionally enriches and enhances the significance of his predicament.[47] In this case, over the course of twelve quatrains, the poet describes the similar experience of the ill-fated Roman grande dame Cornelia, the wife of Pompey ('Corn-élia d'aitheasg rothruagh… do chan cosmhuil mo chomhráidh'). Caesar and Pompey, two great rulers in their day, divided the world between them. Their pride, however, resulted in strife and aggression. As a result, the men of the world gathered in rival camps to face each other in battle in Thessaly. Civil war ensued because of the overweening pride of both great leaders ('D'úabhar na n-airdríogh n-amhra, cuirthear an cath cathardha').[48] Caesar secured victory and Pompey returned to his wife in great sadness which caused her to be overcome with pity and anxiety. Cornelia declares that her love for Pompey and her previous husband Marcus Crassus had resulted in their downfall. She laments aloud that given her experience that she had not been Caesar's wife and then he would have encountered misfortune ("ós dom bhéasaibh ar bhean rum, gan Césair mur fhear agam'). Ó hEódhusa states that he directly resembles Cornelia and like her regrets that he had not taken a lover from among his enemies. In which case, this person would not have been destined for a long life ("s nach sáoghlach an té thoghuim'). By implication, the poet appears to cast himself in a feminine and somewhat passive role and while subject to the malign designs of fortune, he retains sufficient agency, in theory at any rate, to be able to direct the impact of his attendant misfortune. Yet, the poet silently retains his affection for O'Reilly. If he were to give the best of his affection to Máol Mórdha, which he hopes not to have done, he prays that it would not culminate in his destruction as it had for the northern lords.[49]

Reverting once again to the recurrent question of fate, Ó hEódhusa alludes to the danger of disregarding its impact. Then, in a deft and effective display of verbal manoeuvring, the poet apparently admonishes himself aloud for not treating his subject in the classic bardic mode while proceeding directly to laud and flatter O'Reilly. The latter manifests all the conventional traits and virtues of the stereotypical Gaelic lord: handsome, athletic, agile, beloved of women, martial, and generous to poets. In sum, Máol Mórdha is the counterpart of Ioldánach, an epithet for Lugh Lámhfhada, chief of the *Tuatha Dé Danann* ('a aithghin an Ioldánuigh').[50] This composition is binary in terms of its narrative framework. On the one hand, Ó hEódhusa addresses the demise of the seigneurial elite that served as a platform for patronage of praise poetry. Providing a highly personalised reading of this process, the poet locates his predicament in the context of the workings of fate and individual misfortune. On the other hand, while apparently disavowing the continued utility of the bardic template of legitimation, he actually undertakes the prescribed process of poetic validation. This is a nuanced and subtle work that acknowledges political and cultural disruption while concurrently endorsing tradition. The diversity and range of Ó hEódhusa's character is hinted at in his obituary preserved in an early eighteenth-century Fermanagh manuscript (Royal Irish Academy C vi I) but apparently reproduced from an early seventeenth-century source. The obituary records that Ó hEódhusa died on 9 June 1612 and he is acknowledged as a distinguished praise poet, scholar, hospitaller, guest house keeper and as a man held in high esteem by both the Irish and English ('duine mórainmneach ag Gaoidhealaibh 7 ag Gallaibh').[51] As the grant of lands to him under the terms of the plantation indicates, Ó hEódhusa was far from overwhelmed by the new order and successfully negotiated contemporary local political and administrative challenges. Nonetheless, both the poems examined above evidence his perception of a fundamental and negative shift in the authority, influence and dynamism of aristocratic Gaelic society.

Like Ó hEódhusa, Eoghan Ruadh Mac an Bhaird was also born in the 1560s into a learned family traditionally linked to the O'Donnells of Tyrconnell. Possibly the son of Uilliam Óg Mac an Bhaird (d.1576), he was a praise poet and retainer of Rory O'Donnell (d.1608) and Hugh O'Neill (d.1616). As with Ó hEódhusa, Eoghan Ruadh also makes fleeting appearances in extant English administrative records. He is almost certainly the 'Owen roe M'Award' included in a government pardon issued to Rory O'Donnell, his kinsmen and associates in 1603. Mac an Bhaird is also mentioned in the records of an inquisition to ascertain the boundaries of the lands of O'Donnell, O'Doherty and O'Conor Sligo held in Donegal on 26 November 1603. Mac an Bhaird presumably travelled to the continent in the wake of the flight of the earls in 1607. His presence on the European mainland is documented in contemporary Spanish records, initially as a member of the earl of Tyrconnell's retinue in Flanders in 1607–8. Later Spanish references place Eoghan Ruadh or 'the Irish nobleman Don Eugenio Bardeo' in the service of Hugh O'Neill in Flanders and in Rome during the period 1612–14. These references further indicate that he received an allowance from the Spanish monarchy from at least as early as 1608. He maintained his association with the O'Donnells, especially it seems with Nuala, wife of Niall Garbh O'Donnell (d.1625), and the sister of Aodh Ruadh (d.1602), who lived in Louvain during this period. A selection of poems in the 'Book of O'Donnell's daughter' (Bibliothèque royale, Brussels, MS 6131–3), a compendium of work dedicated to various members of the family and possibly compiled in Flanders for Nuala, is ascribed to Eoghan Ruadh. On the basis of a contemporary manuscript reference to 'D. Eugenius Vardeus, nobilis Hibernus', he was possibly still alive and resident in Rome in 1625.[52]

In the poem beginning 'A bhean fuair faill ar an bhfeart' ('My lady who has found the tomb unattended'), Mac an Bhaird addresses Nuala O'Donnell, a solitary mourner at the tomb of her brothers and nephew in Rome: Rory O'Donnell (*d*.1608), Cathbharr O'Donnell (*d*.1608) and Aodh Óg O'Neill (*d*.1609).[53] The first eight quatrains of the composition centre on the poet's graphic delineation of the loneliness of Nuala's vigil in Rome in contradistinction to what surely would have been a communal process of grieving and mourning had this tragic scenario unfolded in Ireland. In fact, in the opening quatrain, Mac an Bhaird suggests that even in Rome, Nuala would not have faced such daunting tribulations unsupported had Gaelic warriors been present ('dá mbeath fian Ghaoidheal ad ghar, do bhiadh gud chaoineadh congnamh').[54] Were this to have happened in Derry, Drumcliff or Armagh, the tomb would not have been unattended for even a day as women would make their way to it.[55] Moreover, the poet asserts that keening women would arrive from across Ireland to assist Nuala. Had they died in Ireland, no house would remain untouched by lamentation ('ní bhiadh an teach gan gháir nguil, dá mbeath láimh re Fiadh bhFionntuin'). In Ireland, their tomb would have been surrounded by mourners deeply affected by grief and sadness. Returning to Nuala's present vigil, Mac an Bhaird depicts her as lying prostrate beside the three bodies, whose untimely demise is described as the 'true culmination of our woe' ('na trí cuirp ré a síneann sibh – fírearr ar n-uilc a n-oighidh!').[56] Mac an Bhaird then proceeds to review the lineage and conventional qualities of each of the deceased. He argues that had they fallen in Ireland their virtues would have occasioned both lamentation and joy – the unexpected latter reaction presumably prompted by their heroism and exemplary demeanour.

Mac an Bhaird considers the potential impact of their deaths had they been killed in any of the successful Gaelic military actions undertaken against the crown during the course of Hugh O'Neill's rebellion (1594–1603). Immediately, he asserts that had they died in action defending their territories it would have constituted a cause of distress for all of Ulster ('dá dtuiteadh duine díobh soin, robadh sníomh uile d'Ulltoibh'). For instance, had Aodh Óg O'Neill fallen at the battle of the Yellow Ford in 1598 when Tyrone, O'Donnell and Maguire inflicted a crushing defeat on Sir Henry Bagenal's force, it would have represented a disaster for the northern party. Likewise, it would have proved a serious setback had Aodh Óg been killed in action during the encounter between Tyrone and Lord Mountjoy at the Moyry Pass in 1600 or in the course of Tyrone's campaign in Munster that same year. In a similar fashion, the death of Rory before the victorious O'Donnell engagement against Sir Conyers Clifford in the Curlew mountains in 1599 or during the latter's march against the O'Donnells via Ballyshannon in 1597 would have been no cause of joy for the Gaelic Irish ('nír fháth gáire ag Gaoidhealaibh'). Likewise with Cathbharr, his death in battle in Ireland would have represented a grave setback. In sum, their passing has proved a source of torment for the north of Ireland ('Nír bheag an léan ar Leith Cuinn'). Certainly, Rory's death in particular presages disaster for Ireland ('sgaradh do Rudhruighe rinn, rabhadh urbhuidhe d'Éirinn').[57] Directly addressing Nuala, the poet asks who of the Gaelic Irish would not weep with her for the loss of the flower of the *Gaoidhil* and he urges her to transcend her sorrow as she will join her kinsmen in due course. God's will is manifest and Nuala must accept the outcome ('do réir thagha an Tí ó bhfuil, go ragha gach ní, a Nualuidh). He advises her to get up from their grave and to reflect on the import of the cross apparently *in situ* beside her.

Turning from Nuala to the Almighty, Mac an Bhaird implores him to relieve the suffering of 'the survivors from war of Conall's blood' and to alleviate the destruction or 'shipwreck' (*longbhriseadh*) which had overtaken them.[58] Addressing the Virgin Mary, traditionally perceived in medieval Christianity as a mediator between the human and divine, Mac an Bhaird continues the theme of maritime disaster, which possibly also alludes to the ship which carried the northern earls into exile. He supplicates her to subdue the waves which engulf them and to rescue the ship of their children ('do chabhair luinge ar leanbán').[59] This latter reference seems to suggest the fate of the next generation of the *Gaoidhil*, or perhaps more narrowly of the O'Donnells, hangs precariously in the balance. Jesus is beseeched to intervene on their behalf. Mac an Bhaird concludes this moving and deceptively complex poem by urging Nuala to seek divine forgiveness in order to assuage the Lord's wrath against her kindred ('nach rabh ní as sia a fherg ret' fhuil, fagh ó Dhia, an Ceard rod chruthuigh').[60] The poet states very simply but effectively that he, like Nuala, had hoped the Gaelic Irish might have been liberated in due course when the deceased men had first departed Ireland.[61] As in the case of Eochaidh Ó hEódhusa, Mac an Bhaird chronicles the impact of a devastating blow to aristocratic Gaelic society. The poet, however, is not acquiescent in the face of death and dislocation. In the final eight quatrains of the poem, Mac an Bhaird urges Nuala to move beyond a sense of fatalistic bereavement by appeasing divine wrath, while concurrently the poet sketches a scenario which posits intervention by God, Jesus and Mary to alleviate the plight of the *Gaoidhil*. In her edition of this poem, Eleanor Knott described it as 'an elegy on the Ulster lords'. It is, however, evidently more than an elegiac and highly artful submission to a *fait accompli*. In lamenting Gaelic misfortunes, Mac an Bhaird, critically, also broaches the possibility of communal redemption.

Significantly, Mac an Bhaird's concern with matters of faith and belief are developed more fully in his elegy on Rory O'Donnell beginning 'Maith an sealad fuair Éire' ('Ireland was long prosperous').[62] This monumental tribute of ninety-one quatrains to the memory of the earl of Tyrconnell essentially combines two narrative strands: one concentrating on the feats and achievements of O'Donnell and the other comprises an elegiac celebration of Gaelic history over three thousand years, culminating contemporaneously in the depredations of the most recent influx of newcomers.[63] The arrival of the Milesians from Spain, progenitors of the *Gaoidhil*, inaugurated 3,000 years of valour and heroism in Ireland. Notwithstanding foreign intervention, the *Gaoidhil* remained steadfast in their control of the island during that period.[64] The influx of Anglo-Norman colonists, however, in the aftermath of 1169 resulted in a lasting cleavage in Gaelic imperium. Their presence in Ireland engendered turmoil and dissension. While many of the Gaelic Irish resigned themselves to the division of the island between *Gaoidhil* and *Gaill*, the O'Donnells had resisted such an accommodation and remained committed to the succour of Ireland.[65] They had managed to retain their territories in Connacht and in Ulster.[66] Moreover, they resisted the encroachment of the followers of Luther and Calvin and also spurned the overtures of what are termed heretical bishops.[67] While the churches of Ireland had been desecrated, ecclesiastical sites in Derry and Raphoe remained intact thanks to the O'Donnells. Such was their resilience, claims Mac an Bhaird, that it prompted the 'Saxon prince' ('Prionnsa Sagsan') to conclude that he could not control Ireland and suppress the indigenous church until the men of Ulster had been subdued. Terrible war ensued and its negative outcome for the Irish side resulted in the departure of O'Donnell from Ireland. Intriguingly, Mac an Bhaird presents Rory's decision to leave for the continent as based on a wish to avoid any further calamity in Ireland which might arise as a result of his continued presence.[68] His death in Rome represents a terrible tragedy for Ireland. Mac an Bhaird adduces a classical comparison for his claim by way of an exemplum outlining the grief endured by the Roman empress Livia at the unexpected death of her son Drusus. Awarded the title Germanicus as a result of his victories over the Teutonic peoples, his triumphal return to Rome was eagerly awaited by Livia. His demise, however, on the homeward journey proved a fearful blow for his mother. Ireland resembles Livia as Drusus is the equivalent of O'Donnell ('An mháthair thruagh Teamhair Chuinn, a mac díleas Ó Domhnuill').[69]

Not unlike other bardic poems composed at this time, this piece combines both tradition and innovation by way of response to unparalleled change. In many respects, Mac an Bhaird celebrates Rory's life in conventional enough terms. Yet, he is also alert to the implications of his passing for the consolidation of English hegemony throughout Ireland and not just in Ulster. The geography of composition differs dramatically from the norm. Rory died in Rome and it seems likely that Eoghan Ruadh composed his elegy while abroad also. Exile resulted in a reformulation of ideological perspective. Additionally, Mac an Bhaird's apparent counter-reformation sensibility is not widely replicated by contemporary Gaelic poets. He reverts to the theme of religious controversy towards the poem's conclusion. At this stage, he laments the straitened circumstances of church organisation, the neglect of worship and the intrusion of ministers of the reformed church into the Raphoe and Derry dioceses.[70] Nonetheless, his dissatisfaction with recent ecclesiastical developments seems informed also by communal awareness and concern. Mac an Bhaird gives the impression that the old faith was esteemed especially for its vibrant social function and not as a result of self-conscious commitment to church dogma. When he bemoans the demise of the uniquely Gaelic designation of *airchinneach* or lay hereditary occupant of church lands, it seems to imply that he valued the medieval church particularly for its temporal contribution to society. In this regard, he viewed the suppression of the church as yet another assault in a concerted programme of attrition against the Gaelic elite. This focus on the communal aspect of the culture of the church is complemented by Mac an Bhaird's earlier overt association of opposition to the established church with resistance to the extension of the crown's writ.[71]

The final poem to be considered is by Fearghal Óg Mac an Bhaird. Born in the late 1540s, Fearghal Óg is remembered as a master poet and retainer of the O'Donnells. One of his earlier works was composed after the inauguration of Toirdhealbhach Luineach O'Neill in 1567 in celebration of that occasion.[72] He was on familiar, if occasionally uneasy terms, with leading members of the O'Donnell family. When Aodh Ruadh acceded to the lordship of the O'Donnells in 1592, Mac an Bhaird composed a carefully constructed inaugural ode illustrating the legitimacy of his succession by means of classical bardic imagery. Fearghal Óg's self-confessed fiery temperament may have contributed to the tension in his relations with Aodh Ruadh evident in 'Ionnmhas ollaimh onóir ríogh' ('A chief poet's wealth is a lord's glory').[73] Indeed, relations between Mac an Bhaird and Rory O'Donnell deteriorated to such an extent at one point during the latter's lordship (1602–7) that the poet withdrew temporarily to Munster.[74] The latest definite date for Mac an Bhaird is approximately 1618, when he composed an elegy on Hugh son of Aodh Dubh O'Donnell who died that year.[75] He addressed two poems lamenting his plight as an impoverished exile to Flaithrí Ó Maolchonaire (d.1629), archbishop of Tuam, while living in Louvain. It is not known when Mac an Bhaird left for the continent nor if he ever returned to Ireland. These compositions are the only certain evidence of his time abroad.[76] Although Ó Maolchonaire had founded the Irish Franciscan college in Louvain in 1607 and often visited the city, he did not live there on a prolonged basis except between November 1623 and October 1626, and for a previous period in 1618. It is possible, therefore, that Mac an Bhaird travelled to the Low Countries in or subsequent to 1618.[77]

In the poem commencing 'Éisd rem égnach, a fhir ghráidh' ('Hear my complaint, dear friend'), Fearghal Óg addressed Ó Maolchonaire firstly by asking him to intercede on his behalf with the Almighty. He proceeds to explain that prosperity had abandoned him in Ireland and he now seeks the assistance of Ó Maolchonaire. He claims to have lost his status, land, cattle and retainers at home ('Do choilleas onóir m'anma...'). Dramatically, the poet states that he has been despoiled and dispossessed by foreigners and this situation obliged him to travel to Louvain ('Goill dom athchur 's dom arguin, rom tug go Labháin tar linn'). He knows that the archbishop has a noble name in Europe ('Ainm uasal san Eoraip ort') and he hopes that Ó Maolchonaire will assist him as appropriate to his clerical status.[78] Indeed, Ó Maolchonaire's honour requires him to alleviate the poet's plight ('is cáir dot onáir m'fhaithchill'). Hopeful of securing the archbishop's assistance, Mac an Bhaird vows optimistically to rise again and observes that when 'the tree sheds one blossom, another appears'.[79] Playing with the imagery of fresh foliage, he claims he will have a second coat before returning to Ireland. As inevitably as heat follows frost and as the sun shines once dark clouds have cleared away, the poet anticipates that similar change will happen in his case.[80] Although composing in a situation which would have been unimaginable to his bardic predecessors, Mac an Bhaird draws on custom to make his case in circumstances that are in no way traditional. Reference is made to a series of mythological and heroic Gaelic figures to endorse his approach to the archbishop. Perhaps more effectively and more convincingly, Fearghal seeks to exploit old family connections in order to elicit Ó Maolchonaire's sympathy. He relates a story in which Ó Maolchonaire's grandfather as a young man became embroiled in an argument with another student poet. Forced to flee, he sought refuge with the Mac an Bhaird family. The latter returned with Muirgheas Ó Maoilchonaire to Connacht and pronounced favourably on the young man's learning ('céim do uaisligh a fhoghluim'). Accordingly, the poet argues that both learned families are united by an enduring link. Mac an Bhaird asks the archbishop to honour this bond as if it were a charter of affection ('mar chairt ar ar gconnailbhe'). The poet exists but does not exist such is his dejection at the loss of prosperity.[81] Clearly, this composition illustrates once again the sense of dislocation and dispossession which characterises poetry composed in the aftermath of the flight of the earls and the Ulster plantation. Like Eochaidh Ó hEódhusa and Eoghan Ruadh Mac an Bhaird, however, Fearghal Óg is not a victim of his circumstances. He demonstrates agency in his choice of exile and recourse to family networks which he seeks to access through the medium of his poetry. Nonetheless,

Mac an Bhaird and Flaithrí Ó Maolchonaire present an instructive contrast and reveal how radically Gaelic intellectual life was being transformed. The old tradition is represented by the distinguished poet exiled and bereft of institutional support and status while new learning and career pathways are symbolised by the fortunes of the archbishop, descendant of a hereditary learned family, counter-reformation polemicist and theologian, advocate of print in Irish and confidant of Philip III of Spain.

In conclusion, the work of the poets discussed in this essay reveals a record of frank and cogent if despondent appraisal of the outcome of political, social, religious, economic, cultural and demographic change in Ulster during the first two decades of the seventeenth century. The poets are acutely cognisant of the implications for elite Gaelic culture of the collapse of aristocratic Gaelic society and leadership. A range of emotions is recurrent in their work: anger, bitterness, sadness, pride, resilience and occasionally optimism. At a personal level, each of these poets manifests a high degree of autonomy and responds to crisis in a manner which exploits available opportunities and resources.[82] They are far from uncomprehending of the turbulent transformation in which they participated. FearFlatha Ó Gnímh in a poem on the downfall of the house of O'Neill ('Gearr bhur ccuairt, a chlanna Néill': 'Your visit is short, descendants of Niall') highlighted a fundamental insight also articulated by these poets. The fall of the O'Neills or the O'Donnells was not a localised phenomenon of provincial significance: once they were overwhelmed, it followed that Gaelic Ireland itself was now conclusively conquered ('an díle air ndol os bhur ccionn, táinic fúibh air fud Éirionn').[83] The evidence of a poem composed to exhort martial valour in 1641 suggests that political and cultural turmoil resulted in a refined sense of collective identity and purpose. The poem beginning 'Dia libh, a uaisle Éireann' ('God be with you, noblemen of Ireland') represents a late manifestation of artistic vigour in the final phase of the cultivation of syllabic poetry. Its author, Uilliam Óg Mac an Bhaird, addressed both historic ethnic communities, Gaelic and Old English, and urged them to mount a unified national campaign of resistance against their common enemy.[84] The enslavement and martyrdom of the Irish was known throughout Europe and the ascendancy of the English in Ireland was well-established.[85] The time has come not only to stand united in defence of Ireland but the Irish must also shield their faith from pernicious attack.[86] A fully-fledged Irish Catholic patriotism now confronted English Protestant authority in Ireland.

It is a cliché of historiography that history is more often than not written by the victors and as such reflects their values and priorities. Like many clichés it contains more than a grain of truth. Moreover, history has traditionally been written from the perspective of archival and literary materials deemed canonical by the victors.[87] Surviving early modern Gaelic documentary sources in no way rival the diversity and extensive volume of contemporary material in English.[88] Indeed, if the English record of the plantation is effectively documentary and bureaucratic, the Gaelic equivalent is purely literary.[89] Praise poetry, therefore, is particularly important for the light it sheds on the cultural priorities, assumptions and mindset of the Gaelic intellectual and dynastic elites. In the case of the distant and silenced Gaelic world of early seventeenth-century Ulster, what survives of their work provides a precious glimpse of the thoughts, emotions and strategies of praise poets during a time of cataclysm and re-invention.

NOTES

1 John O'Donovan (ed.), *Annála ríoghachta Éireann. Annals of the Kingdom of Ireland by the Four Masters* (7 vols, Dublin, 1851), vi, 2359; Bernadette Cunningham, *O'Donnell histories: Donegal and the Annals of the Four Masters* (Rathmullan, 2007), p. 1.
2 Regarding the Gaelic literary tradition locally see Diarmaid Ó Doibhlin, 'Tyrone's Gaelic literary legacy', in Charles Dillon and Henry A. Jefferies (eds), *Tyrone: History & Society* (Dublin, 2000), pp. 403–32. A summary version of the present essay has been previously published as Marc Caball, 'Dispossession and reaction: the Gaelic literati and the plantation of Ulster', *History Ireland*, 17:6 (2009), 24–7. See also Sarah E. McKibben, 'Speaking the unspeakable: male humiliation and female national allegory after Kinsale', *Éire-Ireland*, 43:3–4 (2008), 11–30.
3 Paul Walsh (ed.), *Beatha Aodha Ruaidh Uí Dhomhnaill* (2 parts, London and Dublin, 1948, 1957), ii, pp. 118–25, p. 118, quatrain 6; Douglas Hyde, 'The Book of the O'Conor Don', *Ériu*, 8 (1916), 78–99.
4 Walsh (ed.), *Beatha*, ii, p. 120, quatrains 7–8.
5 William Gillies (ed.), 'A poem on the downfall of the Gaoidhil', *Éigse*, 13 (1969–70), 203–10. Gillies's edition of the poem is based on the earliest extant copy which is in the 'Book of O'Conor Don' and where it is anonymous. However, in other manuscript versions of the text it is attributed to Lochlainn Ó Dálaigh. See Gillies, 'A poem on the downfall', p. 203, note 1. For a previous discussion of this poem see Peter McQuillan, *Native and Natural: Aspects of the Concepts of 'Right' and 'Freedom' in Irish* (Cork, 2004), pp. 67–8.
6 Gillies, 'A poem on the downfall', p. 205, quatrains 4–8.
7 Ibid., pp. 206–7, quatrains 13–18.
8 For a discussion of the widespread use of the Israelite comparison by English Protestants in the sixteenth and seventeenth centuries as a spur for desired moral regeneration among their people see Patrick Collinson, *The Birthpangs of Protestant England: Religious and Cultural Change in the Sixteenth and Seventeenth Centuries* (London, 1988), pp. 4–7, 18.
9 Gillies, 'A poem on the downfall', pp. 208–9, quatrains 19–26.
10 Ibid., p. 203.
11 Thomas F. O'Rahilly (ed.), *Measgra dánta* (2 parts, Dublin and Cork, 1927), ii, no. 54, pp. 144–7, p. 144, lines 1–4 ('Mo thruaighe mar táid Gaoidhil! Annamh

intinn fhorbhaoilidh, ar an uair-se ag duine dhíobh, a n-uaisle uile ar n-imshníomh').

12 Lynn Hunt, 'Introduction: history, culture, and text', in L. Hunt (ed.), *The New Cultural History* (Berkeley and Los Angeles, 1989), p. 14.

13 For example, Ann Dooley has suggested that a specifically 'northern' early seventeenth-century interpretation of Gaelic Irish history has 'prevailed and exerted on our modern national sensibility a disproportionate pressure': 'Literature and society in early seventeenth-century Ireland: the evaluation of change', in Cyril J. Byrne, Margaret Harry and Pádraig Ó Siadhail (eds), *Celtic Languages and Celtic Peoples: Proceedings of the Second North American Congress of Celtic Studies* (Halifax, Nova Scotia, 1992), pp. 513–34, at p. 515.

14 Philip Robinson, *The Plantation of Ulster: British Settlement in an Irish Landscape 1600–1670* (Belfast, 1994), p. 1.

15 Ibid., p. 8.

16 Ibid., p. 1.

17 However, for continuity in settlement patterns in Tyrone see Philip Robinson, 'The Ulster plantation and its impact on the settlement pattern of Co. Tyrone', in Dillon and Jefferies (eds), *Tyrone*, pp. 233–66.

18 By way of background see, for instance, J. E. Caerwyn Williams, 'The court poet in medieval Ireland', *Proceedings of the British Academy*, 57 (1971), 85–135; Brian Ó Cuív, *The Irish Bardic Duanaire or 'Poem-Book'* (Dublin, 1973); James Carney, 'Society and the bardic poet', *Studies*, 62 (1973), 233–61; Brian Ó Cuív, *The Linguistic Training of the Mediaeval Irish Poet* (Dublin, 1973); James Carney, 'Literature in Irish, 1169–1534', in Art Cosgrove (ed.), *A New History of Ireland*, vol. ii: *Medieval Ireland 1169–1534* (Oxford, 1987), pp. 688–707; Damian McManus, 'The bardic poet as teacher, student and critic: a context for the Grammatical Tracts', in Cathal G. Ó Háinle and Donald E. Meek (eds), *Unity in Diversity: Studies in Irish and Scottish Gaelic Language, Literature and History* (Dublin, 2004), pp. 97–123; Marc Caball and Kaarina Hollo, 'The literature of later medieval Ireland, 1200–1600: from the Normans to the Tudors', in Margaret Kelleher and Philip O'Leary (eds), *The Cambridge History of Irish Literature* (2 vols, Cambridge, 2006), i, pp. 74–139. More generally in regard to the common European linguistic and cultural background to praise poetry see M. L. West, *Indo-European Poetry and Myth* (Oxford, 2007), pp. 26–74.

19 Standish Hayes O'Grady (ed.), *Catalogue of Irish Manuscripts in the British Library [formerly British Museum]* (2 vols, Dublin, 1992 reprint), i, 474–6; Pádraig A. Breatnach, 'The chief's poet', *Proceedings of the Royal Irish Academy*, 83:C (1983), pp. 37–79, at p. 56; Marc Caball, *Poets and Politics: Reaction and Continuity in Irish Poetry, 1558–1625* (Cork, 1998), pp. 2–3.

20 Caball, *Poets and Politics*, p. 3; Cf. Daniel MacCarthy (ed.), *The Life and Letters of Florence MacCarthy Reagh* (London, 1867), p. 362; David Beers Quinn, *The Elizabethans and the Irish* (Ithaca, 1966), pp. 44–5.

21 David Greene (ed.), *Duanaire Mhéig Uidhir: The Poembook of Cú Chonnacht Mág Uidhir, lord of Fermanagh 1566–1589* (Dublin, 1972), no. ix, pp. 78–89, p. 82, quatrain 14. More generally in terms of contemporary external perception of the influential communal role of praise poets see Barnaby Rich, *A New Description*

of Ireland (London, 1610), p. 3; Herbert F. Hore, 'Irish bardism in 1561', *Ulster Journal of Archaeology*, 6 (1858), 165–7, 202–12; Charles Hughes (ed.), *Shakespeare's Europe: Unpublished Chapters of Fynes Moryson's Itinerary* (London, 1903), pp. 196–9.
22 Robert Darnton, *The Great Cat Massacre and Other Episodes in French Cultural History* (Harmondsworth, 1985), pp. 24–5; Andrew Pettegree, *Reformation and the Culture of Persuasion* (Cambridge, 2005), p. 8.
23 '... acht ataim ag iarraidh air an leighoir gabhail agam fana danta sa nach fhuair me ughdar leo acht seanndhuine dall do cuir do mheabhair iad ata deich mbliaghna fithceat uadh'. James Carney (ed.), *Poems on the O'Reillys* (Dublin, 1950), p. xiii.
24 Cuthbert McGrath, 'Í Eódhosa', *Clogher Record*, 2:1 (1957), 1–19.
25 Marc Caball, 'Ó hEódhusa (O'Hussey), Eochaidh (*c*.1568–1612)', in James McGuire and James Quinn (eds), *Dictionary of Irish Biography* (9 vols, Cambridge, 2009), vii, 553–5; James Carney, *The Irish Bardic Poet* (Dublin, 1967); Pádraig A. Breatnach, 'Eochaidh Ó hEódhusa (*c*.1560–1612)', *Éigse*, 27 (1993), 127–9.
26 Pádraig A. Breatnach, 'A covenant between Eochaidh Ó hEódhusa and Aodh Mág Uidhir', *Éigse*, 27 (1993), 59–66.
27 O'Grady, *Catalogue of Irish Manuscripts*, pp. 476–8. James Carney translated the first line of this work as 'Ireland is in the mood for love': Carney, *The Irish Bardic Poet*, p. 19.
28 O'Grady, *Catalogue of Irish Manuscripts*, pp. 460–2.
29 Carney, *The Irish Bardic Poet*, p. 31.
30 P. A. Breatnach (ed.), 'Metamorphosis 1603: dán le hEochaidh Ó hEódhusa', *Éigse*, 17 (1977–78), 169–80.
31 Cáit Ní Dhomhnaill (ed.), *Duanaireacht* (Dublin, 1975), no. 12, pp. 99–101. Regarding the date of composition see O'Grady, *Catalogue of Irish Manuscripts*, pp. 471–4.
32 Paul Walsh, *The Will and Family of Hugh O Neill, Earl of Tyrone* (Dublin, 1930), p. 40.
33 Patrick J. Duffy, 'Patterns of landownership in Gaelic Monaghan in the late sixteenth century', *Clogher Record*, 10 (1981), 304–19, at 309; Donald M. Schlegel, 'Sir Brian and Lady Mary MacMahon', *Clogher Record*, 15 (1996), 133–44, at 136, 138.
34 George Hill, *An Historical Account of the Plantation in Ulster at the Commencement of the Seventeenth Century, 1608–1620* (Belfast, 1877; facsimile edn, Shannon, 1970), p. 49, note 63; *Calendar of the State Papers, Relating to Ireland, of the Reign of James I, 1615–1625* (London, 1880), p. 359; Robinson, *The Plantation of Ulster*, p. 54. Sir Henry Dillon wrote to the earl of Salisbury in April 1608 that 'Sir Bryen M'Mahowne, who has been an ancient rebel, he is grown to be every day heavy with surfeit; and albeit he be married to the Lady Mary, daughter of Tyrone, yet if his son, Fert Oge M'Mahowne, be still restrained, he will not stir unless there be a general revolt', *Calendar of the State Papers, Relating to Ireland, of the reign of James I, 1606–1608* (London, 1874), pp. 486–7.
35 Ní Dhomhnaill (ed.), *Duanaireacht*, p. 99, quatrain 1.
36 Ibid., p. 99, quatrains 2–4.
37 Ibid., p. 100, quatrain 6.

38 Ibid., p. 100, quatrains 8–11.
39 Ibid., p. 101, quatrain 15.
40 Ní Dhomhnaill only edited and published the first 15 quatrains of this poem of 41 verses. The poem is extant in Royal Irish Academy manuscript 23 F 16 ('The Book of O'Gara'), compiled by Fearghal Dubh Ó Gadhra, O.S.A., in the Low Countries, particularly in Lille and Brussels, during the years 1655 to 1659. Appreciation of the poem's significance in the seventeenth century is suggested by its presence in Royal Irish Academy manuscript 24 P 12 (40 quatrains), a collection of bardic verse and romantic tales compiled in the early decades of the century, and in a manuscript compendium (11 quatrains) written by the scribe Conchubhar Mhágaodh around 1681. Brian Ó Cuív (ed.), 'A seventeenth-century Irish manuscript', Éigse, 13 (1969–1970), 143–52; Pádraig de Brún, Lámhscríbhinní Gaeilge: treoirliosta (Dublin, 1988), p. 43, item 179.
41 Carney (ed.), Poems on the O'Reillys, no. xxv, pp. 121–7.
42 Robin Flower (ed.), Catalogue of Irish Manuscripts in the British Library (Dublin, 1992), ii, 166.
43 Carney (ed.), Poems on the O'Reillys, p. 228.
44 Ibid., p. 121, quatrain 2.
45 Ibid., pp. 121–2, quatrains 3–7.
46 Ibid., p. 123, quatrains 11–13.
47 Miri Rubin, Gentile Tales: The Narrative Assault on Late Medieval Jews (New Haven and London, 1999), p. 12; Liam P. Ó Caithnia, Apalóga na bhfilí 1200–1650 (Dublin, 1984), p. 126. Cf. James Stewart, 'Stories in poems', Arv: Journal of Scandinavian Folklore, 29–30 (1973–74), 31–5.
48 Carney (ed.), Poems on the O'Reillys, pp. 123–4, quatrains 14–18.
49 Ibid., pp. 125–6, quatrains 19–28.
50 Ibid., p. 127, quatrain 39.
51 Paul Walsh, Irish Chiefs and Leaders (Dublin, 1960), p. 64.
52 Micheline Kerney Walsh (ed.), "Destruction by peace": Hugh O Neill after Kinsale (Armagh, 1986), p. 124; Marc Caball, 'Mac an Bhaird, Eoghan Ruadh (1560s?–1625?)', in McGuire and Quinn (eds), Dictionary of Irish Biography, v, pp. 693–4.
53 Marc Caball, 'Politics and religion in the poetry of Fearghal Óg Mac an Bhaird and Eoghan Ruadh Mac an Bhaird', in Pádraig Ó Riain (ed.), Beatha Aodha Ruaidh: The Life of Red Hugh O'Donnell Historical and Literary Contexts (London, 2002), pp. 74–97, at p. 88. See also Eoghan Ruadh Mac an Bhaird's poem ('Truagh do chor a chroidhe tim') on hearing of Rory O'Donnell's illness, possibly composed in 1608. Osborn Bergin (ed.), Irish Bardic Poetry (Dublin, 1970), no. 4, pp. 35–7. Cf. Marie-Louise Coolahan, Women, Writing, and Language in Early Modern Ireland (Oxford, 2010), pp. 59–61.
54 Eleanor Knott (ed.), 'Mac an Bhaird's elegy on the Ulster lords', Celtica, 5 (1960), 161–71, at 163, quatrain 1. Cf. Tomás Ó Raghallaigh (ed.), Duanta Eoghain Ruaidh Mhic an Bhaird (Galway, 1930), pp. 137–51, no. 11.
55 For an illustration of Armagh cathedral and city in a ruined state in 1602 see J. H. Andrews, The Queen's Last Map-Maker: Richard Bartlett in Ireland, 1600–3 (Dublin, 2008), p. 106, plate 8 (a).

56 Knott (ed.), 'Mac an Bhaird's elegy', pp. 162-4, quatrains 7-9.
57 Ibid., pp. 166-9, quatrains 17-30.
58 Ibid., pp. 168-71, quatrains 31-5.
59 Miri Rubin has argued that 'Mary became the mediator between human and the divine, between the here and the hereafter, and to her were attached yearnings for assistance and consolation on earth and in heaven': *Mother of God: A History of the Virgin Mary* (New Haven and London, 2009), p. xxvi.
60 Knott (ed.), 'Mac an Bhaird's elegy', pp. 170-1, quatrains 36-8.
61 'Do shaoileamar, do shaoil sibh, dál cabhra ag clannuibh Mílidh, tréasan dtriar tarla san uaigh, ag triall ó Bhanbha bheannfhuair', Ibid., p. 170, quatrain 39.
62 Ó Raghallaigh (ed.), *Duanta Eoghain Ruaidh Mhic an Bhaird*, no. 13, pp. 170-205.
63 Caball, *Poets and Politics*, pp. 102-3.
64 'Atá siad ó sin anall, meic Mhíleadh d'aimhdheoin eachtrann, gan tornamh, gan teacht ré a dteann, a neart ar fhonnmhagh nÉireann'. Ó Raghallaigh (ed.), *Duanta Eoghain Ruaidh Mhic an Bhaird*, p. 172, quatrain 7.
65 Ibid., pp. 173-6, quatrains 8-16.
66 Ibid., p. 176, quatrain 17 ('a gcomhroinn do chóigeadh Sreing, 's don chóigeadh orloinn d'Éirinn'). For *cóigeadh Sreing* (Connacht) see Edmund Hogan, *Onomasticon Goedelicum* (Dublin, 1910), p. 315.
67 Ó Raghallaigh (ed.), *Duanta Eoghain Ruaidh Mhic an Bhaird*, p. 178, quatrain 18.
68 Ibid., pp. 178-80, quatrains 19-27.
69 Ibid., pp. 186-8, quatrains 38-47. Ó Caithnia, *Apalóga*, p. 130.
70 Brian Mac Cuarta, *Catholic Revival in the North of Ireland, 1603-41* (Dublin, 2007), pp. 17-36.
71 Ó Raghallaigh (ed.), *Duanta Eoghain Ruaidh Mhic an Bhaird*, pp. 200-2, quatrains 81-6.
72 'Maith do suidhigheadh síol Néill', in Tadhg Ó Donnchadha (ed.), *Leabhar Cloinne Aodha Buidhe* (Dublin, 1931), pp. 101-10.
73 Lambert McKenna (ed.), 'Some Irish bardic poems: C', *Studies*, 41 (1952), 99-104.
74 'Turnamh dóchais díoth muirne' in Láimhbheartach Mac Cionnaith (ed.), *Dioghluim dána* (Dublin, 1938), pp. 419-22.
75 Marc Caball, 'Mac an Bhaird, Fearghal Óg (late 1540s-c.1618)', in McGuire and Quinn (eds), *Dictionary of Irish Biography*, v, pp. 694-6.
76 'Éisd rem égnach, a fhir ghráidh' and 'Fuarus iongnadh, a fhir chumainn' in Cuthbert Mhág Craith (ed.), *Dán na mbráthar mionúr* (2 vols, Dublin, 1967, 1980), i, pp. 117-24. Cf. Pádraig A. Breatnach, 'The aesthetics of Irish bardic composition: an analysis of *Fuaras iongnadh, a fhir chumainn* by Fearghal Óg Mac an Bhaird', *Cambrian Medieval Celtic Studies*, 42 (2001), 51-72; Pádraig Ó Macháin, 'The iconography of exile: Fearghal Óg Mac an Bhaird in Louvain', in Pádraig A. Breatnach, Caoimhín Breatnach and Meidhbhín Ní Úrdail (eds), *Léann Lámhscríbhinní Lobháin/The Louvain Manuscript Heritage* (Dublin, 2007), pp. 76-111.
77 Brendan Jennings, 'Florence Conry archbishop of Tuam: his death, and the transfer of his remains', *Journal of the Galway Archaeological and Historical Society*, 23:3-4 (1949), 83-92, p. 84; Ó Macháin, 'The iconography of exile', p. 83; Pádraig

Ó Macháin, 'The flight of the poets: Eóghan Ruadh and Fearghal Óg Mac an Bhaird in exile', *Seanchas Ard Mhacha*, 21:2/22:1 (2007–8), 39–58; Benjamin Hazard, *Faith and Patronage: The Political Career of Flaithrí Ó Maolchonaire, c.1560–1629* (Dublin, 2009), pp. 127, 130, 170.
78 Mhág Craith (ed.), *Dán na mBráthar*, i, pp. 117–18, quatrains 1–6.
79 Ibid., ii, p. 54, quatrain 8.
80 Ibid., i, p. 118, quatrains 6–9.
81 Ibid., i, pp. 118–20, quatrains 11–25.
82 A similar case for Gaelic intellectual versatility at this juncture is made in Brendan Kane, 'Making the Irish European: Gaelic honor politics and its continental contexts', *Renaissance Quarterly*, 61 (2009), 1139–66; and in Marc Caball, 'Articulating Irish identity in early seventeenth-century Europe: the case of Giolla Brighde Ó hEódhusa (c.1570–1614)', *Archivium Hibernicum*, 62 (2009), 271–93.
83 Brian Ó Cuív (ed.), 'A poem on the Í Néill', *Celtica*, 2 (1954), 245–51, at 246, quatrain 7.
84 Eoin Mac Cárthaigh (ed.), 'Dia libh, a uaisle Éireann (1641)', *Ériu*, 52 (2002), 89–121.
85 Ibid., pp. 96–7, quatrains 8–10.
86 Ibid., pp. 95–6, quatrains 3–4.
87 Eamon Duffy, *Fires of Faith: Catholic England under Mary Tudor* (New Haven and London, 2009), p. 61.
88 In the absence of a Gaelic administrative or cartographic corpus, William J. Smyth has argued persuasively that 'stories of Irish places and peoples were deeply embedded in the world view of the Irish before the New English conquest and colonisation. The ideology and, indeed, central values of the culture were wrapped up in the landscape – its occupation, its use, its names, stories and legends': *Map-Making, Landscapes and Memory: A Geography of Colonial and Early Modern Ireland c.1530–1750* (Cork, 2006), p. 3.
89 Mark Netzloff, 'Forgetting the Ulster plantation: John Speed's *The Theatre of the Empire of Great Britain* (1611) and the colonial archive', *Journal of Medieval and Early Modern Studies*, 31:2 (2001), 313–48, at 315. By way of response to Netzloff see J. H. Andrews, 'Statements and silences in John Speed's map of Ulster', *Journal of the Royal Society of Antiquaries of Ireland*, 138 (2008), 71–9.

The plantation of Ulster: aspects of Gaelic letters

DIARMAID Ó DOIBHLIN

The purpose of this essay is to consider certain aspects of Gaelic letters in the wake of the plantation of Ulster, with particular reference to the area of the six escheated counties (Armagh, Cavan, Donegal, Fermanagh, Londonderry and Tyrone).[1] There are a few basic concepts that should be kept in mind when dealing with the world of pre-plantation Ulster and its indigenous Gaelic-speaking community, the heirs to and keepers of one of the oldest and richest literary traditions in western Europe.[2] It is important to understand that this tradition had been transmitted almost exclusively by way of manuscript, with the notable exceptions of the confessional and grammatical works published by Irish clerics at St Anthony's College (Louvain), Propaganda Fide and St Isidore's College, Rome, the power-houses of the Irish Counter-Reformation.[3] These manuscripts are without question the most important primary source-materials relating to the native Irish.[4] About 4,500 have survived the ravages of time and neglect of which an estimated 600 have their origins in Ulster.[5] According to the early nineteenth-century German traveller Johann Georg Kohl, who visited Ireland on the eve of the famine, manuscripts remained popular and highly-prized in the Gaelic-speaking countryside.[6] They passed from parish to parish, from scribe to scribe and from patron to patron, and were read aloud in the home and workplace.[7] A vast amount of research still needs to be done on these unique sources. Reliable catalogues of the principal collections do exist[8] but little is known about scribal training, their inter-relationships and the role of patrons who ultimately financed their commissions and transcriptions.[9] This is a major lacuna in our understanding of early modern identity formation, given that the language and its manuscripts form the cement that brought cohesion to this discrete confessional and cultural community.

It is also crucial to understand that manuscripts from Munster, Leinster and Connacht circulated freely throughout Ulster's Gaelic-speaking community and one should be careful of over-regionalisation. This does, of course, have its place but the Gaelic world formed a discrete cultural entity and should be regarded and understood as such. Similarly, a tendency among nineteenth- and twentieth-century scholars and writers to romanticise and sentimentalise early modern Irish writing has also muddied the waters. In Paul Walsh's 'The work of a winter 1629/30', which describes Mícheál Ó Cléirigh, the leading luminary of the 'Four Masters', moving from place to place, friary to friary and big house to big house to copy and assemble the basic source materials, one encounters the 'minstrel boy' depicted in later Romantic literature.[10] In fact, Ó Cléirigh should instead be viewed as a Renaissance man going back always 'ad fontes' (to the sources), and listing these along with the time and place where he found them.[11] He did not seek to build a shrine to the past, but rather set down the basic source-materials, which he believed would be at the heart of the new emerging Ireland. A common homeland, a shared Gaelic culture and the realisation that Catholicism comprised an essential component of identity provided the central tenets of a new emerging ideology. It is significant that Ó Cléirigh's fellow friars at St Anthony's College in Louvain chose Coláiste na nÉireannach (the College of the Irish) and not Coláiste na nGael (the College of the Gaels) as a name for their new foundation. It would be an inclusive, not exclusive, foundation.[12] This new development did not escape the attention of Sir William Trumbull, the English ambassador to Brussels (1609–25), who censored the 'perfidious Machiavellian friars of Louvain' who sought:

> by all means to reconcile their countrymen in their affections and to combine those that are descended of the English race and those that are mere Irish into a league of friendship and concurrence against your majesty and the true religion now professed in your kingdom.[13]

The later tendency to 'Ulsterise' the literature of the period is particularly incongruous, and has more to do with the unfortunate and often absurd dialect battles at the beginning of the Gaelic Revival in the late nineteenth and early twentieth century. It seems scarcely credible that Ulster scribes and men of letters, moving around Dublin and throughout the Gaelic world, had not encountered Aogán Ó Rathaille's poem on the effects of the Munster plantations and the subsequent confiscations on the Cromwellian and Williamite periods:

> Do thonnchrith m'intinn, d'imigh mo phríomhdhóchas,
> poll im ionathar, biora nimhe trím dhrólainn,
> ár bhfonn, ár bhfothain, ár monga is ár mínchóngair
> i ngeall le pingin ag foirinn ó chrích Dhóbhair.

Stadfadsa feasta: is gar dom éag gan mhoill
ó treascradh dragain Leamhan, Léin is Laoi:
rachad 'na bhfasc le searc na laoch don chill-
Na flatha fá raibh mo shean roimh éag do Chríost[14]

[My brain shakes as a wave, my chief hope gone,
There's a hole in my gut and foul spikes through my bowels.
Our land, our shelter, our woods, and our level ways,
traded for a penny by a faction from the land of Dover.

I will stop now – my own death closes in
Now that the dragons of the Leamhan, Loch Léin and the Laoi have fallen
In the grave I'll bind myself to those princes
My people served before the death of Christ.[15]

Few poets bring to their craft the passion that one encounters in Ó Rathaile's oeuvre. Gaels all over Ireland in the early modern period faced similarly bleak prospects as a consequence of the English re-conquest of Ireland. Sir John Davies, James VI and I's solicitor-general, determinedly declared that Ireland should be brought to 'civility' with the sword in one hand and the Bible in the other. Queen Elizabeth I, William Bedell and Sir Robert Boyle provided the reading material, Sir Richard Bingham, Sir Humphrey Gilbert, Sir Walter Raleigh, Sir Charles Blount and Sir Arthur Chichester opted for the sword. On his furlough on the shores of Lough Neagh in County Tyrone, the latter chanced upon Patrick O'Quinn and his wife and children whom he speedily put to the sword. He dealt with many of the participants and neutral bystanders of Sir Cathair O Doherty's 1608 rebellion in an equally ruthless fashion.[16]

Nevertheless the policy of the new Jacobean rulers initially tended towards tolerance with regard to the learned professions, seeking as far as possible to win them over. In Scotland the learned classes had embraced Protestantism and there was reason to suspect that the same might happen in Ireland.[17] Eochaidh Ó hEodhasa, poet to the Maguires of Fermanagh and one of the finest poets of his day, received a grant of his lands at the Ulster plantation and Tadhg Óg Ó Cianáin, historian to the Ó Néills, retained his considerable holding in Armagh after the Nine Years War.[18] The Uí Mhealláin, erenagh family in the Parish of Lissan, County Tyrone, also held onto their hereditary townlands at the time of the plantation. Even Mícheál Ó Cléirigh, enjoying the support of the proscribed Franciscan order and the patronage of the Sligo nobleman Fearghal Ó Gadhra, travelled throughout Ireland, despite banishment laws imposed against the Catholic clergy.[19]

The new colonisers appreciated the Gaelic literati's usefulness as a unique repository of historical and legal knowledge. Bishop George Montgomery, newly appointed to the dioceses of Derry, Raphoe and Clogher sought to establish the extent and precise boundaries of the church lands in his Derry diocese. To this end, he assembled a body of men with the local knowledge he required at Limavady in 1609. Seamus Ó Ceallaigh has shown that the participants practically all came from erenagh families, which played a key role in Irish society under the Gaelic system; inhabiting church lands, taking care of buildings and property, as well as overseeing clerical provision.[20] Normally literate and expert in law, local customs and traditions, they often adjudicated in the minor disputes and dissensions that erupted from time to time.[21] They survived as a class in many Catholic parishes right down to the eighteenth century.

Similarly, Sir John Davies describes the jurors who met at Limavady as 'gentlemen of good education and family belonging to the principal septs then inhabiting the country', a stark contrast to 'the wild men of the woods' and the incendiaries encountered in the writings of Edmund Spenser, Barnaby Rich and Fynes Moryson.[22] He reported that 'We had a jury of of [sic] clerkes and scholers for the Jurors', and of '15 in number...13 of them spoke good Latin and that very readily. They conceaved their verdict or praesentment in a similar good forme and methode'.[23] Bishop Montgomery, in turn, lavished particular praise on one Eoghan Mac Cathmhaoil, whom he deemed 'keen witted, artful and crooked'. The 'Black Bishop' planned to win him over, as he 'preferred a fawning dog to a barking one'.[24] The report indicates a reasonably high standard of learning among the Limavady jurors, which, coupled with a similar 'inquisition' held at Dungannon about the same time, diffuses English notions of Irish barbarity, lack of civility and the supposed intransigence of the learned classes.[25] Ironically, in view of their key role in Gaelic Irish society, these erenagh families now functioned as the eyes and ears of the colonisers. Even a cursory perusal of the State Papers, The Fiants of James I, the Rev. George Hill's *Plantation of Ulster* and T. F. O'Rahilly's *Irish poets, historians and judges in English Documents, 1538–1615* provides ample evidence to suggest that many of them had escaped the earliest ravages of the conquest and plantation.[26]

There is little direct or specifically focused contemporary reference to the plantation in Ulster's Gaelic letters, a possible consequence of the immediate and subsequent destruction of manuscripts. It may also be the case that the native Irish set little store by it. Over two decades after its inception Mícheál Ó Cléirigh penned a rather sparse, discursive account in the 'Annals of the Kingdom of Ireland':

> Bá de eiccin, & do imtheacht na n-iarladh att-rubhramar, tainicc a n-domhnus & an-dúthaigh, a b-forba, & a b-fearann, a n-dúine, & a-diongnadha, a c-cuanta caomhthurcarthacha, & a n-inbeara iaiscc iomdha do bhein do Gaoidhelaibh Chóiccidh Uladh, & a t-tabhairt ina f-fiadhnaisi do eachtair chenélaibh & a c-cor-somh for athchur, & for ionnarbadh in aile críochaibh comhaigtibh go ro éccsat a n-ermhór.[27]

[It was indeed from it, and from the departure of the Earls we have mentioned, it came to pass that their principalities, their territories, their estates, their lands, their forts, their fortresses, their fruitful harbours, and their fishful bays, were taken from the Irish of the province of Ulster and given in their presence to foreign tribes; and they were expelled and banished into other countries, where most of them died.][28]

He provides scant detail on the provincial or regional impact of the plantation: not a mention of the barony of Raphoe, the precints of Portlough, Boylagh and Bannagh, Doe, Fanad, Kilmacrennan or Inishowen. Furthermore, he makes little reference to the new landlords or settlers.[29] Other contemporary writers, however, were more specific. A fine poem by Lochlainn Ó Dálaigh, a native of Bréifne (modern County Cavan), captures the upheaval experienced by the Gaelic Irish. In his 'Cá háit ar ghabh na Gaeil?' ('Where have the Gaels gone?'), he notes that they can no longer be seen and that the whole landscape has been transformed:

Atá againn ina n-ionad
Díorma uaibhreach eisíon
D'fhuil Ghall, de ghasra Mhonaidh,
Saxain ann is Albanaigh.

Roinnid í eatorra féin,
An chríoch seo Chlanna saor Néill
Gan puinn de Mháigh lachtmhar Fhlainn.
Nach bhfuil ina n-acraí againn.

Tharla dúinn –trom an dua
Port oireachtais dá bhfású.
An torchaire ag searg ina sreabh
Dorch-fhoithre sealg ina sráidibh.

Comhthionól tuata i dteach naomh,
Seirbhís Dé faoi dhíon fionnchraobh,
Cuilt cléire ina gcuilce tána
Sliabh ina ghortaibh gabhála.

Aonaigh ina n-áitibh sealga
Sealga ar shlite suaitheanta,
Creasa thar faiche ina bhfál air.
Gan dáil ghraifne fána ghruannna[30]

[We have in their place, a proud and impure swarm of strangers of English and Scottish extraction. Saxons are there and Scotsmen.
The land of noble Niall's posterity they divide out among themselves, and there is not a jot of Flann's milk yielding plain that is not divided up into acres,
We have lived to see the tribal convention places emptied, the wealth perished away in a stream, dark thickets of the chase turned into streets.

A boorish congregation in the house of God, while God's service is
performed under the shelter of simple boughs, the clerical robes are litter for
cattle, the mountain is all fenced in fields, Fairs are held in places of the
chase, hunting there is upon the plain highways; the green is crossed by
girdles (i.e enclosures) of twisting fences, and they (the strangers) practise
not to gather together their horses for the races.][31]

The poem is ultimately about alienation, the uprooting and estrangement of
a people in their own land, and the ostracism of their customs and pieties.
Ó Dálaigh's stark message calls to mind a passage from a timeless, early
eighteenth-century Irish sermon from County Down in which Dr Séamus
Mac Póilín, one-time Vicar-General of the Diocese of Dromore, reflects on
the surrounding landscape and the state of religion.[32] He subsequently reproduced this particular passage in the introduction to a small Catechism published in 1742:[33]

Féucuigidh anois air fhíneamhuin na hÉirion annair ar cuireadh an síol maith
ó chianaibh agus a thug na tórtha céadach uaidhche, aniodh'na fásach
d'easbhuidh gníomha agus glanshaothar a dhéanamh innte! Ant-úrghairdín se
darbh ainm anallad Oilean na Naomh agus a bhí líonmhar an árdscoltaibh agus
a ndaoine diaga fóghlamtha eagnuidh, atá sí aniodh lán ainbhfis, dorchadais
agus eirceacht. An tír a bhí roimhe seo líonmhar an easbuigaibh, an abuibh, a
mbráthairibh, agus a mnáibh riaghalta, a mainistreachaibh agus a gcealluibh
coisreaga, ionta a raibh creidiomh Chríosda go laetheamhail 'ga mhíniúghadh
agus 'ga theagasc don phobal ainbhfiosach; féach mar atá an tírse aniodh, gan
ab, gan chill, gan chléir, gan altóir, gan adhradh, gan aifreann, gan iodhbuirt,
gan le fághail air lorg na naomh, ach áiteacha uaigneacha, agus ionadadh falmha,
balladh dorcha dubha airna gcaitheamh le haois agus le haimsir ag titim agus
ag tuirneamh go talam: Dún, Sabhal agus Fachairt áras na naomh, longphoirt
Phadraic, Bhrithide agus Cholaim Cille, gan díon, gan chrann, gan chraoibh,
muna bhfuil dreas no eidhnean ag déanamh fasgaidh do bheathachuibh állta
nó d'éunlaidh uaigneacha an aeir. Broic agus sioncha ag tamhan agus ag scréachaidh, feithideadh agus préachain ag gragarnaigh an áit na n-organ, na psailm
agus na n-iomnadh agus na gcainticeadh ceolmhara a bhí 'nallad 'ga sinm sna
hárasaibh naomha sa.[34]

[Let us look now at the vineyard of Ireland where the good seed had been sown
long ago and which had yielded fruit a hundredfold, today it is a desert through
lack of action and pure effort. This fresh garden which was formerly known as
the Island of Saints and where high quality schools and learned wise people
were plentiful, today it is full of ignorance and, darkness and heresy. This
tender garden, once called the Island of Saints, where previously there were in
plenty excellent schools, and a learned wise religious people is today full of
ignorance, darkness, and heresy. This land which previously had in abundance
bishops and abbots, brothers and nuns, monasteries and blessed churches, and
where the faith of Christ was being daily explained and taught; look at this

country today as it is, without an abbot or church, without clergy or altar, without worship or mass, without sacrifice, without any trace of the saints, but only lonely and empty places, dark black walls worn by age and time, falling to the ground: Down, Saul and Faughart, the dwelling place of the saints, the stronghold of Patrick, Briget and Colam Cille, without roof or branch or tree, apart from brambles or ivy which provides cover for wild beasts or the lonely birds of the sky. Badger and fox barking and howling, wild creatures and crows raucous there in the place of the organs and psalms, the hymns and melodious canticles which formerly were played in those holy places.]

A number of observations could be made about the confessional, cultural and historical content of this particular passage but it is crucial to understand that its language register is almost entirely modern. Indeed, one could deliver this sermon in any Gaelic-speaking community today and be perfectly understood. This shows that the Gaelic literati at the time of the plantation had begun to adapt their language and their letters to the evolving world. The European Renaissance, Protestant Reformation and Catholic Counter-Reformation continually stressed the importance of 'national' languages and Martin Luther's exploitation – if exploitation is the proper word – of the national language in his theological and philosophical writings reinforced the need to touch the minds and hearts of people through a medium which they readily understood and used in their everyday lives.[35] Language too played a crucial part in the ongoing search for an Irish national identity, a point succinctly driven home by Mícheál Ó Cléirigh in a passage which should have a wider prominence:

> De bhrí go n-áilíosaíonn nó go santaíonn gach tír a bheith freacraithe go maith ina dteangacha máthartha agus dúchais, agus a bheith eolach foirfe go beacht agus caingne nó cúise a dtíre a thuiscint go maith iomlán ionas go dtuilchéimnidís nó go dtéidís ar a n-aghaidh gan tuisliú gan teibeadh chomh maith i neithe diadha agus daonna in éinfheacht; gonadh aire sin do shantaíos féin, de réir m'acfuinne cuidiú le mo choimhthíoracha le cruinniú agus le cur síos ár dteanga nádúrtha, an Ghaeilge, beagnach, nó go mórmhór ionnarbtha óna tír, tréigthe óna dúchasacha, agus dísbeagaithe óna haois leanúna, i bhfoirm fhoclóra ó choinne cháich, dochum go n-aithnidís a saine, nó a n-idirdhealú féin ó chiníocha eile rann an domhainse.[36]

> [Because every country seeks and desires to be contemporary in their mother tongues, and to be knowledgeable and accurate and to understand fully and well the cause of their country so that they would progress without trip up or failure in divine and human matters: and it is for that reason I myself desired according to my abilities to assist my fellow countrymen in assembling and describing our native language, Irish, in the form of a dictionary, a language which is now almost totally driven from our land, abandoned by the natives, and reviled by its followers; so that they might recognise their diversity, and their distinctiveness from all the other races of the world].

That 'diversity' and 'distinctiveness' remained a central tenet of Irishness in the succeeding centuries.

Ó Cléirigh and the Irish Franciscans at Louvain knew that they would have to simplify the language to touch the hearts and intellects of the native Irish. To that end, they penned historical, hagiographical and religious works in a manner comprehensible to those in a position to read their texts. It is reasonable to assume, therefore, that the question of language formed a key item on the agenda of an important meeting in Paris when Thomas Messingham, the head of the city's Franciscans, sat down with Hugh Ward and Patrick Fleming to discuss the collecting and collating of ancient Irish source-materials.[37] Ó Cléirigh, who would play a pivotal role in the process, provided the *Sanasán* or Dictionary while Giolla Brighde/Bonabhentúra Ó hEodhasa's *Rudimenta Grammaticae*, which set down the grammatical basics of the Irish Language, made the most important contribution.[38] A native of the Clogher diocese, descended from the bardic family of Ó hEodhasa, which traditionally provided poets to the Maguires of Fermanagh, Bonabhentúra received his training in an official bardic school and was regarded by his contemporaries and colleagues as an authority on language.[39] Aodh Mac Cathmhaoil (Aodh Mac Aingil/Hugh Cavellus), Franciscan scholar, Roman Catholic Archbishop of Armagh and Primate of All-Ireland acknowledged both his piety and mastery of Irish:

> Cé go bhfuilid morán daoine foghlamtha ag ár naision ní raibh le fada d'aimsir duine coimhdheas agus coimhoirdhirc le Bonabhentura a léighionn, a nGaoidhilg, a ccrábhadh.[40]

> [Although our nation can claim many learned people, it is a long time since there was any person as accomplished and as illustrious as Bonabhentura, in learning, in Irish, in piety.]

Furthermore, Mac Aingil suggested that Bonabhentura had planned to write much more

> do rachadh a leas anma agus a n-onóir shaoghalta don náision...ach faraor do goireadh air a n-am a thoraidh agus a dtosach a shaothar do chur a gcló.[41]

> [...which would bring spiritual wellbeing and worldy distinction to the nation, but alas he was called in the time of his fruitfulness and as he was beginning to publish his work.]

Ó hEodhasa had been a student at Douai in September 1605, and successfully completed an MA in the Humanities. He subsequently joined the Franciscan Order in 1607 and took holy orders in 1609. He died quite suddenly in November 1614 at St Anthony's College, Louvain where he had risen to the rank of guardian. Three years before his death, he published a small Catechism at Antwerp, which laid out in more simple Irish the basic tenets of the Catholic faith as outlined at the Council of Trent.⁴² Copied again and again into manuscripts in the north and south of Ireland, one could argue that it formed the catechetical basis of Irish Catholicism down to the late nineteenth century. Ó hEodhasa simplified the discourse, and drawing on direct metaphors and images, he set out the teaching of the Catholic Church at the time. The imaginative fluvial images, which he adopted when confronting the idea of the Blessed Trinity, provide an effective example of his methodology:

> Cuirid daoine foghlamtha an tAthair i gcosúlacht le tobar a mbiadh sruth ag sileadh dochum locha, an Mac i gcosúlacht leis an sruth féin, agus an Spiorad Naomh leis an loch. Oir de reir mar is e an t-éanuisce amháin bhíos an tobar shíleas chum an tsrutha agus on tsruth chun an locha, is mar sin is aoindiacht amháin, énnádúr agus énsubstaint atá sanAthair sa Mhac agus sa Spiorad Naomh.⁴³

> [Learned people compare the Father to a well, from which a stream flowed into a lough, and the Son to the stream itself, and the Holy Spirit to the lough. And just as it is the same water which is the well which flows to the stream and the stream to the lough, so there is one divinity, one nature, and one substance which is in the Father, and the Son and the Holy Spirit.]

Ó hEodhasa undoubtedly plundered contemporary sources, but the metaphor of lough and stream would have resonated with audiences in his native Fermanagh. By simplifying the language and abandoning manuscript for the printed, published text, Ó hEodhasa and his fellow friars had pioneered a whole new and significant approach to Gaelic letters. Later writers of devotional texts would follow this lead.⁴⁴

Aodh Mac Aingil, Ó hEodhasa's contemporary, also had a distinguished career as a theologian and philosopher.⁴⁵ Indeed his work on Duns Scotus, published in the seventeenth century, still featured on the syllabi of European Departments of Philosophy up to the end of the nineteenth century. Active in south Derry in the early seventeenth century, Pádraig Modartha McGlone's contemporary deposition located Mac Aingil in the Barony of Loughinsholin in the company of Henry Mellon, guardian of the Franciscan Friary in Armagh. Both clerics urged the people to stay with the old faith.⁴⁶ His Irish-language defence of the Sacrament of Penance, *Scathán Shacramuinte na h-Aithrithe*/Mirror of the Sacrament of Confession (1618), adopts Ó hEodhasa's approach in advocating and promoting a simpler more direct form of language in the prose medium.⁴⁷ Ó hEodhasa and Mac Aingil also wrote poems of considerable grace and merit, which are worthy of further consideration.⁴⁸

Toirealach Ó Mealláin, another Franciscan and possibly a close relation of the aforementioned Henry, was also a Franciscan friar and the author of the first real military account of the push by the old Irish against the recent settlements in Ulster. His text survives in two manuscripts in the Boole

Library, Cork and the Huntington Library, San Marino, CA, and the former may well be the original copy.[49] The author identified himself among the Clann Laochthréan (O'Loughrans), a distinguished erenagh family in the Parish of Domhnach Mór in County Tyrone.[50] The Uí Mheallâin or O'Mellons served as erenaghs of the nearby parish of Lissan and their territory, clearly marked on seventeenth-century maps as Meallanaght, runs into present day Cookstown.[51] This unique account of the 1641 Rebellion is clearly worthy of further analysis, in conjunction with other under-utilised sources such as 'The Aphorismical Discovery of the Treasonable Faction', 'The Commentarius Rinnucianus', 'The Historia', 'Philopater' and the five seventeenth-century political poems edited by Cecile O Rahilly, not least because they all give a distinctly Irish take on the war and conquest of the 1640s and 1650s.[52]

A local document, Ó Mealláin's journal provides a fascinating and original perspective from Gaelic Ulster. It has been described as a diary, but in many places it seems to be little more than jottings taken down perhaps in the field, with very few continuous prose passages. The account details how the 1641 Rebellion initially comprised a largely bloodless seizure of new towns and planter garrisons in south Derry and east Tyrone by the Gaelic Irish. Ó Mealláin's text suggests his close proximity at times to the action.[53] This unusual type of personal witness is also found in Niall Mac Muireadhaigh's contemporary 'Caithréim Chlainne Raghnaill'.[54] Sir Feilim Ó Néill, who initially led the old Irish forces to some early military successes, believed that the plantation had been undone and the old world re-established which prompted him to lift his family's traditional tribute of a fatted cow, a measure of butter and four shillings from every balliboe from Glasdrummond to Tullyhogue.[55] As the rebellion began to disintegrate, however, the day of the 'meadur ime' and the 'mart ramhar' was clearly over.

General Eoghan Rua Ó Néill, the hero of Ó Mealláin's narrative, had enjoyed a distinguished military career in Spanish Flanders, before returning to lead the Ulster Irish forces in 1642. His most successful thrust came at Benburb, County Armagh, in 1646, where he effectively destroyed a Scottish covenanting army, which had been sent to Ulster under the command of General Robert Monroe. Ó Néill addressed his troops before the battle and urged them to fight bravely. Ó Mealláin purportedly records his actual words:

Ansúd chugaibh eascairde Dé agus naimhde bhur n-anma: agus déanaidh calmacht na n-aghaidh aniu. Oir is iad do bhen dhibh ar dtighearnaidh, bhur gclann agus bhur mbeatha spiradálta agus temporálta agus bhen bhur ndúthaigh dibh is do chuir ar deoraigheacht sibh.[56]

[There before you are God's enemies and the enemies of your souls; and fight bravely against them today. Because it is they who took your lordships from you and took your families and your spiritual and temporal life. And they took your lands from you and sent you into exile.]

Whether Ó Mealláin heard those words himself or had them reported to him, they reflect a native angst at the Ulster plantation, which would, in turn, be massively intensified by the subsequent Cromwell conquest.

Little else is known of Ó Mealláin's other writings or subsequent career. A graveyard not far from Benburb, County Tyrone, contains a burial place called the Friar's Grave and locals consider it to be the final resting-place of the author. The friar's own account, however, suggests that his community had been forced to abandon Armagh Friary and flee to the Fews in the south of the county.[57] Furthermore, the 1665 Hearth Money Rolls for Cornevannaghan, in the townland of Carricklehane and Barony of the Lower Fews, County Armagh, list Henry O Mellon and Turlagh O Mellon.[58] It is possible that the two friars had simply slipped underground and moved into the Fews district in the South Armagh/North Louth area, which became a haven for the dispossessed Gaels and their literary traditions.[59]

A poem entitled 'An Dibirt go Connachta' ('The expulsion to Connacht') ascribed to one Fear Dorcha Ó Mealláin, is one of the great Irish 'caointe' (laments) associated with the Cromwellian period. Edited by Rose Maud Young (Rois Uí Ógáin) of Galgorm House, Ballymena, one of the most gifted scholars of her generation, she made reference in the introduction to Cromwell's so-called 'Hell or Connaght' Proclamation:

> In August 1652, Parliament passed an Act to dispose of the Irish. All Catholics (and many Protestant royalists above the rank of tradesman or labourer) were to remove themselves and their families to Connacht and Clare where they were given small allotments. Any of those ordered found east of the Shannon after May 1st 1654 might be killed by whoever met them. The move had to be made in Winter. The season was very severe and the roads almost impassable. Hundreds perished on the way.[60]

Subsequent editors associated the author, Fear Dorcha Ó Mealláin, with modern County Down,[61] the likely consequence of the poet's reference to Colmán Mac Aoidh, an early saint associated with Newry. The present author, however, has suggested elsewhere that the saint alluded to is Colman 'Mucaí' or Colmán the swineherd who founded the old monastery at Arboe in Munterevlin,[62] and that the poet's family served as erenaghs in the adjacent parish of Lissan. In 1608, one 'Feardorcha O Mellon gentleman' received a pardon and the grant of a townland. Our poet almost certainly descended from him.[63]

The poem begins with a traditional prayer-*an iterarium*, a blessing at the start of a journey, and thereafter the poet seeks to guide and direct his people in the face of the new threat to their lives and their values. He urges them to be patient in these difficult times, to remain steadfast in their faith and put their trust in God, a God also worshipped by their persecutors:

Ionann Dia dúinn is dóibh
Aon Dia fós a bhí is atá
Ionann Dia a bhfus is thiar
Aon Dia riamh is a bheidh go brách.[64]

[Our God and their God is identical
One God there was and still is
Here and in the West God is the same
One God ever and shall be]

Their stark situation compares to the sufferings of the enslaved children of Israel in Egypt, whom God would ultimately lead through the Red Sea to freedom. He also urges the Irish, however, not to concern themselves with worldly possessions, which will matter little when they are in the grave. This strong moral tone runs throughout the poem and the parallels between the children of Israel and the native Irish would become a recurring trope in Gaelic literature.

The plantation of Ulster, copper-fastened and extended by the Cromwellian Settlement after 1652, led to the confiscation of lands, and the arrival of new landlords and masters. Initially quite small in extent, its expansion brought catastrophic political, socio-economic, religious and cultural upheavals upon the native population and the exile of many political figures, soldiers and clergy. They established an early modern diaspora on mainland Europe, which would also make a major contribution to Irish and Hiberno-Latin letters.[65] According to the contemporary state papers, Somhairle Buí Mac Dónaill, son of James Mac Dónaill of the Glens of Antrim, spoke no language but 'Erse'.[66] Forced to flee Ireland in the early seventeenth century, he hijacked a ship and made his way over to the Low Countries where he became a captain in the O'Neill regiment of the Spanish Army.[67] He later fought his way across Europe as far as Bohemia. Although described in official English documentation as a pirate and a bandit he earned a very high reputation in Prague for the discipline and good order of the Irish troops under his command. This, in turn, may help to explain the cordial reception the Irish Franciscans received when they arrived to found a monastery in the Bohemian capital in 1631.[68]

While stationed in Ostend, Somhairle had engaged two scribes Niall Ó Catháin, presumably from O'Cahan' Country (County Derry/Londonderry), and Aodh Ó Dochartaigh of Inis Eoghain (County Donegal) to assemble all the Fiannaigheacht material they could get access to. The Ó Catháin/ Ó Dochartaigh manuscript (A20, Franciscan Collection), housed in the Ó Cléirigh Institute in University College Dublin,[69] contains a copy of the 'Agallamh na Seanórach' ('Tales of the Elders of Ireland') and the most delicate poems of 'Duanaire Finn'.[70] Some scholars have questioned whether Somhairle could have read and understood the texts.[71] Nevertheless, a colophon in the manuscript from Niall Gruama Ó Catháin gives some sparse information on the undertaking and reminds us of our debt to his patron:

> crích an lebuir seo conuici seo agus go ndena Dia trocairi ar in bhfer rosgribh i.e. Niall Gruama Ó Cathain agus ar fhir an lebuir mar in cedna i.e. Somhairle Mac Domhnaill.[72]

> [the end of the book until this point and may God have mercy on the writer Niall Gruama Ó Cathain and on the book's owner Somhairle Mac Domhnaill]

Similarly, Aodh Ó Dochartaigh's dedication acknowledges Somhairle's role in preserving one of the most precious gems of the Irish literary tradition:

> Ag sin dhuitsi a chaiptin Somhairle et da bhfaghuinn ni badh mo ina sin do Dhuanaire Finn re na sgríobhadh do dhénainn dhaoibhse é. Mise Aodh O Dochartaigh do sgriobh a nOisdin 12 Februari 1627.[73]

> [Here is this for you Captain Somhairle and if I could find more of Duanaire Finn to write I would do it for you, I am Aodh Ó Dochartaigh who wrote this in Ostend 12 February 1627]

Somhairle Buí also arranged for the compilation of the manuscript book now known as the Book of O'Conor Don, which contains over three hundred poems.[74] It is impossible to judge the level of Somhairle's literary accomplishments but his willingness to fund two scribes from a paltry captain's salary to carry through this important literary task has, at the very least, something to do with that 'saine' (diversity) and 'idirdhealú' (distinctiveness) emphasised by Mícheál Ó Cléirigh in the introduction to his Sanasán.[75]

The anonymous 'An Síogaí Rómhánach', a long narrative poem composed in the early 1650s, catalogues the cataclysmic transformation of Gaelic Ireland and the Irish literary tradition by the second half of the seventeenth century. While prostrate on the tombs of Hugh O'Neill second earl of Tyrone and Rory, first earl of Tyrconnell, in the Church of St Pietro in Montorio, Rome, the poet encounters and engages a beautiful young woman who bemoans Ireland's sufferings from the Reformation to the execution of Charles I. The plantation of Ulster concludes a process that began with the Tudor conquest. James VI and I, the penultimate villain in a rogues' gallery containing the incestuous Henry and monstrous Elizabeth, is ultimately responsible for seizing the lands of the Gaelic Irish and allocating them to Saxons.[76] The poem's simple stress metre (caoineadh) bears testimony to a democratisation of Irish literature in the first half of the seventeenth century. The decline of the bardic schools system and collapse of the tradition of hereditary literary families heralded the arrival of a new literary caste, some of whom sprang from the ranks of the Old English descendants of the original Anglo-Norman colonisers. Seathrún Céitinn,[77] Piaras Feiritéar,[78] Pádraigín Haicéad,[79] Muiris Mac Dháibhí Dubh Mac Gearailt,[80] Dáibhí Ó Bruadair,[81] and the named and anonymous

Gaelic Irish authors of five long narrative political poems subsequently edited by Cecile O'Rahilly,[82] composed their verse for a wide audience. They vividly catalogued the wars, massacres, transportations and transplantations inflicted upon the Catholic Irish and noted the impact on the Irish language, as well as the sufferings of the aristocracy, gentry and Catholic clergy.

This literary transformation also witnessed a proliferation of prose texts, hagiographical works, biographies, diaries and social commentaries, a good number of which had their origins in post-Plantation Ulster. Tadhg Ó Cianáin, author of 'Imeacht na nIarlaí ('The departure of the earls'), accompanied the fugitives from Lough Swilly and provides an invaluable eye-witness account of their travels through France and Flanders to Rome.[83] Lughaigh Ó Cléirigh's 'Beatha Aodha Rua Uí Dhomhnaill' ('The Life of Red Hugh O'Donnell') comprises a hagiographical, heroic biography of the warrior-prince of Tír Chonaill.[84] 'Pairlement Chloinne Tomáis' ('Parliament of Clann Thomas'), a bitter invective on the effects of the social revolution which occurred in the aftermath the Cromwellian wars, attacked Oliver Cromwell as the king of churls who had championed every boor and upstart at the expense of the native aristocracy. It proceeded to scurrilously lampoon his main benefactors, the vile progeny of 'Clann Tomáis', who aped the manners, dress and language of the English settlers.[85]

The Irish Catholic religious orders, operating from Louvain, Salamanca and Rome and utilising the scions of the tradition learned families (Uí Gnímh, Uí Chléirigh, Uí Eodhasa, Uí Mhaoil Chonaire and Uí Dhuibhgheannáin), would eagerly embrace the scholarly tenets of European Renaissance and the print-culture of the Reformation and Counter-Reformation. Their two-fold project, 'Docum Glóire Dé agus Onóra na hÉireann' ('For the Glory of God and the Honour of Ireland'), inscribed above the gates of Coláiste na nÉireannach in Louvain, would inform the religious and secular writings emerging from behind its walls. A stream of confessional, theological and secular works, as well as religious primers, catechisms, dictionaries, annals, saints lives and heroic biographies emanated from these colleges, all penned in a simplified Irish idiom they reflect both the continental and native bardic training of many of their authors.

The Four Masters' compilation of 'Annála Ríoghachta Éireann' ('Annals of the Kingdom of Ireland') was pivotal to these ongoing efforts to promote Catholicism and preserve the Irish Catholic nation's literary heritage. This was not, as Roy Foster has claimed, a 'monument to a dead civilisation', but rather the first book of an emerging Irish Catholic Nationalist identity. Seathrún Céitinn's 'Fóras Feasa ar Éirinn' ('Foundation of Knowledge on Ireland') attacked those anti-Irish writers such as Giraldus Cambrensis, Edmund Spenser, William Camden, Richard Stanihurst, Edmund Campion, Meredith Hamner and Fynes Moryson who he deemed to be 'dall aineolach i dteangaibh na tíre' (blind and ignorant in the language of the country). He also justified his extensive use of poetry, 'Cnáimh agus smior an tSeanchusa' ('the bones and marrow of history') as a source for his work, a pronouncement which should not be lost on historians who wish to assess either the importance of Irish literary sources or Gaelic Ireland's reaction to the plantation of Ulster.

NOTES

1 For a wider perspective on Irish literature in the seventeenth century see B. Ó Cuív, 'The Irish language in the early modern period', in T. W. Moody, F. X. Martin and F. J. Byrne (eds), *The New History of Ireland*, iii: *Early Modern Ireland, 1534-1691* (Oxford, 1976), pp. 509-42; C. Ó Maonaigh, 'Scríbhneoirí Gaeilge an Seachtú haois Déag', *Studia Hibernica*, 2 (1962), 182-208; B. Ó Doibhlin, *Manuail de Litríocht na Gaeilge*, fascicle 3 (2007) and fascicle 4 (2008); B. Cunningham 'Native culture and political change in Ireland', in C. Brady and R. Gillespie (eds) *Natives and Newcomers* (Dublin, 1986), pp. 156-9; B. Ó Buachalla, *Aisling Ghéar: Na Stíobhartaigh agus an tAos Léinn, 1603-1788* (Dublin, 1996). For the bardic period in particular see M. Caball, *Poets and Politics: Reaction and Continuity in Irish Poetry, 1558-1625* (Cork, 1998) and his essay in this present volume. W. J. Smyth's *Map Making, Landscapes and Memory: A Geography of Colonial and Early Modern Ireland c.1530-1750* (Cork, 2006) has perceptive and valuable readings of texts from this era.
2 We too readily forget that viable Irish language communities survived in the so called escheated counties until our own times – in south County Londonderry and Tyrone, in south Down and south Armagh, in Fermanagh and in Cavan.
3 E. W. Lynam 'The Irish character in print 1571-1923', *Transactions of the Bibliographical Society*, 4th series, 4 (1924), 286-325.
4 B. Ó Cuív. 'Ireland's manuscript heritage', *Éire/Ireland*, 11 (1984), 87-107. See also R. Ó hUiginn (ed.), 'Oidhreacht na Lámhscríbhinní', *Léachtaí Cholm Cille*, 35 (Maynooth, 2004).
5 P. de Brún, *Lámhscríbhinní Gaeilge: Treoirliosta* (Dublin, 1988). On the wanton destruction of manuscripts, see E. Ó Muirgheasa, *Abhráin Airt mhic Chubhthaigh agus Abhráin eile* (Dundalk, 1916) p. 198; B. Ó Buachalla 'Art Mac Bionaid, Scríobaí', *Seanchas Ardmhacha*, 60 (1979), 338-49.
6 J. G. Kohl, *Travels in Ireland* (London, 1840), pp. 31-2.
7 P. de Brún, 'Gan teannta buird na binse: scríobhaithe na Gaeilge 1650-1850', *Comhar*, Samhain (1972), 15-20.
8 P. de Brún, *Lamhscríbhinní Gaeilge: Treoirliosta*, pp. 1-37.
9 N. Ní Shéaghdha, *Collectors of Irish Manuscripts: Motives and Methods* (Dublin Institute of Advanced Studies, 1999); B. Ó Buachalla, *I mBéal Feirste Cois Cuain* (Dublin, 1968); B. Ó Conchúir, *Scríobhaithe Chorcaí 1700-1850* (Dublin, 1982).
10 Paul Walsh, 'The work of a winter, 1629-30', in N. Ó Muraíle (ed.), *Mícheál Ó Cléirigh, His Associates and St Anthony's College, Louvain* (Dublin, 2008), pp. 146-53.

11 B. Ó Buachalla, 'Annála Rioghachta Éireann is Foras Feasa ar Éirinn: An Comhthéacs Comhaimseartha' *Studia Hibernica*, 22–3 (1982-83), 59–105; B. Cunningham, *The Annals of the Four Masters: Irish History, Kingship and Society in the Early Seventeenth Century* (Dublin, 2010).
12 Ó Buachalla, *Aisling Ghéar*, pp. 90–8.
13 Cited in C. P. Meehan, *The Fate and Fortunes of Hugh O'Neill, Earl of Tyrone and Rory O'Donnell, Earl of Tyrconnell* (Dublin, 1868), p. 328. See also J. Leerssen, *Mere Irish and Fíor-Ghael: Studies in the Idea of Irish Nationality, its Development and Literary Expression prior to the Nineteenth Century* (Notre Dame, 1997), pp. 151–220; T. O'Connor, '"Perfidious machiavellian friar": Florence Conroy's campaign for a Catholic restoration in Ireland, 1592–1616', *Seanchas Ardmhacha*, 19 (2002), 91–105.
14 B. Ó Buachalla, *Aogán Ó Rathaille* (Dublin, 2007), p. 43. Translations are placed in square brackets and, unless otherwise stated in the footnotes, are the author's own.
15 S. Ó Tuama and T. Kinsella, *An Duanaire 1600–1900: Poems of the Dispossessed* (Dublin, 1994), pp. 165–6.
16 Chichester wrote 'we have burned and destroyed along the Lough, even within four miles of Dungannon where we killed man, woman, child, horse, beast and whatever we found. The last service, from which we returned yesterday was upon Patrick O'Quinn, one of the chief men of Tyrone, dwelling within four miles of Dungannon fearing nothing, but we lighted upon him and killed him, his wife, son, daughters, servants and followers being many and burned all to the ground', *Calendar of the State Papers Relating to Ireland* (24 vols, London, 1860–1911) (*CSPI*) 1601, p. 334. Such massacres, as well as the sustained use of martial law in the planted areas of Ulster, featured among the grievances of the Ulster rebels in 1641: see 'The Remonstrance of the Irish of Ulster', in J. T. Gilbert (ed.), *The History of the Irish Confederation and the War in Ireland* (7 vols, London, 1882–91), i, 450–60.
17 S. Meigs, *The Reformations in Ireland* (Dublin, 1997), pp. 80–9 and J. Dawson 'Calvinism and the Gaidhealtachd in Scotland', in A. Pettegree, A. Duke and G. Lewis (eds), *Calvinism in Europe 1540–1620* (Cambridge, 1994), pp. 231–53.
18 T. F. O'Rahilly, 'Irish poets, historians and judges in English documents 1538–1615', *Royal Irish Academy Proceedings*, 36 (1942), 86–120, at 117–18; Marc Caball, 'Eochaidh Ó hEodhusa', in J. McGuire and J. Quinn (eds), *Dictionary of Irish Biography* (9 vols, Cambridge, 2009), vii, pp. 553–5.
19 Walsh, *Irish Leaders and Learning through the Ages*, pp. 350–61.
20 S. Ó Ceallaigh, *Gleanings from Ulster History* (Cork, 1951), pp. 109–18.
21 J. Watt, *The Church in Medieval Ireland* (Dublin, 1972), p. 209; Ó Ceallaigh, *Gleanings from Ulster History*, pp. 111–12; K. Nicholls *Gaelic and Gaelicised Ireland in the Middle Ages* (Dublin, 1972), pp. 111–13; E. Ó Doibhlin *Domhnach Mór* (Omagh, 1969), pp. 26–42.
22 Ó Ceallaigh, *Gleanings from Ulster History*, p. 113.
23 Ibid., p. 114.
24 Ibid. See also H. A. Jefferies, 'Bishop George Montgomery's survey of the parishes of Derry diocese: a complete text from c.1609', *Seanchas Ardmhacha*, 17 (1996), 44–76.

25 B. Rich, *A New Description of Ireland* (London, 1610), p. 2; C. P. Meehan, *The Fate and Fortune of the Earls of Tyrone and Tyrconnell* (Dublin, 1886), pp. 109–11. For English colonial attitudes towards Ireland, and the Irish, in particularly the literati, see A. Hadfield, *Strangers to That Land: British Perceptions of Ireland from the Reformation to the Famine* (Buckinghamshire, 1994); idem, *Literature, Politics, National Identity: Reformation to Renaissance* (Cambridge, 1994); idem, *Literature, Travel and Colonial Writing in the English Renaissance, 1545–1625* (Oxford, 1998); idem, *Edmund Spenser's Irish Experience* (Oxford, 1977); W. Maley, *Salvaging Spenser: Colonialism, Culture and* Identity (London, 1997); W. Maley, B. Bradshaw and A. Hadfield, *Representing Ireland: Literature and the Origins of Conflict* (Cambridge, 1993); R. McCabe, *Spenser's Monstrous Regiment: Elizabethan Ireland and the Poetics of Difference* (Oxford, 2002); P. Palmer, *Language and Conquest in Early Modern Ireland* (Oxford, 2001).

26 T. F. O'Rahilly, 'Irish poets, historians and judges in English documents 1538–1615'; G. Hill, *An Historical Account of the Plantation of Ulster at the Commencement of the Seventeenth Century 1608–20* (Belfast, 1877).

27 J. O'Donovan, *Annals of the Four Masters* (2nd edn, 7 vols, Dublin, 1856), vi, 2353–9.

28 Ibid. The translation is by O'Donovan.

29 Braidwood has an excellent and detailed statement of the origins of the new settlers in the escheated counties; see J. Braidwood 'Ulster and Elizabethan English', in G. B. Adams (ed.), *Ulster Dialects* (Belfast, 1964), pp. 5–109.

30 B. Ó Doibhlin, *Manuail de Litríocht na Gaeilge*, fascicle 2 (Dublin, 2006), pp. 245–7.

31 Standish Hayes O'Grady (ed.), *Catalogue of Irish Manuscripts in the British Library [formerly British Museum]* (2 vols, Dublin, 1992 repr.), i, pp. 374–5.

32 D. Ó Doibhlin, *The Ulster Irish Translation of the De Imitatione Christi* (Monaghan, 2000), pp. 23–31; P. Kearns, 'James Pulleinee, an 18th century dean of Dromore', *Seanchas Ardmhacha*, 11 (1983/4), pp. 70–9.

33 J. Pulleine, *An Teagasg Críosdaidhe Criosdaidhe anGOIDHELG nGoidhleigHELG* (n.p. 1782).

34 C. Ó Maonaigh, *Seanmónta Chúige Uladh* (Dublin, 1965), pp. 36–7.

35 D. MacCulloch, *Reformation Europe's House Divided 1490–1700* (London, 2003), pp. 115–32; A. J. Krailsheimer, *The Continental Renaissance* (London, 1971), pp. 103–4.

36 D. Ó Floinn, *An Éigse Ilcheasach*, p. 1, no. 1, text taken from A. W. K. Miller, 'Foclóir no sanasan nua', *Revue Celtique*, 4 (1879–80), 349–428, v (1881–83), 1–69.

37 B. Ó Doibhlin *Manuail de Litríocht na Gaeilge*, fascicle iii (Dublin, 2007), pp. 17–18. See also M. Mac Craith, 'Printing in the vernacular: the Louvain project', *History Ireland*, 15:4 (2007), 27–32.

38 T. Ó Cléirigh, *Aodh mac Aingil agus Scoil Nua-Ghaeilge i Lobháin* (Dublin, 1936), pp. 18–34, at 28–34; P. Mac Aogáin, O.F.M., *Graiméir Ghaeilge na mBráthar Mionúr* (1968) and S. de Napier, *Coimhréir agus Gramadach Ghaeilge Uladh i 1600* (Dublin, 2001).

39 C. McGrath, 'Í Eodhusa', *Clogher Record*, 2 (1957), 1–19.

40 C. Ó Maonaigh (ed.), *Scáthán Shacramuinte na hAithridhe* (Dublin, 1952), p. 95. See also Salvador Ryan, 'Bonaventura Ó hEódhusa's An Teagasg Críosdaidhe (1611/14): a reassessment', *Archivium Hibernicum*, 58 (2004); Salvador Ryan, 'A wooden key to open heaven's door: lessons in practical Catholicism from St Anthony's College, Louvain', in Edel Bhreatnach, Joseph McMahon and John McCafferty (eds), *The Irish Franciscans, 1534–1990* (Dublin, 2009), pp. 221–32.

41 Ó Maonaigh (ed.), *Scáthán Shacramuinte na hAithridhe*, p. 94. See also M. Mac Craith, 'The political and religious thought of Florence Conroy and Hugh MacCaughwell', in A. Ford and J. Mc Cafferty (eds), *The Origins of Sectarianism in Early Modern Ireland* (Cambridge, 2006), pp. 183–202; idem, '*Scáthán Shacramuinte na hAithridhe*: scáthán na sacraiminte ceanna', *Léachtaí Cholm Cille*, 30 (Maynooth, 2000), 28–64; B. Hazard, *Faith and Patronage: The Political Career of Flaithrí Ó Maolchonaire, c.1560–1629* (Dublin, 2009).

42 F. Mac Raghnaill (ed.), *An Teagasg Críosdaidhe* (Dublin, 1976).

43 Ibid., pp. 10–11.

44 Theobald Stapleton in his *Catechismus, seu Doctrina Christiana Latino-Hibernica* (Brussels, 1639; facsimile reprint Dublin, 1945), article 31, makes a forceful plea for the use of the vernacular and for the simplification of Irish 'as coir agus is iomchuibhe dhuinne na Herenuig bheith ceanamhail gradhach onorach air ar tteangain nadurtha fein, an ghaoilag noch ata chomh fulethach, chomh muchta soin nach mor na deacha si as cuimhn na nduinne, a mhillean is feidir a chur ar an Aois Elaghain dhon Teangain do chuir fa fordhoireacht agus cruos focal, dha scibha i nodaibh agus foclaibh deamhaire dorcha do thuicseanta, agus ni foilid soar moran dar nduinibh uaisle do bheir a tteanga dhuchais nadurtha (noch ata fortill fuirithe, onorach folamtha gearchuiseach inti fein ttarcaisne agus a neamhchionn agus a chaitheas an aimsir a saorthudh agus a foghlaim teangtha coimhtheacha ele'.

45 T. Ó Cléirigh, *Aodh Mac Aingil agus an Scoil Nua-Ghaedhilge i Lobháin*, pp. 41–90.

46 *CSPI*, vol 4, pp. 428–30. The contemporary state papers contain numerous other accounts of sermons preached to thousands of people by other Louvain-educated friars such as Turlough MacCrodyn, Hugh McKale, and Henry O'Mellan who warned the faithful against Protestantism, attributed their political and religious trials to sin and lack of devotion and promised speedy deliverance from their sufferings; The National Archives, London (TNA), SP 63/227/96, Chichester to the Privy Council, 4 July 1609; TNA, SP 63/232/21, Examination of Tadhg Modder Mac Glone, 21 Oct. 1613; British Library, Cotton Mss, Titus BX, Examination of Shane MacPhelim O'Donnelly, 22 Oct. [1613], fols 236–7.

47 C. Ó Maonaigh (ed.), *Scáthán Shacramuinte na hAithridhe*, p. 5. An leabhrán so do sgríobhadh go simplidhe [to write this book in a simple plain manner].

48 C. Mhag Craith (ed.), *Dán na mBráthar Mionúr*, i, 23–71 and 154–77.

49 T. Ó Donnchadha, 'Cín Lae O Mealláin', *Analecta Hibernica*, 3 (1931), 1–62. See also C. Dillon, 'Cín Lae Uí Mheallain Friar O Meallan journal', in C. Dillon and H. A. Jefferies (ed.), *Tyrone: History and Society* (Dublin, 2000), pp. 327–401; M. Níc Cathmhaoil, 'Cín Lae Uí Mhealláin', in *Seanchas Ardmhacha*, 21 (2006), pp. 35–54.

50 'Bamar ceathrar bráthar san mBreantur Toirealach Ó Mealláin sagart agus triar bráthar de chlann Laochthréan, Eoghan, Benedictus agus Tadhg a n-anmonna,' [There were four of us, friars, in Branty. Toirealach Ó Mealláin, a priest, and three friars of the O 'Loughran family, Eoghan, Benedictus and Tadhg were their names], in T. Ó Donnchadha, 'Cín Lae O Mealláin', p. 21. See also E. Ó Doibhlin, *Domhnach Mór*, pp. 26–42.
51 S. Ó Ceallaigh, *Gleanings from Ulster History*, pp. 66–72.
52 E. Ó Doibhlin, *Domhnach Mór*, pp. 111–35; M. Nic Cathmhaoil 'Cín Lae Uí Mheallain', pp. 35–54. The *Commentarius Rinuccianus*, complied in Florence by two Capuchin friars (Richard O'Ferrall and Robert O'Connell) between 1661 and 1666, comprises the history of Archbishop Giovanni Battista Rinuccini's embassy as Papal Nuncio the Confederate Catholics of Ireland. The work is complemented by O'Connell's substantial *Historia Missionis Hiberniae Fratrum Minorum Capuchinorum*. It is arguably O'Connell's own voice and temperament which resound in the *Commentarius*, as can be seen by comparison with the tenor of the *Historia*. John Callaghan, *Vindiciarum Catholicorum Hiberniae, Authore Philopatre Irenaeo ad Alithophilum Libri Duo* (Paris, 1650), which defended the Confederate peace with Ormond and the Inchiquin truce, raised the hackles of the Irish prelates at Louvain and was subsequently placed on the Index Librorum Prohibitorum in June 1654. See also Cecile O'Rahilly (ed.), *Five Seventeenth-Century Political Poems* (Dublin, 1977).
53 'Tangadar na hAlbonnaigh móra go Dún Geanainnis do sgaoileadar frais lamhaigh in aon uaim; is do chualas féin an mhórfhuaim sin' [The great Scotch army reached Dungannon and discharged a volley of shots in unison. I myself heard that great noise], in T. Ó Donnchadha, 'Cín Lae O Mealláin', p. 15.
54 'acht nach sgríobhtar annso acht na daoine do chonaic mé féin agus feas coda dá ngníomharthaibh ar cuimhne agam' [but I write here of the people I saw myself and had knowledge of some of their deeds in my memory], in S. Laoide and E. Mac Néill (eds), N. Mac Muireadhaigh, *Alasdair Mac Colla* (Dublin, 1916), pp. 12–13.
55 'd'ordaigh mart ramhar, meadur ime, agus ceithre scillinge d'airgiod san mbaile bhó ón Ghlas Dromuinn go Tulaigh Óg' [He ordered a fatted cow, a measure of butter and four silver shillings per ballyboe from Glasdromman to Tullyhogue], in T. Ó Donnchadha, 'Cín Lae O Mealláin', p. 7.
56 Ibid., p. 41.
57 Ibid., p. 27.
58 L. P. Murray, 'The County Armagh hearth money rolls A.D 1664', *Archivium Hibernicum*, 8 (1941), 167–76.
59 T. Ó Fiaich (ed.), *Art Mac Cumhaigh: Dánta* (Dublin, 1973), p. 11.
60 R. Ní Ógáin (ed.), *Duanaire Gaedhilge* (Dublin, no date), ii, 77–9, 134.
61 E. Ó Muirgheasa (ed.), *Dánta Diadha Uladh* (Dublin, 1936), pp. 184–7; S. Ó Tuama and T. Kinsella, *An Duanaire, 1600–1900: Poems of the Dispossessed* (Dublin, 1994), pp. 103–9.
62 J. B. Leslie, *Armagh Clergy and Parishes* (Dundalk, 1911), p. 95; J. H. Todd (ed.), *Felire na naomh nerennach Martyrolgium Dungallense seu Calendarium Sanctorum Hiberniae collegit et digessit Fr. Michael O'Clery* (Dublin, 1864), p. 55.

63 S. Ó Ceallaigh, *Gleanings from Ulster History*, p. 68.
64 E. Ó Muirgheasa (ed.), *Dánta Diadha Uladh*, p. 186. The poem is discussed in B. Ó Doibhlin, *Manuail de Litríocht na Gaeilge*, fascicle 4 (Dublin, 2008), pp. 118–20.
65 J. J. Silke, 'The Irish abroad, 1534–1691', in Moody, Martin and Byrne (eds), *The New History of Ireland*, pp. 587–632; B. Millet, 'Irish Literature in Latin, 1550–1700', in ibid., pp. 561–86.
66 S. Ó Ceallaigh *Gleanings from Ulster History*, p. 101.
67 R. Gillespie, *Conspiracy: Ulster Plots and Plotters in 1615*, pp. 33–5.
68 P. Walsh, *Irish Chiefs and Leaders* (Dublin, 1960), pp. 110–40.
69 M. Dillon, C. Mooney and P. de Brún, *Catalogue of Irish Manuscripts in the Franciscan Library Killiney* (Dublin, 1969), pp. 39–45.
70 E. Mac Neill and G. Murphy (eds), *Duanaire Finn* (3 vols, Irish Texts Society, London, 1908–4).
71 R. Gillespie, *Conspiracy: Ulster Plots and Plotters in 1615*, p. 34.
72 M. Dillon, C. Mooney, and P. de Brún (eds), *Catalogue of Irish Manuscripts in the Franciscan Library Killiney*, p. 40. See also H. MacDonnell, *The Wild Geese of the Antrim MacDonnells* (Dublin, 1996).
73 Dillon, Mooney and de Brún, *Catalogue of Irish Manuscripts*, p. 43.
74 P. Walsh, *Irish Chiefs and Leaders*, pp. 139–40.
75 See note 37 above.
76 'An Síogaí Rómhánach', in C. O'Rahilly (ed.), *Five Seventeenth-Century Political Poems*, pp. 17–32, at 21.
77 E. Mac Giolla Eáin, *Dánta, amhráin is caointe Sheathrúin Céitinn* (Dublin, 1900); B. Cunningham, 'Keating, Geoffrey [Seathrún Céitinn] (b. c. 1580, d. in or before 1644)', in *Oxford Dictionary of National Biography*, online edition, ed. Lawrence Goldman (Oxford, 2004), www.oxforddnb.com/view/article/15224 (accessed 30 September 2011).
78 P. Ua Duinnin (ed.), *Dánta Phiarais Feiritéir* (Dublin, 1934).
79 M. Ní Cheallacháin (ed.), *Filíocht Phádraigín Haicéad* (Dublin, 1986); M. Hartnett (trans.), *Haiceád* (Oldcastle, 1993).
80 N. Williams (ed.), *Dánta Mhuiris Mhic Dháibhí Dhuibh Mhic Gearailt* (Dublin, 1979).
81 J, MacErlean (ed.), *Duanaire Dháibhidh Uí Bhruadair* (3 vols, London, 1910, 1913, 1917).
82 O'Rahilly (ed.), *Five Seventeenth-Century Political Poems*.
83 N. Ó Muraíle (ed.), *Turas na dTaoiseach nUltach as Éirinn: From Ráth Maoláin to Rome: Tadhg Ó Cianáin's contemporary narrative of the so-called 'Flight of the Earls'* (Dublin, 2008).
84 P. Walsh (ed.), *Beatha Aodha Ruaidh Uí Dhomhnaill* (2 vols, London, 1948–57).
85 N. J. A. Williams (ed.), *Pairlement Chloinne Tomáis* (Dublin, 1980).

12

Angling for Ulster: Ireland and plantation in Jacobean literature

WILLY MALEY

My title risks creating a ghost topic if by 'literature' we understand imaginative writing – plays, poetry, prose fiction – since many critics doubt that events in Ireland from 1603 to 1625 impacted sufficiently on English literary culture to feature more than fleetingly in the writing of the period. Three main lines of thought emerge. First, a moratorium on public exploration of Irish affairs triggered by the outbreak of the Nine Years War in 1594 had a knock-on effect in the drama of the early seventeenth century, in the post-war peace and plantation years. Thus Ireland featured in texts of the time only obliquely, allegorically or furtively. As Michael Neill notes: 'Given the amount of political, military, and intellectual energy it absorbed, and the moneys it consumed, Ireland can seem to constitute...one of the great and unexplained lacunae in the drama of the period.'[1] Andrew Murphy discerns 'a failure, or unwillingness, on the part of English dramatists to engage with one of the most urgent and important political crises of the close of the sixteenth century: the war in Ireland'.[2] Murphy, like Andrew Hadfield, attributes this silence to censorship arising from ongoing conflict.[3] For Tristan Marshall: 'Theatrical events located or concentrating on Ireland were in fact extremely rare in the early Jacobean period, with the most obvious, Jonson's *Irish Masque at Court* (1613) being notable only for the fact that it was performed twice, and that it unsurprisingly deals with the idea of the Irish giving way to English civility thanks to the presence of King James.'[4] Joel Altman cites one contemporary, George Fenner, writing on 30 June 1599, in the midst of Essex's unsuccessful expedition: 'it is forbidden, on pain of death, to write or speak of Irish affairs'.[5] Yet Altman observes that the patron of Shakespeare's company was Sir George Carey, 'whose signature appeared on most of the orders commanding the lords lieutenant to levy soldiers for Ireland', so Ireland loomed large as the elephant in the room.[6]

The second line on literary representations of Jacobean Ireland is that it was business as usual. England – now 'Britain' – was as preoccupied with Ireland as ever, from the moment James I, inspecting the documents stored in Whitehall upon his accession, exclaimed 'We had more ado with Ireland than all the world besides.'[7] The Ulster plantation was both crowning achievement and consolation for stalled plans for fuller union. Why, though, is this pervasive presence in the state papers such a puzzling absence from the London stage and the licensed page?[8]

A third line is that a shift occurred in the kind of writing that dealt with Ireland. Thus according to David Armitage: 'The New World remained largely in the realm of fiction and fancy for Britons until at least the early seventeenth century, when Francis Bacon compared the solid success of the Ulster Plantation with the risks of the new venture in Virginia, "an enterprise in my opinion differing as much from [Ulster], as *Amadis de Gaul* differs from Caesar's *Commentaries*".'[9] Mark Netzloff shares Armitage's view that Ulster was all about non-fiction, arguing that this particular plantation 'becomes textually located in a documentary form of writing, an accumulation of records that constitute the majority of documents within the archive'.[10] The implication that Ulster was a workaday world far removed from the wonders of the 'New World' would chime with some historical evidence. 'Ulster', we are told, 'was a rough-hewn frontier, which meant that the Celtic Scots were already geographically and physically preconditioned to live where English and Lowland Scots, for the most part, could not. Most English settlers found a physical environment much too harsh and foreboding for their tastes... Eventually it became clear to the commissioners that only certain of the planters were fit for the rigors of Ulster life.'[11] The argument that Ulster lacked the exoticism of other colonies and was therefore less literary is backed up by Captain Arthur Chichester, later Lord Chichester of Belfast, who informed King James in 1610: 'I had rather labour with my hands in the plantation of Ulster, than dance or play in that of Virginia.'[12] Ulster had long had a reputation as a hostile environment. Camden's *Remaines* tells of '*Savage* a Gentleman which amongst the first English, had planted himselfe in *Vlster* in *Ireland*, advised his sonne for to builde a castle for his better defence against the Irish enemy, who valiantly answered; *that hee woulde not trust to a castle of stones, but to his castle of bones*, Meaning his body'.[13]

There are problems, though, with Armitage and Netzloff's perspectives. Bacon had grave misgivings about plantation; amidst the debate on Anglo-Scottish Union, 'Britons' themselves were more fantasy than reality; and he was anticipating the success of the Ulster plantation, not reporting on it. As for unfavourable comparisons with Virginia, as Sarah Irving reminds us, 'Bacon's name appears on the second charter of Virginia, dated May 23, 1609'.[14] Irving notes that 'evidence of Bacon's misgivings...about colonisation...occur...

throughout his writing and correspondence'.[15] In 'Of Plantations', Bacon wrote: 'I like a Plantation in a pure Soyl, that is, where People are not Displanted, to the end, to Plant others; for else it is rather an Extirpation, than a Plantation.'[16]

Denise Albanese initially appears to echo Armitage's assertion that Bacon signals a shift from literary to political discourse in his treatise on the Ulster plantation. Commenting on the classical comparison of Ulster and Virginia, Albanese writes: 'The advice here is pragmatic and direct – but takes a strange detour by likening Jacobean imperialism to written, indeed, literary texts. As the reference to Caesar suggests, Ireland is "another Britain", and the British may take Caesar's part in a war represented as discursive strategy, whose impetus, and hence ontology, are linear.'[17] Since this line of argument radically underestimates the richly metaphorical nature of Bacon's text, Albanese goes on to qualify this claim, getting behind Bacon's stated purpose. His treatise, 'written to persuade the king that Ireland can be subjugated to English needs... produces the literary both as a warrant for settlement and as a blind for the real acts of violence such settlement necessitates... Virginia, too, figures as a fiction, as the romantic narrative of colonialist ease, to whose plot one more episode, one more encounter, can always be added. The open-ended form of romance at once authorizes and mystifies the work of domination Bacon proposes, and provides aesthetic distance from the world of colonialist practices. This disavowal of the need for closure, this pre-emptive use of fictional form, permeates the *New Atlantis*: it is an ideological move that enables the scientific to be produced under the ostensible sign of the literary, and results in an act of discursive colonization that tropes the cultural agendas of Jacobean imperialism'.[18] Anne Fogarty has shown that John Davies is as contradictory and as literary as Bacon in his depiction of Ireland.[19] Thus the claim for a shift from fantasy to fact also overlooks the blurred boundary between fiction and history in the period.

Jacobean writers engaged with Ireland in various ways. The translator of Montaigne, John Florio, was in Ireland in the summer of 1609 on some business connected with Ulster, collecting £500 on behalf of King James, forfeited by Cathair O'Doherty after his abortive rebellion in 1608.[20] John Donne almost went to Ireland from France in November 1597 to rebuff a potential Spanish attack at Waterford.[21] He wrote an epistle to his friend Henry Wotton who served there in 1599 under Essex,[22] applied for a position as a secretary in Ireland in 1608, using the influence of the Countess of Bedford,[23] preached the virtues of colonization in 1622, and gave the oration at the funeral of Sir John Davies.[24] In his sermon on the Virginia Plantation, Donne stressed the social and economic advantages of colonisation: 'if the whole Countrey were but such a Bridewell, to force idle persons to work, it had a good vse... alreadie the imployment breedes Marriners; alreadie the place giues Essayes, nay, Fraights of Marchantable Commodities'.[25]

Donne's epistle to Wotton is concerned not merely with the safety but the selfhood of his friend:

> Went you to conquer? and have so much lost
> your self, that what in you was best and most
> Respective frendship should so quickly dye?[26]

In Elegie XX, 'Loves Warre', Donne writes:

> Sick Ireland is with a strange warr possest
> Like to an Ague; now raging, now at rest;
> Which time will cure: yet it must doe her good
> If she were purg'd, and her head vayne let blood.[27]

But as Tom Cain suggests, empire's impact on Donne goes deeper than such explicit verses.[28] Holy Sonnet XIV, 'Batter my Heart', contains siege imagery with specific Irish resonances:

> I, like an usurpt towne, to'another due,
> Labour to admit you, but Oh, to no end,
> Reason your viceroy in mee, mee should defend,
> But is captiv'd, and proves weake or untrue.[29]

The word 'viceroy' had a particular Irish context as the designation given to the office of the Lord Deputy.

A siege mentality is evident in Francis Bacon's 'A Speech in Parliament touching the Naturalization of the Scottish Nation' (1608), where he speaks of two entrances into England used by its enemies, France and Spain: '*France* had *Scotland*, and *Spain* had *Ireland*: For these were the two Accesses, which did comfort, and encourage, both these Enemies, to assail, and trouble us. We see, that of *Scotland*, is cut off, by the *Vnion*, of both these *Kingdoms*; If that, it shall be now made constant, and permanent. That of *Ireland*, is likewise cut off, by the convenient situation, of the *North* of *Scotland*, toward the *North* of *Ireland*, where the Sore was: Which, we see, being suddainly closed, hath continued closed, by means of this Salve; So as now, there are no Parts, of this State, exposed to Danger, to be a Temptation, to the Ambition of *Forrainers*, but their approaches, and Avenues, are taken away.'[30] Plantation, made possible by Union, provided a shield against invasion, the paling of Britain.

In Jonson's *Irish Masque*, first performed on 29 December 1613, Donnell introduces 'Ty good shubshects of Ireland, and pleash ty mayesty', to which Dennise adds: 'Of Connough, Leymster, Vlster, Munster. I mine one shelfe vash borne in te English payle and pleash ty Mayesty.'[31] Which pale does Dennise allude to? Is the joke that, as Davies claimed in his *Discovery*, under James, and in the wake of the Ulster plantation, there is no pale, no limit to the king's writ, as all Ireland is under British (Anglo-Scottish) control?[32]

According to Margaret Rose Jaster, Jonson's *Irish Masque* 'activates convenient Irish stereotypes to legitimate whatever policy the savior-king will employ'.[33] David Lindley suggests Jonson's *Masque* shares its tropes with Davies's *Discovery*: 'Sir John Davies, indeed, in expressing his hopes for the future, comments that "these ciuil assemblies and assizes and sessions haue reclaimed the Irish from their wildness, caused them to cut off their glibs and long hair, to conuert their mantles into cloaks, to conform themselves to the manner of England in all their behaviour and outward form." Perhaps this work, published in 1612, actually inspired Jonson's device, but in any case it makes clear the force of the central symbol of *The Irish Masque*.'[34]

For James Smith, the casting of Jonson's *Masque*, with the roles of the Irish ambassadors taken by 'five English and five Scottish Jacobean courtiers... reflects precisely the bifurcated nature of King James's sovereignty... and parallels exactly the ethnic make-up of the New English colonizers just then enforcing the Ulster Plantation in Ireland'.[35] For Smith, 'Jonson's decision to fashion his epithalamion for the Somerset wedding around Irish visitors to court, suggests the extent to which Ireland served as an alternative arena into which Jacobean society could conveniently displace compromising realities.'[36] Furthermore, he contends that colonial texts were in conversation with one another in the period. Davies's imagery certainly anticipates Jonson's dramatic metaphors of transformation and 'provides the masque with a contemporary analogue, using identical language to venerate James's adoption of Ireland into the nation state'.[37] Finally, Smith concludes that while 'Jonson proposes a translated and idealized Ulster', his 'masque did not silence the native alternative any more than it provided an idealized solution to Old English discord'.[38]

Jonson's mantled aristocrats conceal another truth, namely the creation of a new colonial aristocracy. Bacon had advised James:

> For *Honour*, or *Countenance*, if I shall mention to your *Majesty*, whether, in wisdome, you shall think convenient, the better to express your Affection to the *Enterprise*, and for a Pledge there of, to adde the *Earldome* of *Vlster*, to the Princes Titles; I shall, but learn it, out of the practice, of *King Edward* the First; Who first used the like course, as a mean, the better to restrain, the *Countrey* of *Wales*: And, I take it, the *Prince of Spain*, hath the Addition, of a *Province*, in the *Kingdome* of *Naples*; And other *Presidents*, I think, there are, and it is like, to put more life, and Encouragement, into the *Undertakers*. Also, considering the large *Territories*, which are to be Planted, it is not unlike, your *Majesty* will think, of raising some *Nobility* there; which, if it be done meerly, upon new Titles, of *Dignity*, having no manner of Reference to the Old; And if it be done also, without putting, to many *Portions*, into one *Hand*; And lastly, if it be done, without any great *Franchises*, or *Commands*; I do not see, any Perill, can ensue thereof: As, on the other side, it may draw some *Persons*, of great

Estate, and *Means*, into the Action, to the great Furtherance, and Supply, of the charges thereof.'³⁹ John Selden observed in *Titles of honor* (1614), 'that of BARONET became a new erected distinct Title vnder our present Soueraigne, who, for certain disbursments toward the Plantation in *Vlster*, created diuers into this Dignitle, and made it hereditarie.⁴⁰

Colonial metamorphosis through upward social mobility is also key to *Epicoene: or, the silent woman* (1609), where Jonson's Sir Amorous La-Foole declares: 'I have been a mad wag in my time, and have spent some crowns since I was a page in court, to my lord *Lofty*, and after, my lady's gentleman-usher, who got me knighted in *Ireland*, since it pleased my elder brother to die.'⁴¹ In addition to his *Irish Masque* and *Epicoene*, Jonson used Irish characters or topics in *Prince Henry's Barriers* (1610), *The Devil Is An Ass* (1613), *Bartholomew Fair* (1614), and *The New Inn* (1629). Rebecca Ann Bach notes that: 'Both the inn and the fair house a stage-Irish character, the prototypical other to England; *Bartholomew Fair* has the Irish bawd Captain Whit, and *The New Inn* has the Irish nurse Shelee-nien.'⁴² Jonson's unmasking trick is recycled in *The New Inn*: 'The drunken Irish woman Shelee-nien Thomas, is not Irish at all but instead the Lady Frampul, an English noblewoman.'⁴³ In *Prince Henry's Barriers* (1610), Jonson's claim for James – '*Ireland* that more in title, than in fact/Before was conquer'd, is his *Lawrels* act' – anticipates Davies' *Discovery*.⁴⁴

If its colonisers gendered Ireland female – Spenser's Irena being the most notable example – then Ulster was also a site of sexualised depictions of imperial power. Critics have shown the extent to which the language of ostensibly non-literary texts is saturated with metaphor. Anna Suranyi notes of Thomas Blenerhasset's 1610 promotional pamphlet that 'he began by urging "Fayre England" to take control over her niece, Ulster. Blenerhasset emphasised Ulster's potential fertility despite the "lamentable" warfare that had "bereaved" her and "defaced whatsoever was beautiful in her to behold". Ulster, he stressed, had been left stripped; she was "naked", desolate and bereft of resources; "she presents her-selfe...in a ragged sad sable robe...with a very little showe of any humanitie". The [male] inhabitants had done little to redeem the "depopulated" land and had "dispoyled" Ulster. For Blenerhasset, the elder aunt England should take a nurturing role by endeavouring to "cover her nakedness" so that "naked Ulster may be relieved, deckt, and richly adorned"...English colonisers would take on the male roles of governance, colonisation, warfare, as well as husbandry, in both senses of the term, to create a relationship that, for Blenerhasset, was both necessary and beneficial'.⁴⁵ Ireland was gendered too by Gaelic poets lamenting Ulster's loss. Speaking of 'the humiliation of plantation', Sarah McKibben says: 'In his twenty-six-quatrain lament, 'Cáit ar ghabhadar Gaoidhil?' (*Where have the Gaels gone?*), Lochlainn Ó Dálaigh conveys the desolation and shame surrounding the departure of Ireland's noblemen and the loss of their hereditary property to colonial upstarts... To capture the affront, Ó Dálaigh offers a telling analogy for Ireland: "clár óir fá fhoirinn tacoir" ("a golden [chess] board under a base [or artificial/ersatz] set of men"...). This image of a noblewoman beneath or below boorish men expresses class outrage and sexualized revulsion at ignoble men trespassing on and "taking" the precious land of Ireland. Through this coded image of rape... Ó Dálaigh both amplifies and deflects the national horror by again displacing it onto a gender-normative figure of suffering and weakness.'[46]

Ulster occupied a unique position in early modern Ireland, unique in not being occupied and mapped in the way that other parts of the country were. The Ulster plantation had its roots in the sixteenth century, with proposals put forward from as early as the 1560s. Moves to settle that unruly region were brought into sharp relief by the death of Shane O'Neill in 1567.[47] The queen immediately wrote to the Lord Deputy, Sir Henry Sidney, urging that Ulster be properly mapped, in a letter reproduced by John Hooker in the second edition of Holinshed's *Chronicles* in 1586.[48] In 1568, Sidney informed Cecil that a private venture would defray the costs of garrisons or settlements borne by the crown.[49] Robert Lythe's surveys of 1567–70 were followed by the ill-fated efforts at settlement of Sir Thomas Smith and Walter Devereux, first earl of Essex, in the 1570s.[50] Smith, author of *De Republica Anglorum*, initiated a failed venture in the Ards Peninsula in the early 1570s, and Essex was part of another doomed project to colonise Ulster in 1573–75, and died in Dublin in 1576.[51] Philip Sidney, whose father, Sir Henry, had made the subduing of Ulster his special mission, posed the following question in Sonnet 30 of his *Astrophil and Stella*: 'How Ulster likes of that same golden bit,/ Wherewith my father once made it half tame.'[52]

John Davies later decried such ventures, arguing that 'when Priuate men attempt the Conquest of Countries at their own charge, commonly their enterprizes doe perrish without successe: as when, in the time of Queene *Elizabeth*, Sir *Thomas Smith* vndertooke to recouer the *Ardes*'.[53] This did not prevent Davies profiting directly from the later settlement. He 'obtained grants of land as an undertaker not only in Tyrone but also in counties Fermanagh and Armagh'.[54] Presenting the Ulster Plantation as some thing entirely new, despite the earlier attempts of Smith and Essex, Davies wrote to the earl of Salisbury on 24 August 1609:

> The use and fruit of this survey and description will not only consist in this, that his majesty shall hereby know what land he hath here, and how to distribute it to undertakers; but in this also, that it will discourage and disable the natives henceforth to rebel against the crown of England; and be a special means hereafter of preventing and suppressing rebellions in this country. For this country (wherein there were never any cities or towns to draw commerce

or trade, and wherein the crown of England never appointed magistrates or visitations of justice till within these five years past) was heretofore so obscure and unknown to the English here as the most inland part of Virginia is yet unknown to our English colony there; so as our ignorance of their places of retreat and fastnesses made them confident in their rebellions and was their only advantage; and made us diffident in our prosecutions and was our only disadvantage, whereas now we know all the passages, have penetrated every thicket and fast place, have taken notice of every notorious tree or bush; all which will not only remain in our knowledge and memory during this age; but being found by inquisitions of record, and drawn into cards and maps are discovered and laid open to all posterities.[55]

But there were risks in surveying hostile territory, as Davies explains in a letter: 'Wee pursue our first course in describing and destinguishing the land. Our Geography hath had the speedier dispath heer, for that the County is but little, consisting only of three Baronies; and for that wee sent 2 surveyors before, to perambulate the cuntry, and to prepare the Busines by gathering noates of the names, scites, and extents of the Townelands; wch they performed well and readily, being accompanied but wth a slender guard. I speake of a guard, as of a necessary circumstance; for though the cuntry bee now quiet, and the heades of greatnes gone; yit our Gographers do not forgett what entertaynement the Irish of Tirconnell gave to a map-maker about the end of the late great Rebellion: for one Barkeley being appoynted by the late earle of Devonshire to draw a trew and perfect mapp of the North parts of Vlster (for that the old mappes were false and defective) when hee came into Tirconnell, the enhabitants tooke of his head, bycause they would not have their cuntrey discovered.'[56]

Mark Netzloff notes that if Scotland goes off the map in accounts of Ireland, Ulster itself also disappears: 'The most puzzling aspect of Speed's map of Ulster lies in the fact that, although engraved in 1610, its details do not reflect any awareness of the Jacobean Ulster plantation. Instead, Speed anachronistically attributes much of Ulster to regional Irish chiefs, demarcating the land along the lines of sixteenth-century divisions that reflect Ulster's earlier status as a region largely resistant to English colonial infiltration. There were empirical reasons for Speed's inability to map contemporary Ulster, as there was no authoritative map that he could use as a model.'[57] For Netzloff: 'Speed's map of Ulster is thus profoundly ambivalent: it commemorates the unsettled status of the region, marking the necessity of fortifications and military rule in the province, while also emphasizing its pacification, eliding signs of the recent conquest...Yet Speed's map nonetheless divides the Ulster landscape primarily along the lines of regional Gaelic Irish lordships, memorializing a social hierarchy that had been effectively displaced from Ulster by 1610.'[58]

Speed 'follows other early modern historiographers in justifying the conquest of Ireland by analogy with the Roman conquest of Britain', while 'William Camden concluded, "a blessed and happy turne had it beene for Ireland, if it had at any time been under [Roman] subjection"'.[59] David Scott Wilson-Okamura points out that both Sir Thomas Smith and Sir John Davies 'took Virgil's description of the colony at Carthage as a model for' the settlement of Ulster.[60] David Beers Quinn suggests that Smith 'proposed to engage himself in the creation of a model "Roman" colony, equipping his illegitimate son, Thomas, with a patent and some corporate backing to found the city of "Elizabetha" in the Ards in Ulster, and issuing a printed pamphlet, a prospectus and a map to draw in men and money. Curiously, he avoided in this printed publicity any mention of Rome and even of the word "colony" and spoke only of "inhabiting" the Ards, but in his correspondence he reiterated time after time for some four years the classical jargon about colonies which had become for him a living directive to action.'[61] Quinn notes that this continued through the Munster Plantation of the 1580s and into Ulster's successful colonisation in 1609: 'The habit of reference to the Roman past continued in Ireland during the growing pains of the great Ulster plantation, which was to surpass Munster and Virginia alike as a new colony in the period before 1630. Thomas Blenerhasset in 1610 thought James I should preen himself like a Roman emperor on the strength of his colonising achievements in Ulster, while Sir John Davies, making still another review of the mistakes of the initial Norman conquest, before going on to praise the Ulster plantation, could not refrain from pointing out how the Romans would have done much better than Henry II and the rest.'[62]

Not all those who journeyed to Ulster in this period were settlers or soldiers. The Scottish traveller William Lithgow, visiting Ireland from August 1619 to February 1620, wrote: 'I dare avow, there are more Rivers, Lakes, Brooks, Strands, Quagmires, Bogs and Marishes, in this Countrey, then in all Christendome besides.'[63] Lithgow mentions 'two intolerable abuses of protections in that Kingdome: The one of Theeves and Wood-carnes, the other of Priests and Papists... The first is prejudicalll to all Christian civilnesse, tranquill government, and a great discouragement for our colloniz'd plantators there, belonging to both soyles of this Island; being daily molested, and nightly incombered with these blood-sucking Rebells. And notwithstanding of their barbarous cruelty, ever executed at all advantages, with slaughter and murder upon the *Scots* and *English* dwellers there; yet they have and find at their owne wills Symonicall protections, for lesser or longer times; ever as the confused disposers, have their law-sold hands, filled with the bloody bribes of slaughtered lives, high-way, and house robbed people'.[64] One of Lithgow's anecdotes concerns a trip to Ulster:

And now I call to memory (not without derision) though I conceal the particular place and prelate, it was my Fortune in the County of *Dunagale*, to bee joviall with a bishop at his Table, whereafter diverse Discourses, my ghostly Father grew offended with mee, for terming of his wife Mistresse: which when understood, I both called her Madam, and Lady Bishop: Whereupon he grew more incensed; and leaving him unsatisfied: resolve me Reader, if it be the custome here or not? and amends shall repay over-sight, a ghostly wife shall be still Madam Lady with me; if not, mine observed manner shall be Mistresse.[65]

Lithgow denounces Irish barbarity in a passage whose language has a familiar feel for anyone acquainted with writers from Gerald of Wales to Spenser, though in a style all his own:

But leaving this, and observing my Method, I remember I saw in *Irelands* North-parts, two remarkable sights: The one was their manner of Tillage, Ploughes drawne by Horse-tayles, wanting harnesse, they are onely fastned with straw, or wooden Ropes to their bare Rumps, marching all side for side, three or foure in a ranke, and as many men hanging by the ends of that untoward Labour. It is as bad a husbandry, I say, as ever I found among the wildest Savages alive; for the *Caramins*, who understand not the civill forme of Agriculture, yet they delve, hollow, and turn over the ground with manuall and wooden instruments: But they the *Irish*, have thousands of both Kingdomes daily labouring beside them, yet they cannot learne, because they will not learne to use harnesse, as they doe in *England*, so obstinate and perverse they are in their Barbarous consuetude, unlesse punishment and penalties were inflicted; and yet most of them are content to pay twenty shillings a yeare, before they will change their custome. The other as goodly sight I saw, was women travelling the way, or toyling at home, carry their infants about their necks, and laying the Dugges over their shoulders, would give sucke to the Babes behinde their backs, without taking them in their armes: Such kind of breasts me thinketh, were very fit to bee made money-bags for East or West-*Indian* Merchants, being more than halfe a yard long, and as well wrought as any Tanner, in the like charge, could ever mollifie such Leather.[66]

Here, the depiction of Irish husbandry and women not only justifies plantation as a civilising process but leads to a crude quip about the general profitability of colonialism.

Barnaby Rich's Irish career under Elizabeth is bookended by service under both earls of Essex. Arriving on the ill-fated expedition to colonise Ulster in the company of the first earl in July 1573, Rich participated in his son Robert's equally disastrous campaign in 1599. Irish historians value Rich as a chronicler of Dublin life under James, exemplified in his best-known treatise on Irish affairs, *A new description of Ireland* (1610). Rich's prose dialogue, 'The Anothomy of Irelande' (1615), whose form echoes Spenser's *View* and whose subtitle, 'truly dyscoverynge the state of the cuntry', mimics Davies' *Discovery*, leaves no doubt as to the cause of Ireland's ills: 'they say that a dyscease once knowne. is halfe cured. but I thynke the dysceases of *Irelande* are not knowne as they shuld be, they are but superfycyally lookt into: and for thys *Canker*. that is crept in by popery, it hath so spred it self. through Cytty towne & cuntry, that it hath left hys Ma^ti allmost neuer a sounde subiecte'.⁶⁷ In 1621, Robert Burton, in his *Anatomy of Melancholy*, provided a more subtle use of medical metaphors in colonial discourse. Burton's *Anatomy* is rarely read as part of the corpus of texts on Jacobean Ireland, yet it contains several revealing references to that country.⁶⁸ Burton had read Davies, as he acknowledges in a key passage:

> This Island among the rest, our next neighbours the *French* and *Germanes* may be a sufficient witnesse, that in a short time by that prudent policy of the *Romanes* was brought from barbarisme; see but what *Caesar* reports of vs, and *Tacitus* of those old *Germanes*, they were once as vnciuill as they in *Virginia*, yet by planting of Colonies, & good lawes, they became from babarous outlawes, to be full of rich and populous cities, as now they are, & most florishing kingdomes; and so might *Virginia*, and those wild *Irish* haue beene ciuilised long since, if that order had beene heretofore taken, which now begins of planting Colonies &c. I haue read a Discourse printed A°. 1612. *Discouering the true causes, why Ireland was neuer intirely subdued or brought vnder obedience to the crowne of England, vntill the beginning of his Maiesties happie raigne.* But if his reasons were throughly scanned by a Iuditious Politition, I am afraid he would not altogether be approued, but that it would turn to the dishonour of our nation, to suffer it to lie so long wast.⁶⁹

Later, discoursing on discontents and cares, Burton declares: 'Our whole life is like an *Irish Sea*, wherein there is nought to be expected but tempestuous stormes, and troublesome waues, no *Halcyonian* times, wherein no man can hold himselfe secure, or agree with his present estate.'⁷⁰ The Irish denote desperation in Burton's *Anatomy*: 'many men are affected like *Irishme[n]* in this behalfe, that if they haue a good skimiter, had rather haue a blow on their arme, then their weapo[n] hurt, they had rather loose their liues, then their goods'.⁷¹ Burton says that 'all feares, griefes, suspitious, discontents are swallowed vp & drowned in this *Euripus*, this Irish sea, this Ocean of misery, as so many small brookes'.⁷² The Irish are wheeled out repeatedly as illustrations of extreme emotions: 'And many generous spirits, and graue staid wise men otherwise are so tender in this, that at the losse of a deare friend they will cry out, houle and roare, and teare their haire, lamenting many months after, houling as those *Irish* women and *Greekes* at their graues, & commit many vndecent actions, and almost goe besides themselues.'⁷³ We are told that 'An *Irish* Sea is not so turbulent & raging as a litigious wife',⁷⁴ that there's an affinity between 'poligamy of *Turkes*, or *Irish* deuorcement',⁷⁵ that 'the *Muscouites*, if they suspect their wiues, will beat them till they confesse, & if this will not auaile, like those wild *Irish*, be diuorced at their plesures, or else knock them on the heads'.[76]

So, reflecting on the representation of Jacobean Ireland, and specifically Ulster, is it a matter of 'much ado'? Was there, as some suggest, a moratorium or a conspiracy of silence?[77] Thomas Wilson wrote to James on 10 March 1619, ten years into the Ulster Plantation, reminding him about that earlier visit to the record office on his accession when he had been overwhelmed by the Irish archive.[78] How much had changed in the intervening years? Ireland had featured intermittently in the drama of the period. Irish characters, colloquialisms or costumes appear in the anonymous *Captain Thomas Stukeley* (c.1605), George Chapman's *Bussy D'Ambois* (1607), Thomas Dekker and John Webster's *Northward hoe* (1607), Thomas Middleton's *Puritan Widow* (1607), Dekker's *The Honest Whore, Part Two* (1608), Beaumont and Fletcher's *The Coxcomb* (c.1609), Nathan Field's *Amends for Ladies* (c.1611), Dekker's, *If This be not a Good Play, the Divell is in It* (1612), Webster's *The White Devil* (1612), John Cooke's *Greenes Tu quoque* (1614), Thomas Heywood's *The Four Prentices of London* (1615), Beaumont and Fletcher's *The scornful ladie* (1616), Middleton and William Rowley's *A Fair Quarrel* (1617), Middleton's *The Triumphs of Honour and Industry* (1617), Chapman's *Two wise men and all the rest fooles* (1619), Middleton and Rowley's *The Welsh Embassador* (c.1623), John Ford's *Perkin Warbeck* (1633), and Thomas Randolph's, *Hey for Honesty, Down with Knavery* (before 1635).[79]

Chapman's *Bussy D'Ambois* alludes to the ending of 'The nine yeeres warre'.[80] In Beaumont and Fletcher's *The scornful ladie* (1616), Welford says of Abigail, whose deliberately dropped glove he is expected to retrieve: 'This is the strangest pamperd peece of flesh towards fiftie, that euer frailty cop't withall, what a trim *Lenuoy* heere she has put vpon me: these woemen are a proud kinde of cattell, and loue this whorson doing so directly, that they wil not sticke to make their very skinnes Bawdes to their flesh. Here's dogskin and storax sufficient to kill a Hauke: what to do with it, beside nayling it vp amongst Irish heads of Teere, to shew the mightines of her palme, I know not: there she is, I must enter into Dialogue. Lady you haue lost your gloue.'[81] In Dekker's *The Honest Whore, Part Two*, written around 1607 but not published till 1630, a play laced with Irish references, Hipolito echoes Shakespeare's Richard II, saying of Irish footman Brian: 'that Irish Judas,/Bred in a country where no venom prospers,/But in the nation's blood, hath thus betray'd me'.[82] Irish immigration rather than Anglo-Scottish settlement of Ulster preoccupies Lodovico, who declares: '*England* they count a warme chimny corner, and there they swarme like Crickets to the creuice of a Brew-house...why Sir, there all Costermongers are Irishmen.'[83]

Many of these dramatic depictions masqueraded as entertainment rather than propaganda, but had the insidious effect of sustaining the Elizabethan image of Irish barbarity well into the seventeenth century. Still, despite this scattershot of scurrilous stereotypes on stage, non-fiction remained the prime site for discussions of plantation in James's reign. Promotional discourse on the Ulster plantation is a recurrent feature in Jacobean literature on a range of subjects. A marginal note to the 1611 edition of George Turberville's *The booke of falconrie*, first published in 1575, reads: 'But truely there is no Goshawke more excellent then that which is bread in Ireland in the north parts, as in *Vlster*, and in the Country of *Tyrone*.'[84]

A section of Ranulf Higden's *Polychronicon*, published as *The descrypcyon of Englonde* (1498), declared: 'Ther is a lake in Vlster & moche fysshe therin.'[85] England had been angling for Ulster for a very long time. Discussing Irish rivers in *Mikrokosmos* (1625), Peter Heylyn linked Ulster and Virginia through Spenser's river schemata in Book IV of *The Faerie Queene*: 'The chiefe riuers are 1 Shennin or Sinei, which beginning in *Vlster*, runneth the course of 200 miles, to the Verginian Sea, & is nauigable 60 miles. 2 The Slane. 3 Awiduff, called by the English, Blackwater. 4 Showre. These and the other riuers of principall note, take along with you, according as I find them registred by that excellent Poet M. Spencer, in his Canto of the mariage of Thames and Medwaie'.[86] For Thomas Gainsford, who acquired land in Ulster around 1610–14, authored *The True Exemplary, and Remarkable History of the Earle of Tirone* (1619), and whose *True and Wonderfull History of Perkin Warbeck* (1619) informed John Ford's 1633 play, in which Perkin's base is Ireland,[87] Ulster was one of the glories of England, a catch fit for a king: 'The Prouince of VLSTER, and called the *North* is very large, and withall mountanous, full of great Loughs of fresh water, except Lough Cone, which ebbeth and floweth, as the Sea shouldreth aside the streites at *Strangford*, and with that violence at the ebb, that a ship vnder saile with a reasonable gale of winde cannot enter against the tide. These lakes nature hath appointed in steed of riuers, and stored with fish, especially Trowt, and Pike, of such strange proportion, that if I should tell you of a Trowt taken vp in *Tyrone* 46 inches long, and presented to the *L: Montioy*, then Deputie: you would demand, whether I was *oculatus testis*, and I answer, I eat my part of it, and as I take it both my *L: Dauers*, and *Sir William Goodolphin* were at the table, and worthy *Sir Iosias Bodley* hath the portraiture depicted *in plano*.'[88] Ulster's reputation as a prime site for anglers was well established, and as a colony it was quite a catch.

Of course, plantation remained economically driven, whether or not its leading figures were merchants, adventurers, or both. Gerard Malynes, in *The ancient law-merchant* (1622), picking up on the 'unfinished business' rhetoric of Sir John Davies in his *Discovery*, wrote:

> And here I remember a good obseruation heretofore made touching the kingdome of Ireland, Why the same was not brought vnto perfect obedience to

their soueraigne these 400 yeares, but vnder our most gratious king *Iames*? which is attributed to the mistaking of the place of the plantation of the first aduenturers, that were deceiued in their choice; for they sate downe and erected their castles and habitations in the plaines and open countries, where they found most fruitfull and profitable lands, & turned the Irish into the woods and mountaines, which, as they were proper places for Outlawes and Theeues, so were they their naturall castles and fortifications: thither they draue their preys and stealths; they lurked there, & waited to do euill and mischiefe; for these places they kept vnknowne, by making the waies and entries thereunto impassible; there they kept their cattle, liuing by the milke of the cow, without husbandrie or tillage; there they increased and multiplyed vnto infinit numbers by promiscuous generation among themselues; there they made their assemblies and conspiracies without discouerie; but they discouered the weaknesse of the English dwelling in the open plaines, and thereupon made their fallies and retraits with great aduantage. Whereas on the other side, if the English had builded their castles and townes in those places of fastnes, & had driuen the Irish into the plaines and open countries, where they might haue had an eye and obseruation vpon them, the Irish had beene easily kept in order, and in short time reclaimed from their wildnesse, and would haue vsed tillage, and by dwelling together in towneships learned mechanicall Arts and Sciences. This discourse may seeme strange to the Law of Merchants: but when Merchants vndertake Plantations (as we see they do) no man will hold the same to be impertinent.[89]

In *A View of the State of Ireland*, Spenser had mocked Richard Stanyhurst's conjecture that the Irish descended from the Egyptians on the grounds that an Irish battlecry and Ulster name, 'Ferragh', sounded like 'Pharaoh', and for suggesting that the ancient Irish, or Scots, derived from *scotos* ('darkness').[90] But the impression of a dark chapter in Anglo-Scottish history, a thing of darkness acknowledged by a new state prospering through plantation persists. The obscurity of Ulster in Jacobean literature – an obscurity that can be exaggerated – is bound up with the shift from England to Britain and from the history play to a more domestic drama. Although an Irish context has been claimed for *The Tempest* (1611), it is another Shakespeare play that surfaces in Ulster.[91]

Surviving evidence would suggest that English theatre companies visited the south of Ireland in the early seventeenth century, including playwrights William Fennor and Robert Daborne.[92] Contemporary records also point to some Ulster productions. Alan Fletcher indicates an 'aborted performance of *Much Ado About Nothing*', set for some time before 28 May 1628 at Coleraine in County Londonderry, which he conjectures is 'probably Shakespeare's play of that name', citing 'A coppie of a certificate touching abuses offered to his Maiesties Commissioners', wherein 'Edward Harfleite of Coleraine told me that they had provided themselues to haue entertained ye Commissioners at their coming to Coleraine with a play, the title [of] it was. Much adoe about nothing which play he purposed to haue Plaid but that they heard ye Commissioners... Tooke a Song that was sung (they are come from seeing ye buildings soe much to hart... by them that they durst not play there play for feare of offending the Commissioners.'[93] That a proposed production of Shakespeare's comedy in the newly planted province of Ulster was barred is intriguing, suggestive of a vexed site of censorship and secrecy, prompting subtle allegorical depiction as well as provocative political representation. Fletcher finds it 'hard to see how any of the songs in *Much Ado* might have caused... offence, unless it be Balthasar's song on the faithlessness of men (*Much Ado*, Act II, scene III). But if so, the commissioners must have been touchy indeed. Perhaps more likely they had been the butt of some lampoon'.[94] Fletcher assumes the dispute revolved around the commissioners' role and colonists' rights. He guesses the settlers were staging the play, and 'it is probable that the song to which the commissioners took exception was some sort of jibe, however mild, aimed at them by persons unknown'.[95]

Perhaps the play caught the conscience of the commissioners? Balthasar's song could credibly offend crown agents come to police a plantation:

Men were deceivers ever,
One foot in sea, and one on shore
To one thing constant never. (2.3.57–9)

Fletcher observes that in choosing *Much Ado* 'the planters... showed themselves aware of what was currently popular in London theatrical circles, and thus likely to please their guests'.[96] Or perhaps displease them. They may also have shown themselves aware of the political uses of drama and the possibilities of oblique critique.[97] Either way, despite the amnesia of empire and a tendency to forget its past, as Mark Netzloff observes, 'England quite literally had "more to do" with Ireland than with any other site, and in Ulster, still does.'[98]

NOTES

I would like to thank Andrew Hadfield and John Kerrigan for helpful comments on an earlier draft of this essay.

1 Michael Neill, 'Broken English and broken Irish: nation, language, and the optic of power in Shakespeare's Histories', *Shakespeare Quarterly*, 45 (1994), 11.
2 Andrew Murphy, 'Shakespeare's Irish history', *Literature & History*, 5 (1996), 38.
3 Andrew Hadfield, '"Hitherto she ne're could fancy him": Shakespeare's "British" plays and the exclusion of Ireland', in Mark Thornton Burnett and Ramona Wray (eds), *Shakespeare and Ireland: History, Politics, Culture* (London, 1997), pp. 46–67.

4 Tristan Marshall, 'The Tempest and the British Imperium in 1611', *Historical Journal*, 41 (1998), 378.
5 Joel B. Altman, '"Vile Participation": the amplification of violence in the theatre of *Henry V*', *Shakespeare Quarterly*, 42 (1991), 12.
6 Altman, '"Vile Participation"', 15.
7 J. H. Andrews, 'Appendix: the beginnings of the surveying profession in Ireland – Abstract', in Sarah Tyacke (ed.), *English Map-Making 1500–1650: Historical Essays* (London, 1983), p. 20.
8 I say 'the licensed page' because Andrew Hadfield and others have argued that censorship determined what was published on Ireland and explains why treatises like Edmund Spenser's prose dialogue were published well after their date of composition. In Spenser's case *A View of the Present State of Ireland*, written in 1596, was entered in the stationer's Register in 1598 but not did not appear in print until 1633. Manuscript copies did circulate, and there was clearly a considerable archive of state papers on Ireland, as Thomas Wilson's anecdote indicates. See Andrew Hadfield, 'Another case of censorship? The riddle of Edmund Spenser's *A View of the Present State of Ireland* (c.1596)', *History Ireland*, 4:2 (1996), 26–30. My focus in this essay is on printed material.
9 David Armitage, 'Literature and empire', in Nicholas Canny and Alaine Low (eds), *The Oxford History of the British Empire, Volume I: The Origins of Empire: British Overseas Enterprise to the Close of the Seventeenth Century* (Oxford, 1998), p. 111.
10 Mark Netzloff, 'Forgetting the Ulster plantation: John Speed's *The Theatre of the Empire of Great Britain* (1611) and the colonial archive', *Journal of Medieval and Early Modern Studies*, 31:2 (2001), 314–15.
11 J. Michael Hill, 'The origins of the Scottish plantations in Ulster to 1625: a reinterpretation', *Journal of British Studies*, 32 (1993), 37.
12 Cited in George Hill, *An Historical Account of the MacDonnells of Antrim* (Belfast, 1873), pp. 57–58; Howard Mumford Jones, 'Origins of the Colonial Idea in England', *Proceedings of the American Philosophical Society*, 85 (1942), 464.
13 William Camden, *Remaines of a greater worke, concerning Britaine, the inhabitants thereof, their languages, names, surnames, empreses, wise speeches, poësies, and epitaphes* (London, 1605), p. 198.
14 Sarah Irving, '"In a pure soil": colonial anxieties in the work of Francis Bacon', *History of European Ideas*, 32 (2006), 253, n. 21.
15 Irving, '"In a pure soil"', 251.
16 Francis Bacon, *The essays, or councils, civil and moral, of Sir Francis Bacon* (London, 1696), p. 92.
17 Denise Albanese, 'The *New Atlantis* and the uses of Utopia', *English Literary History*, 57 (1990), 503.
18 Albanese, 'The *New Atlantis* and the uses of Utopia', p. 504.
19 Anne Fogarty, '"This Inconstant Sea-Nimph": history and the limitations of knowledge in John Davies' writings about Ireland', in Timothy P. Foley, Lionel Pilkington, Sean Ryder, and Elizabeth Tilley (eds), *Gender and Colonialism* (Galway, 1995), pp. 23–34.

20 Arundell del Re, 'References to Florio in the Irish state papers', *Review of English Studies*, 12 (1936), 194–7.
21 E. K. Chambers, 'John Donne, diplomatist and soldier', *Modern Language Review*, 5 (1910), 492.
22 Andrew Murphy, 'Gold lace and a frozen snake: Donne, Wotton and the Nine Years War', *Irish Studies Review*, 2 (1994), 9–11; Ted-Larry Pebworth and Claude J. Summers, '"Thus Friends Absent Speake": the exchange of verse letters between John Donne and Henry Wotton', *Modern Philology*, 81 (1984), 365–6, 372.
23 Tom Cain, 'John Donne and the ideology of colonization', *English Literary Renaissance*, 31 (2001), 443, n. 5; P. Thomson, 'John Donne and the Countess of Bedford', *Modern Language Review*, 44 (1949), 330.
24 Hans S. Pawlisch, *Sir John Davies and the Conquest of Ireland: A Study in Legal Imperialism* (Cambridge, 1985), p. 33.
25 John Donne, *Sermon vpon The Eight Ver[s]e of the First Chapter of the Acts of The Apostles. Preached To the Honourable Company of the Virginian Plantation, 13. Nouemb. 1622* (London, 1624), pp. 21–3, cited in Howard Mumford Jones, 'The colonial impulse: An analysis of the "Promotion" literature of colonization', *Proceedings of the American Philosophical Society*, 90 (1946), 147.
26 'Henrico Wottoni in Hibernia belligeranti', in C. A. Patrides, *John Donne: The Complete English Poems* (1985; repr. London, 1994), p. 321.
27 Patrides, *John Donne*, p. 186.
28 Cain, 'John Donne and the ideology of colonization', passim.
29 Patrides, *John Donne*, p. 443.
30 Francis Bacon, *Resuscitatio, or, Bringing into publick light severall pieces of the works, civil, historical, philosophical, & theological, hitherto sleeping, of the Right Honourable Francis Bacon, Baron of Verulam, Viscount Saint Alban according to the best corrected coppies: together with His Lordships life* (London, 1657), p. 22.
31 Ben Jonson, *The workes of Beniamin Ionson* (London, 1616), p. 1001.
32 Sir John Davies, *A Discovery of the True Causes why Ireland was never entirely Subdued, nor brought under Obedience of the Crowne of ENGLAND, untill the Beginning of his Majesties happie Raigne* (London, 1612).
33 Margaret Rose Jaster, 'Staging a stereotype in Gaelic garb: Ben Jonson's *Irish Masque*, 1613', *New Hibernia Review*, 2 (1998), 96–7.
34 David Lindley, 'Embarrassing Ben: the masques for Frances Howard', *English Literary Renaissance*, 16 (1986), 353.
35 James M. Smith, 'Effaced history: facing the colonial contexts of Ben Jonson's *Irish Masque at Court*', *English Literary History*, 65 (1998), 297–8.
36 Ibid., 297.
37 Ibid., 310.
38 Ibid., 312.
39 Bacon, *Resuscitatio*, p. 260.
40 John Selden, *Titles of honor* (London, 1614), p. 356.
41 Ben Jonson, *Epicoene, or the silent woman A comedie. Acted in the yeare 1609. By the children of her majesties revels* (London, 1620), sig. C3.
42 Rebecca Ann Bach, 'Ben Jonson's "Civil Savages"', *Studies in English Literature*, 37:2 (1997), 284.

43 Bach, 'Ben Jonson's "Civil Savages"', 285.
44 Cited in Rebecca Ann Bach, '"Ty Good Shubshects": the Jacobean masque as colonial discourse', in Leeds Barroll (ed.), *Medieval and Renaissance Drama in England* (London: Associated University Presses, 1995), p. 206. Bach points out that: 'Many of the prominent dancers in James's masques were veterans of Essex's disastrous (for the English) 1599 campaign in Ireland' (p. 206).
45 Anna Suranyi, 'Virile Turks and maiden Ireland: gender and national identity in early modern English travel literature', *Gender & History*, 21 (2009), 248-9.
46 Sarah E. McKibben, 'Speaking the unspeakable: male humiliation and female national allegory after Kinsale', *Éire-Ireland*, 43 (2008), 23.
47 Ciaran Brady, 'The killing of Shane O'Neill: some new evidence', *Irish Sword*, 15 (1982-3), 116-23.
48 Bernhard Klein, 'The lie of the land: English surveyors, Irish rebels and *The Faerie Queene*', *Irish University Review*, 26 (1996), 207.
49 D. B. Quinn, 'Sir Thomas Smith (1513-1577) and the beginnings of English colonial theory', *Proceedings of the American Philosophical Society*, 89 (1945), 544.
50 J. H. Andrews, 'The Irish surveys of Robert Lythe', *Imago Mundi*, 19 (1965), 22-31.
51 See Hiram Morgan, 'The colonial venture of Sir Thomas Smith in Ulster, 1571-1575', *Historical Journal*, 28 (1985), 261-78.
52 Sir Philip Sidney, *Astrophil and Stella*, in William E. Ringler, Jr (ed.), *The Poems of Sir Philip Sidney* (Oxford, 1962), p. 180, sonnet 30: 9-10.
53 Davies, *A discouerie of the true causes why Ireland was neuer entirely subdued*, p. 159.
54 D. M. Waterman, 'Sir John Davies and his Ulster Buildings: Castlederg and Castle Curlews, Co. Tyrone', *Ulster Journal of Archaeology*, 3rd ser., 23 (1960), 89.
55 J. H. Andrews, 'The maps of the escheated counties of Ulster, 1609-10', *Proceedings of the Royal Irish Academy*, 74 (1974), 140-1.
56 John Davies, 'Plantation of Ulster: original letter from Sir John Davys, giving an account of the Inquisition on the forfeited lands in the County of Colerain, A.D. 1609', *Ulster Journal of Archaeology*, 1st ser., 4 (1856), 194-5.
57 Netzloff, 'Forgetting the Ulster plantation', 321.
58 Ibid.
59 Ibid., 317-19.
60 David Scott Wilson-Okamura, 'Virgilian models of colonization in Shakespeare's *Tempest*', *ELH*, 70 (2003), 716-18.
61 D. B. Quinn, 'Renaissance influences in English colonization', *Transactions of the Royal Historical Society*, 5th ser., 26 (1976), 73-93, at 80.
62 Quinn, 'Renaissance influences in English colonization', 88.
63 William Lithgow, *The totall discourse, of the rare adventures, and painefull peregrinations of long nineteene yeares travailes from Scotland, to the most famous kingdomes in Europe, Asia, and Affrica* (London, 1640), p. 432. On Lithgow's Irish sojourn see Clifford Edmund Bosworth, *An Intrepid Scot: William Lithgow of Lanark's Travels in the Ottoman Lands, North Africa and Central Europe, 1609-21* (Aldershot, 2006), pp. 151-6.
64 Lithgow, *The totall discourse*, pp. 434-5.

65 Ibid., pp. 428-9.
66 Ibid., p. 436.
67 Edward M. Hinton, 'Rych's *Anothomy of Ireland* [1615], with an account of the author', *PMLA*, 55 (1940), 84.
68 Edward Snyder cites two of these, but not the most interesting ones. See Snyder, 'The Wild Irish', p. 699.
69 Burton, *Anatomy of Melancholy*, p. 51.
70 Ibid., p. 145.
71 Ibid., p. 213.
72 Ibid., p. 274.
73 Ibid., p. 412.
74 Ibid., p. 647.
75 Ibid., p. 648.
76 Ibid., p. 686.
77 According to Mark Netzloff, 'The withholding of cartographical information regarding the Ulster plantation reveals the important commercial advantages sought by both state officials and private investors'. Netzloff, 'Forgetting the Ulster plantation', 327.
78 Ibid., 313-14.
79 J. O. Bartley, 'The development of a stock character, I: the stage Irishman to 1800', *Modern Language Review*, 37 (1942), 440-1; H. Macaulay FitzGibbon, 'Ireland and the Irish in the Elizabethan Drama', *Irish Monthly*, 56:665 (1928), 593.
80 George Chapman, *Bussy D'Ambois* (London, 1607), p. 15.
81 Francis Beaumont and John Fletcher, *The scornful ladie* (London, 1616), sig. E2.
82 Thomas Dekker, *The second part of The honest whore* (London, 1630), sig. E4.
83 Dekker, *The second part of The honest whore*, sig. A2v.
84 George Turberville, *The booke of falconrie* (London, 1611), p. 60. Turberville had served in Ireland, but this marginal note is absent from the original publication.
85 Ranulf Higden, *The descrypcyon of Englonde Here foloweth a lytell treatyse the whiche treateth of the descrypcyon of this londe which of olde tyme was named Albyon and after Brytayne and now is called Englonde and speketh of the noblesse and worthynesse of the same* (Westminster, 1498), sig. E1.
86 Peter Heylyn, *Mikrokosmos A little description of the great world. Augmented and reuised* (Oxford, 1625), pp. 516-17.
87 John J. O'Connor, 'A lost play of Perkin Warbeck', *Modern Language Notes*, 70 (1955), 567.
88 Thomas Gainsford, *The glory of England* (London, 1618), p. 147.
89 Gerard Malynes, *Consuetudo, vel lex mercatoria, or The ancient law-merchant* (London, 1622), pp. 235-6.
90 Edmund Spenser, *A View of the State of Ireland*, ed. Andrew Hadfield and Willy Maley (Oxford, 1997), pp. 60-1.
91 See David J. Baker, 'Where is Ireland in *The Tempest*?', in Mark Thornton Burnett and Ramona Wray (eds), *Shakespeare and Ireland: History, Politics, Culture*, pp. 68-88; Dympna Callaghan, 'Irish Memories in *The Tempest*', in *Shakespeare Without Women: Representing Gender and Race on the Renaissance Stage* (London and New York, 2001), pp. 97-138.

92 W. H. Grattan Flood, 'Fennor and Daborne in Youghal in 1618', *Modern Language Review*, 20 (1925), 321–2. Daborne settled in Ireland, becoming dean of Lismore. The prologue of one of Daborne's plays touches on 'the Irish Shore'. See *A Christian turn'd Turke: or, The tragicall liues and deaths of the two famous pyrates, Ward and Dansiker* (London, 1612), sig. A. Fennor, the same year, published a prose pamphlet dedicated to Prince Henry that treated the Irish with familiar contempt: 'There are of all Nations that professe this needy life, but of all, there are most *Irish*, and I can not blame them, for by nature they are borne to be lowsy, straglers and stirrers vp of rebellion; And since it hath pleased God of his mercy to plant his peaceable word, and by the meanes of his Seruant, our most gratious Soueraigne Lord, King *Iames* (whome God long preserue) to make a ciuill Countrey, of the most barbarous parts in *Ireland*: There are many that haue harkened to ye Diuels whispering, & sworne their true seruice to his adopted Sonne, the Romish Priest'. William Fennor, *Pluto his trauailes, or, The Diuels pilgrimage to the Colledge of Iesuites. Lately discouered by an English gentleman* (London, 1612), p. 10.

93 Alan J. Fletcher, *Drama, Performance, and Polity in Pre-Cromwellian Ireland* (Cork, 2000), pp. 238–9.

94 Ibid., p. 430, n. 159.

95 Ibid., pp. 239–40.

96 Ibid., p. 431, n. 160.

97 Jean E. Howard, 'Renaissance antitheatricality and the politics of gender and rank in *Much Ado About Nothing*', in Jean E. Howard and Marion F. O'Connor (eds), *Shakespeare Reproduced: The Text in History and Ideology* (London, 1987), pp. 163–87.

98 Netzloff, 'Forgetting the Ulster plantation', 313–14.

13

'The Scottish inhabitants of that Province are actually revolted': John Milton on the failure of the Ulster plantation

NICHOLAS MCDOWELL

On 28 March 1649, only two weeks after John Milton (1608–74) had been employed as Secretary for Foreign Tongues by the newly formed Council of State of the English republic, he was instructed to 'make some observations upon the Complicacion of interest which is now amongst the severall designers against the peace of this Commonwealth, and they to be made ready to be printed with the papers out of Ireland, which the House has ordered to be printed'.[1] The 'Complicacion of interest' refers to the 'Articles of Peace' signed on 17 January 1649 between the Catholic Confederate Association, made up of Gaelic Irish and 'Old English' settlers, and Charles I's lord lieutenant in Ireland, James Butler, Marquis of Ormond (1610–88). Ormond's army was joined in the opening months of 1649 by Cavaliers fleeing England after defeat in the second civil war and the execution of Charles I on 30 January. A further 'complication' was the horror of the mainly Scottish Presbyterian settlers in Ulster at the regicide and their antagonism to an English republican regime dominated by Independents, generally more tolerant of sectarianism and opposed to a compulsory Presbyterian church government in England. By March–April 1649 Ireland had become the site of what looked to the republic's Council of State like an unholy alliance of Irish Catholics, Scottish Presbyterians and English Cavaliers. This alliance threatened to scupper the Council's urgent efforts to establish the authority and stability of the new kingless state by plunging it into a fresh 'British' conflict.[2]

On or around 16 May a tract appeared, published anonymously but with the imprint 'By Authority', entitled *Articles of Peace Made and Concluded with the Irish Rebels, and Papists, by James, Earle of Ormond, For and in behalfe of the late King, and by vertue of his Autoritie. Also a Letter sent by Ormond to Col. Jones, Governour of Dublin, with his Answer thereunto. And A Representation of the Scotch Presbytery at Belfast in Ireland.* Upon which *all are added Observations*.³ Milton's *Observations* is a piece of official propaganda, in other words, and its appearance anticipated the Cromwellian conquest of Ireland in August–September 1649. On 23 March 1649 Cromwell had used the same language of interest and multiple threat as found in the instructions to Milton, telling his fellow Army officers: 'I had rather be overrun with a cavalierish interest than a Scotch interest; I had rather be overrun with a Scotch interest than an Irish interest; and I think of all this is the most dangerous. If they [the Irish] shall be able to carry on their work they will make this [the English] the most miserable people in the earth, for all the world knows their barbarism.'⁴ Milton agreed with Cromwell on the barbarity of the Irish if not on their world-wide notoriety: in the *Observations* he cites Irish farming practices as evidence of 'a disposition not only sottish but inducible and averse from all Civility and amendment, and what hopes they give for the future, who rejecting the ingenuity of all other Nations to improve and waxe more civill by a civilizing Conquest, though all these many yeares better taught and shown, preferre their own absurd and savage Customes before the most convincing evidence of reason and demonstration: a testimony of their true Barbarisme'. The best-known of the various insults in the tract, however, is probably the dismissal of Belfast as 'a barbarous nook of *Ireland*', 'whose obscurity till now never came to our hearing' – a comment which looks ironic only in the light of the unfortunate celebrity of the city since the 1960s.⁵

It is not quite a joke to say that, for all these crude and conventional attitudes, some of Milton's best friends were from Ireland: when Katherine Jones (née Boyle), Viscountess Ranelagh (1615–91), whose nephew and son Milton tutored in the late 1640s, had to return to her homeland in 1656 he expressed his sorrow since 'to me…she has stood in the place of all relations'.⁶ The impressive Lady Ranelagh, who apparently knew German and Hebrew and who extended some patronage to educational and reforming projects such as the proposed natural history of Ireland by Robert Wood (1621/22–85), might have given him a more nuanced picture of the Irish than we find in the *Observations*.⁷ But, even taking account of the work's polemical purpose, there is no indication of any more subtle understanding of the land or its people. For all his advanced thinking about divorce, the free circulation of information and political rights, Milton subscribed to a conventional Aristotelian taxonomy of sexual and racial difference, according to which women are not only naturally inferior and subject to men, as slaves are subject to their masters, but certain races such as the Irish (and Indians and Turks) are naturally inferior and subject to those who are possessed, or at least potentially possessed in the case of the English, of a higher form of rationality.⁸ As a result of the stereotypical characterisation of the Irish as barbarous and savage in the *Observations* and the associations of the pamphlet with the notorious Cromwellian campaign, Milton scholars, many of whom like to claim a proto-liberal position for their author on matters of religious toleration, tended until really quite recently to avoid the work. With a renewed taste, however, both for historical and political readings of Milton and for understanding the Civil Wars in terms of the problems of governing the three Stuart kingdoms, critical interest in the tract has risen considerably since the 1990s. As Joad Raymond archly puts it: '[Milton's] first duty as a state servant was to pen a tract attacking the factions struggling in Ireland, a tract to which scant attention has been paid, perhaps because of the political offence it gives to the image of John Milton as a liberal thinker. It has recently undergone, however, a revival of interest, perhaps because of the political offence it gives to the image of John Milton as a liberal thinker.'[9] The recent enthusiasm to enlist Milton as the ideologist of Cromwellian conquest can sometimes race ahead of the facts. A substantial recent collection of essays on the subject of Milton and toleration twice describes the *Observations* as 'defending Cromwell's re-conquest of the rebellious Irish in 1649' when Cromwell of course did not set sail for another three months.[10] Nonetheless, as Gordon Campbell and Thomas N. Corns comment in their recent biography, with an eye on twenty-first-century geopolitics: 'Milton produced a tendentious dossier designed to launch and excuse a dubious war of aggression. He would not be the last public servant to do so; though he may, perhaps, have been the first.'[11]

In fact most of the tract published in May 1649 and which we call the *Observations* is not made up of Milton's words at all. Forty-four of the sixty-five quarto pages consist of the four texts to which Milton responded, the most substantial of which are the *Articles of Peace* and the *Necessary Representation*, a condemnation of the regicide and of the alleged tolerationist policies of the new republic issued on 15 February by the Belfast Presbytery (and which survives in full only as printed in the *Observations*).[12] Whereas Cromwell in his speech to his fellow officers placed the royalist, Scottish and Irish interests in a hierarchy of degeneracy with the Irish coming out on top, recent scholarship has rather seen Milton as more concerned in his Irish tract about the threat to the security and liberty of the republic from the Presbyterians and the Scots than the Irish Catholic alliance with the Cavaliers. This point seems to be born out by the form of the *Observations*. While Milton devoted four pages to answering the thirty-three pages of the *Articles of Peace*, he spent eleven pages refuting the four pages of the *Representation* by the Belfast Presbyterians. Willy Maley puts it bluntly: 'Milton's greatest concern in the *Observations* is not with Ireland at all, but with Scotland.'[13] And John Kerrigan has recently suggested that Milton exploited the 'overdetermined position of Ulster, as a seat both of the 1641 rebellion by Irish Catholics and of Scottish encroachment (yet again) into a kingdom under English sovereignty, to shift the political agenda from a country (Ireland) that he regarded as merely troublesome towards one that in his view posed a threat to the revolution. He secured this by blurring the distinction that has so often been made between the Scots and the 'mere' Irish in the province'.[14]

This final point is exemplified when Milton heard news, even while writing, that 'the *Scottish* inhabitants of that Province are actually revolted, and have not only besieged in *London-Derry* those forces which were to have fought against *Ormond*, and the Irish Rebels; but have in a manner declar'd with them, and begun op'n war against the Parliament; and all this by the incitement and illusions of that unchristian Synagogue at *Belfast*' (*CPW*, iii. 322). The fortified city of Derry in the west of Ulster, which had been taken by the Parliamentarian Sir Charles Coote in late 1648, was in May 1649 under siege by the uneasy and *ad hoc* alliance of confederate, royalist and covenanter forces. The Parliamentary commander in Ulster, Colonel George Monck, who six months earlier had taken Belfast and Carrickfergus in the east of Ulster, had been forced to retreat south to Dundalk. With that word 'synagogue' Milton managed not only to collapse Irish Catholics with Scottish Presbyterians, but both of them with Jews as well – his point was to emphasise how Presbyterianism was as foreign to true English values as Catholicism or Judaism. Ironically, propagandists for the Jacobean Church of England also adopted Milton's polemical strategy of confusing Jewish legalism and Presbyterian preciseness.[15] But the reference to Presbyterian 'judaism' also suggests that we should not place overwhelming emphasis on the importance of particular ethnic or national difference, of Irishness and Scottishness, in the argument of the tract. As Thomas Corns observes, for Milton the 'aversion to Catholicism or Presbyterianism overrides ethnic considerations'.[16]

The Belfast Presbytery, largely composed of Scottish planter stock, became a proxy in the *Observations* not just for the Scottish threat to the English republic's borders but of the Presbyterian clericalism consistently excoriated by Milton in both prose and verse since the mid-1640s, when Presbyterian ministers in the Westminster Assembly had charged his arguments for divorce on the grounds of incompatibility with heresy. In 1644 Presbyterian outrage at Milton's views led to his divorce tracts being denounced in a sermon before Parliament and to Milton being called to explain himself before the House of Lords. The first reaction to *The Tenure of Kings and Magistrates*, Milton's apology for the trial and execution of the king published in February 1649, was from a Presbyterian MP, Clement Walker (*d*.1651), who brushed Milton aside as 'a Libertine, that thinketh his Wife a Manacle...one that (after the Independent fashion) will be tied by no obligation to God or man'. It is probably no coincidence that within weeks Walker was arrested and Milton himself seized Walker's papers to prepare a case of treason against him.[17] Milton's personal and political animus against the Presbyterians was the driving force of his writing between 1644 and 1649. 'New Presbyter is but old

Priest writ large', Milton declared in the final line of his devastating 1646 poem 'On the New Forcers of Conscience under the Long Parliament'. The Presbyterian will to power and to tyranny over the conscience, manifested most acutely for Milton in their reaction to his ideas about divorce, had shown them to be another clerical manifestation of the popish spirit of persecution.[18] The conclusion of the poem was echoed in the *Observations* in the attack on Presbyterianism, exemplified by the Belfast clergy, as 'aspiring to be a compulsive power upon all without exception...to punish and amerce by any corporall infliction those whose consciences cannot be edifi'd...we hold it no more to be *the hedg and bulwark of Religion*, then the Popish and Prelaticall Courts, or the *Spanish Inquisition*' (*CPW*, iii. 326). But this Presbyterian threat to true English liberty was also inside England, and inside parliament, as well as outside in Scotland and Belfast, as had been starkly demonstrated by the need for the New Model Army to purge parliament of Presbyterian MPs at the end of 1648 in order to pave the way for the trial of the king.

Indeed, while writing the *Observations*, Milton was also at work on his *History of Britain*, although this would not be published until 1670; and while dating the composition of the six-book *History* remains a matter of educated conjecture, the consensus in recent years is that Milton composed most of the first four books, and probably also the 'Digression' to Book III (which was not published until 1681 as the *Character of the Long Parliament*) in the months after the regicide.[19] Book III deals with the condition of Britain after the Romans departed, a period during which the British showed themselves, in Milton's opinion, to be unable to govern themselves and take their chance of liberty, resulting in their fatal decision to try to stave off the incursions of the Danes and the Picts by inviting in the pagan Saxons. The 'Digression' explicitly compares the 'confused anarchy' after the Roman withdrawal with the behaviour of the Presbyterian Long Parliament after the victory over the king's forces in the civil wars, and contains long passages of vitriolic attack on the moral and spiritual degeneracy of the Presbyterian clergy during the 1640s: 'thir intents were cleere to be no other then to have set up a spir[i]tual tyrannie by a secular power to the advancing of thir owne authorit[ie] above the magist[r]ate' (*CPW*, v. 447). The analogy suggested that Presbyterian dominance would lead England into a condition of spiritual tyranny, paving the way for conquest by and enslavement to foreign, pagan powers – and for the Saxons in post-Roman Britain Milton read Roman Catholicism in post-war England. Milton repeatedly identified this cycle in Britain's history: the sloth and vice of the Saxons in turn forced them to succumb to the Danes, and the lack of godly virtue in English hearts prepared the way for the Norman conquest and the imposition of corrupt French values. Slavish behaviour, religious, political, and moral, on the part of a people leads to enslavement of their nation. This is Milton on the lapse of the once fearsome Saxons into vice:

> when God hath decreed servitude on a sinful Nation, fitted by thir own vices for no condition but servile, all Estates of Government are alike unable to avoid it. God had purpos'd to punish our instrumental punishers, though now Christians, by other Heathen, according to his Divine retaliation; invasion for invasion, spoil for spoil, destruction for destruction. The *Saxons* were now full as wicked as the *Britans* were at their arrival, brok'n with luxury and sloth, either secular or superstitious; for laying aside the exercise of Arms, and the study of all vertuous knowledge, some betook them to over-worldly or vitious practice, others to religious Idleness and Solitude, which brought forth nothing but vain and delusive visions; easily perceav'd such, by thir commanding of things, either not belonging to the Gospel, or utterly forbidden, Ceremonies, Reliques, Monasteries, Masses, Idols, add to these ostentation of Alms, got ofttimes by rapine and oppression, or intermixt with violent and lustfull deeds, sometimes prodigally bestow'd as the expiation of cruelty and bloodshed. What longer suffering could there be, when Religion it self grew so void of sincerity, and the greatest shews of purity were impur'd? (*CPW*, v. 259)

The Saxons turned in their degeneracy to the idolatrous devotional practices of Roman Catholicism: in the narrative pattern of Milton's *History*, as Andrew Hadfield observes, 'each [physical] invasion is preceded by a spiritual invasion of the false principles of Rome'.[20]

The representation of Ulster and the Ulster planters in the *Observations* can be placed into this wider context of Milton's theory of conquest and slavery in the *History of Britain* and the 'Digression', according to which slavish behaviour or a slavish disposition leads to conquest, and thus to literal enslavement. The Catholicism of the Irish unquestionably made them, for Milton, a barbarous and slavish race deserving of conquest.[21] But Milton's representation of the Belfast Presbyterians in the *Observations* also made very clear his opinion that the plantation of Ulster had failed: James I may have sought to pacify an unruly Ulster through the migration of Calvinist planters, but in Milton's view the planters, or rather the faction of Presbyterianism for which they stood in the *Observations*, presented as great a threat to the liberties of the English state as the Gaelic Irish. The Ulster Presbyters sought to dictate, Milton declared in the *Observations*, to the 'sovran Magistracy of England, by whose authoritie and whose right they inhabit there', just in the same way as Scottish ministers who, having been 'neighbourly admitted, not as the Saxons by merit of thir warfare against our enemies, but by the courtesie of *England* to hold possessions in our Province, a Countrey better than thir own, have, with worse faith than those Heathen, prov'd ingratefull and treacherous guests to thir best friends and entertainers' (*CPW*, iii. 333–4). Milton was concerned that Presbyterian tyrants would invade from Scotland

and Ireland to make a conquest of England ('a Countrey better than thir own'). The real anxiety, however, was that the English would invite tyranny upon themselves, as they had done repeatedly in their history, and that God would consequently punish them as an apostate people.

Here a connection can be made to Milton's disillusionment with the rapturous popular reception in England of the *Eikon Basilike* (February 1649), the book published within days of the regicide, which purportedly recorded the thoughts of Charles I in the weeks before his death. It went through 35 editions within the year. In *Eikonoklastes*, first published in October or November 1649, Milton lambasted the English people for displaying their 'besotted and degenerate baseness of spirit' in their devotion to the false, idolatrous pieties of the king's book, 'except some few, who yet retain in them the old English fortitude and love of Freedom, and have testifi'd it by thir matchless deeds, the rest, imbastardized from the ancient nobleness of thir Ancestors, are ready to fall flat and give adoration to the Image and Memory of this Man' (*CPW*, iii. 344). For Milton, only the republican 'rump' of MPs who put Charles Stuart to death preserved the natural English propensity to valour in the cause of liberty, which presumably he traces back to the original fighting spirit of the Saxons before they succumbed to vice and then bowed their heads to the Norman yoke. Milton turned again to the familiar anti-Catholic language of idolatry, as he did throughout *Eikonoklastes*: the English were turning Catholic in their idolatry of the dead king and his book, and in the cycle of British history the spiritual invasion of Rome precedes literal invasion and conquest:

> And how much their intent, who publish'd these overlate Apologies and Meditations of the dead king, drives to the same end of stirring up the people to bring him that honour, that affection, and by consequence, that revenge to his dead Corps, which hee himself living could never gain to his person, it appears both by the conceited portraiture before his Book, drawn out to the full measure of a Masking Scene, and set there to catch fools and silly gazers... for though the Picture sett in Front would Martyr him and Saint him to befool the people, yet the Latin motto in the end, which they understand not, leaves him, as it were a politic contriver to bring about that interest by faire and plausible words, which the force of Armes deny'd him. But quaint Emblems and devices begg'd from old Pageantry of some Twelf-nights entertainment at *Whitehall*, will doe but ill to make a Saint or Martyr: and if the People resolve to take him Sainted at the rate of such a Canonizing, I shall suspect thir Calender more than the *Gregorian*. (*CPW*, iii. 342–3)

Milton called upon the language of Puritan anti-theatricalism to identify the king's book, and in particular the famous frontispiece showing Charles in a pose of Davidic, penitential meditation, exchanging his earthly for a heavenly crown, as a form of popish idol. The *Eikon Basilike* was presented as a decorous textual statue of the king which, like a model of a saint or the Virgin Mary in Catholic practice, turned devotion into passive, 'silly' gazing on a meaningless visual sign, when it should instead take the form of active intellectual engagement with the Word of God in the scriptures. Milton's emphasis on the Latin motto continued the association with Catholicism but the reference to popular ignorance ('which they understand not') also introduced the disdain that resurfaced throughout *Eikonoklastes* for 'an inconstant, irrational, and Image-doting rabble' (*CPW*, iii. 601), whose appetite for the King's Book showed their reason to be enslaved to their passions, skilfully 'stirr[ed] up' within them by the Machiavellian clerics behind its publication.

There may be a revealing Shakespearean allusion in Milton's charge that the English were 'ready to fall flat and give adoration to the Image and Memory of this Man' – one that can lead us back in the direction of what Milton regarded as the failed Protestant plantation of Ulster. 'Fall flatt' is a common phrase in descriptions of idolatrous behaviour in the period: for instance John Foxe, attacking the survival of Catholicism in Scotland, declared that although 'God commaundeth not to worship any Grauen Images, the Scots fall flat to them, and offer them: Incense'.[22] But there are grounds to believe that Milton had in mind one of the most memorable episodes in *The Tempest* (1611), the Shakespeare play that influenced Milton's early poetry more than any other – particularly the *Masque Performed at Ludlow Castle*, or 'Comus', which contains a series of analogies and echoes, associated mostly with the tempter Comus and his demonic power.[23] When Caliban encounters Trinculo and Stephano, he initially presumes them to be minions of Prospero: 'Here comes a spirit of his, and to torment me/For bringing wood in slowly. I'll fall flat;/Perchance he will not mind me' (II. ii. 15–17).[24] While Caliban's posture here is one of self-protection it soon turns into one of self-debasement after Stephano pours wine into his mouth: 'I do adore thee...And I will kiss thy foot. I prithree, be my god...I'll swear myself thy subject' (ll. 138, 147, 150). The blasphemy of Caliban's adoration of these ridiculous false gods is encapsulated in Stephano's repeated injunctions that Caliban must 'kiss the book' (ll. 129, 140, 155). The drinking of the wine becomes a parodic, idolatrous version of swearing an oath on the Bible, with a parodic echo also of sacramental ceremony that Milton the religious anti-formalist might have particularly appreciated, if not necessarily for the reasons Shakespeare intended. In *Eikonoklastes* the 'besotted' English fall flat like savage Calibans before the image of the king in the *Eikon Basilike*, idolatrously kissing his book, which blasphemously masquerades as a form of scripture in its numerous analogies between the regicide and the Psalms and the Passion narratives. In Milton's *Masque* Comus is the off-spring of Bacchus and Circe, 'Whose charmed cup/ Whoever tasted, lost his upright shape,/And downward fell into a grovelling swine' (ll. 51–3); in *Eikonoklastes* Milton laments that 'so many sober

Englishmen... like men enchanted with the *Circean* cup of servitude, will not be held back from running their own heads into Yokes of Bondage' (*CPW*, iii. 488).

From his earliest major writings, the Circe episode in Homer, in which Odysseus's men are transformed into swine by drinking the wine offered by Circe, was for Milton a mythical lesson in the workings of idolatry, showing how idolatry originated in men's susceptibility to sensual ease and had an essential link to slavery.[25] The scene of Caliban's voluntary enslavement in *The Tempest* is a retelling of the Circe myth, except Caliban is already only half a man ('What have we here, a man or a fish?' (ll. 24–5)), and Stephano and Trinculo are pathetic parodies of the seductive Circe. And in *The Tempest* as in the *Observations* and *Eikonoklastes*, it is hard to tell whether the idolater's monstrosity ('A most poor credulous monster! (l. 144)) leads to his idolatry or vice versa. When Milton condemns an Irish 'disposition not onely sottish but inducible and averse from all Civility and amendment' he sounds rather like Prospero and Miranda damning Caliban as 'Abhorred slave,/which any print of goodness wilt not take' (I. ii. 352–3). Several critics have indeed argued persuasively that 'Ireland provides the richest historical analogue for [*The Tempest*'s] colonial theme'.[26] In the *Observations*, Milton ran together Scottish Presbyterians and English royalists with Irish Catholics as a monstrous and foreign threat to English Protestant values: Ormond led 'a mixt Rabble, part Papists, part Fugitives, and part Savages' (*CPW*, iii. 315). If the memory of *The Tempest* did indeed shape Milton's bitter attacks on the alacrity with which the English were willing to remain enslaved to Stuart tyranny, then he identified Caliban's brutish susceptibility to idolatry with an un-English 'baseness of spirit' among the English Presbyterians and royalists who adored the King's book that only a few months earlier he had linked in the *Observations* with the savage 'inducible' nature of the Irish.

Sir John Davies (1569–1626), in *A Discoverie of the True Causes why Ireland was never entirely Subdued* (1612), had called on the Homeric Circe myth to convey his disgust at how the 'old' English settlers who had come to Ireland seeking to 'make a perfect conquest of the Irish, were by them perfectly and absolutely conquered': having left behind 'the Ciuill and Honorable Lawes and Customes of *England*...they became degenerate and metamorphosed like *Nabuchadnezzar*: who although he had the face of a man, had the heart of a Beast; or like those who had drunke of *Circes* Cuppe, and were turned into very Beasts; and yet tooke such pleasure in their beastly manner of life, as they would not returne to their shape of men againe' (164, 182). In *Eikonoklastes* Milton accused Charles I of fomenting the Irish rebellion of 1641 in his efforts to obtain Irish Catholic troops, and thus of being directly responsible for the attacks on the Protestant population. Milton's ludicrously inflated figure of Protestant deaths in 1641 is '154000 in the Province of *Ulster* onely... which added to the other three, makes the total sum of that slaughter in all likelihood fowr times as great' (*CPW*, iii. 470). Milton's wider point is that in the idolatry of their worship of the king the Scottish Presbyterians and English Cavaliers are no more civil than, and just as un-Protestant as, the Catholic Irish. But the comparison with Davies on the degeneration of the 'old' English settlers also illuminates an important but subtle distinction between Milton's representation of the planter in Ireland and earlier representations such as those of Davies and before that Edmund Spenser. For Milton, the Belfast Presbyterians had become accomplices of the Catholic confederates, 'their Copartning Rebels in the South, driving on the same Interest to loose us that Kingdome that they may gaine it themselves, or at least share in the spoile: though the other be op'n enemies, these pretended Brethren' (*CPW*, iii. 317). But these 'pretended Brethren' had not been conquered and metamorphosed by Irish and popish vice in the manner of the old English settlers in Davies's account. Rather this was a natural turn of events because, after their initial support for the civil wars, the Presbyterians had begun to act like Catholics in their religious intolerance and persecutory will to power, manifested most personally for Milton in the controversy over his arguments for divorce. The planters had not been conquered by the Irish; rather in sliding back from true religion they turned themselves Irish-like in their adoption of the principles of Rome. They allowed themselves to be conquered by their own idolatrous passions, and this left them open to actual conquest by Catholic powers.

Milton did not cite Edmund Spenser's *A View of the Present State of Ireland* in the *Observations*, but we know from his commonplace book that he had read it closely, probably in the early 1640s.[27] In the *Observations* Milton maintained that the Irish 'by their endlesse treasons and revolts have deserv'd to hold no Parlament at all, but to be govern'd by Edicts and Garrisons, as absolute and supreme in that Assembly as the People of *England* in their own Land' (*CPW*, iii. 303). This echoed the suggestion in Spenser's dialogue that Ireland might best be controlled by a form of martial law imposed by four garrisons placed at strategic points across the island. Indeed there may be an allusion to the *View* at the end of the *Observations* which condemned the Ulster Presbyterians as tainted by the same Gaelic degeneracy as the native Irish: 'By thir actions we might rather judge them [the Belfast Presbyterians] to be a generation of High-land theevs and Red-shanks' recalls Spenser on how the 'O-Neales are neerlye allyed... to the Earl of Argile, from whom they use to have all theyr succours of those Scots and Reddshankes' (*CPW*, iii. 333 n. 96).

In *Eikonoklastes* Milton did not refer directly to Spenser but to *The Faerie Queene* (1590–1609), in a moment revealing of his frustration with the legal process of the trial of the king and the opportunity it gave Charles to adopt the pose of martyr, and also of his identification of the popular idolatry of the King's Book with the sort of religious and moral degeneracy stereotypically ascribed to the Irish: 'If there were a man of iron, such as *Talus*, by our Poet *Spencer*, is fain'd to be page of Justice, who with his iron flaile could doe all this, and expeditiously, without those deceitfull formes and circumstances of Law, worse than those ceremonies of Religion; I say God send it don, whether by one *Talus*, or by a thousand' (*CPW*, iii. 390). In Book V of *The Faerie Queene* Artegall, 'Champion of true justice', is helped by the ferociously iconoclastic 'yron man' Talus – 'Immoveable, resistlesse, without end./Who in his hand an yron flaile did hould,/With which he threst out falsehood, and did truth unfold' – to save the virgin Irena, a figure of Elizabeth's rule in Ireland, from the tyrant Grantorto. At one point Artegall has to restrain Talus from 'slaughter' of their defeated enemies but nonetheless in his determination to 'reforme that ragged commonweale' he sends Talus to search out those 'who did rebell gainst lawfull government;/On whome he did inflict most grievous punishment'.[28] Milton goes on to remind his readers how it was 'this iron flaile the People' that 'threw down' all 'those Papistical innovations' in the Church as well as tyrannous organs of government such as the Star Chamber (391). But the 'iron flaile' that had so recently purged Parliament and paved the way for the execution of the king was not 'the People' but the New Model Army. The approving reference to Talus evoked violent Miltonic fantasies of the annihilation of the idolatrous enemies of liberty, without regard to the traps and snares of the law. In *Eikonoklastes* those enemies included those Scottish and English Protestants who had lost their humanity in their adoration of the king's book and so become like the degenerate, 'inducible', slavish Irish.[29] The Spenserian reference is chilling in its suggestion that Milton, disillusioned with the reaction of English and Scottish Protestants to the regicide, contemplated mowing down, 'whether by one *Talus*, or by a thousand', those that resisted republican efforts to 'reforme that ragged commonweale'.

It may be that there were good tactical reasons for Milton's polemical concentration on the Belfast Presbyterians rather than the native Irish. Given his place in the Commonwealth government Milton would probably have known something about the negotiations for a truce taking place in May 1649 between the Ulster Catholic military leader Owen Roe O'Neill, who had become estranged from the Confederate hierarchy, and the Commonwealth's Colonel Monck. O'Neill had come to the conclusion that toleration of Catholicism might be more likely secured under the Commonwealth rather than Stuart rule, and while Westminster predictably refused his demands for various religious and political concessions, in the meantime his assistance helped to scatter the royalist siege of Derry in the summer of 1649.[30] But Milton's attitude towards Presbyterianism in the *Observations* is absolutely consistent with what he had been arguing since 1644, targeting them because of their failure to maintain the virtue that they had previously displayed: 'the Scottish inhabitants of that Province are actually revolted'. 'Revolt' is a key term both in Milton's depiction of the Presbyterians in the 1640s and in the theory of conquest he outlined in the *De Doctrina Christiana*, the work of systematic theology dating from 1650s but not published until the nineteenth century. 'Revolt' is a verb that can mislead us today because we associate it with revolution in the sense of an uprising against established authority. But Milton used it in the sense of revolve or back-slide rather than rebel, as in the attack on the Presbyterian clergy in the opening of *The Tenure of Kings and Magistrates*:

> after they [the Presbyterians] juggl'd and palter'd with the world, bandied and borne arms against the King, devested, disanointed him, nay curs'd him all over in thir Pulpit & thir Pamphlets, to the ingaging of sincere and real men beyond what is possible and honest to retreat from, not only turne revolters from these principles, which only could at first move them, but lay the straine of disloyaltie, and worse, on these proceedings, which are the necessary consequence of thir own former actions. (*CPW*, iii. 191)

The Presbyterian clerics were implicitly compared to the Scottish witches in Shakespeare's *Macbeth* (c.1606), 'these juggling fiends...That palter with us in a double sense' (V. ix. 19–20) – a further example of how Shakespearean drama shaped Milton's reading of the 'British problem' in 1649. Milton instinctively turned to Shakespeare to articulate the threat to the English republic from the Presbyterian and Catholic barbarism that dominated Scotland and Ireland, almost as though he saw Shakespeare (as he is seen today) as a defining voice of English nationhood.[31]

What makes Milton so bitter about the Presbyterians was that they 'revolted' from the very Reformation principles that in part inspired the Jacobean plantation. The strategy of the second edition of *The Tenure of Kings and Magistrates* (c.September–October 1649), to which Milton appended a list of impeccable Reformation authorities in an effort to undermine Presbyterian condemnation of the regicide as un-Protestant, was anticipated in the *Observations*, where Milton reminded the Ulster Presbyterians, or rather 'those blockish Presbyters of *Clandeboy*', that their own kinsman '*John Knox*, who was the first founder of Presbytery in *Scotland*, taught professedly the doctrine of deposing and of killing kings' (*CPW*, iii. 329). Milton feared that the English would follow the Scottish Presbyterians in revolting from their republican principles, spurning the opportunity to seize their liberty just as they had several times before in their history.

In the *De Doctrina Christiana*, which he may have begun to compile by 1649, Milton quoted a series of biblical texts to show the terrible consequences for Israel of its national apostasy. When discussing issues of free will and divine foreknowledge, Milton considered the question of God's foreknowing that the Israelites would lapse from their true religion to idolatrous worship of alien gods, and rejected the notion that their fall was predestined: 'the Israelites did not revolt', he wrote, 'because God knew that they would: rather, God knew that they would because he knew the causes of their revolt' (*CPW*, vi. 165–6). Israel suffered for its apostasy more painfully than pagan nations suffered for their idolatry because, as Milton declared at the end of Book III of the *History of Britain*, 'so much more tolerable in the eye of Heav'n is infidelity profess't, than Christian Faith and Religion dishonoured by unchristian works' (*CPW*, v. 183).[32] The Catholic Irish were not infidels or pagans, but Milton more or less depicted them as such in their irremediable barbarity. The Presbyterians, however, stood accused of the worse crime of religious hypocrisy; and in consorting and allying with them, the English risked repeating their history and allowing the spiritual invasion of Rome to precede the physical conquest of the nation. In the same passage in the *History of Britain* Milton declared that he has shown 'the many miseries and desolations brought on by divine hand on a perverse Nation; driv'n, when nothing else would reform them, out of a fair Country, into a Mountanous and Barren Corner, by Strangers and Pagans' (v. 183). If the English paid heed to the protests against regicide of the Ulster Presbyterians, then eventually the whole of England would become, like Ireland and Scotland, a savage and unchristian land, 'a Mountanous and Barren Corner'; indeed, a 'barbarous nook'.

NOTES

1 D. F. McKenzie and Maureen Bell (eds), *A Chronology and Calendar of Documents Relating to the London Book Trade 1641–1700* (3 vols, Oxford, 2005), i, p. 252.
2 Among the many books which offer a narrative of the complicated events and alliances in Ireland in the opening months of 1649, see the cogent narrative in Micheál Ó Siochrú, *God's Executioner: Oliver Cromwell and the Conquest of Ireland* (London, 2008), pp. 52–76. See also Jane Ohlmeyer, 'The civil wars in Ireland', in John Kenyon and Jane Ohlmeyer (eds), *The Civil Wars: A Military History of England, Scotland, and Ireland, 1638–1660* (Oxford, 1998), pp. 73–102; Jane Ohlmeyer (ed.), *Ireland from Independence to Occupation, 1641–1660* (Cambridge, 1995); Ian Gentles, *The English Revolution and the Wars in the Three Kingdoms, 1638–52* (Harlow, 2007).
3 The London bookseller George Thomason inscribed 16 May on his copy of the *Observations*. There was only one issue of the *Observations* and copies are rare: in his standard biography Parker lists only nine libraries with copies. See W. R. Parker, *Milton: A Biography*, revised Gordon Campbell (2 vols, 1968; Oxford, 1996), ii, p. 960.

4 'Cromwell's speech to the General Council of the Army at Whitehall, 23 March 1649', in W. C. Abbot (ed.), *The Writings and Speeches of Oliver Cromwell* (4 vols, Cambridge, 1937–47), ii, pp. 36–9.
5 D. M. Wolfe (general ed.), *Complete Prose Works of John Milton* (8 vols in 10, New Haven, 1953–82) (hereafter *CPW*), iii, pp. 304, 317, 327. All references to Milton's prose works are to this edition unless otherwise noted. Milton's charge against Belfast appears, for example, as an epigraph in Patricia Craig (ed.), *The Belfast Anthology* (Belfast, 1999).
6 Milton's (Latin) letter of 21 September 1656 is printed in Frank A. Patterson (general ed.), *Works of John Milton* (18 vols in 21, New York, 1931–8), xii, pp. 78–83.
7 On Ranelagh's circle and interests, see Sarah Hutton, 'Jones [*née* Boyle], Katherine, Viscountess Ranelagh (1615–1691)', in *Oxford Dictionary of National Biography*, online ed. Lawrence Goldman (Oxford, 2008), www.oxforddnb.com/view/article/66365 (accessed 30 September 2011).
8 On Milton and slavery, see further the bold essay by Martin Dzelzainis, 'Conquest and slavery in Milton's *History of Britain*', in Nicholas McDowell and Nigel Smith (eds), *The Oxford Handbook of Milton* (Oxford, 2009), pp. 547–68, at p. 557. On the common imagery applied by the English to the Irish and other races supposedly open to conquest and enslavement due to their natural incapacity to rule themselves, see e.g. Nicholas Canny, 'The ideology of English colonization: from Ireland to America', *William and Mary Quarterly*, 30 (1973), 575–98.
9 Joad Raymond, 'Complications of interest: Milton, Scotland, Ireland, and National Identity in 1649', *Review of English Studies*, 55 (2004), 315–45, at 316.
10 Elizabeth Sauer and Sharon Achinstein, 'Introduction'; Elizabeth Sauer, 'Toleration and nationhood in the 1650s: "Sonnet XV" and the case of Ireland', both in Sharon Achinstein and Elizabeth Sauer (eds), *Milton and Toleration* (Oxford, 2007), pp. 1–19 (2); pp. 203–23, at p. 211.
11 Gordon Campbell and Thomas N. Corns, *John Milton: Life, Work, and Thought* (Oxford, 2008), p. 218.
12 In John MacBride's *A Sample of Jet-Black Prelatick Calumny*, printed in Glasgow in 1713, MacBride says that not everyone has a copy of Milton's tract so he has copied the *Necessary Representation* from the 'original copy' (p. 106). After quoting it, he says that it was 'publickly read' in congregations in Belfast and then sent to Derry, to Sir Charles Coote, 'who then held Derry for the Parliamentary party' (p. 109). Coote responded aggressively: his letter of March 7 is printed in Edmund Borlase, *History of the Excrebable Irish Rebellion* (1680), pp. 205–8. Coote may have passed the *Neccessary Representation* on to London, where Milton was shown it. An anonymous attack on the *Necessary Representation* had already been published in London before Milton's tract appeared, and had printed sections of it to refute them according to the standard polemical technique of animadversion: *A necessary examination of a dangerous design and practice against the interest and soveraignty of the nation and Common-wealth of England, by the Presbytery at Belfast in the province of Ulster in Ireland; in their scandalous, malicious and treasonable libel* was collected by George Thomason on 17 April.

13 Willy Maley, *Nation, State and Empire in English Renaissance Literature* (Basingstoke, 2003), pp. 135–6. The point about the form of the *Observations* is emphasised in the essay which probably did most to provoke new interest in the pamphlet: Thomas N. Corns, 'Milton's *Observations upon the Articles of Peace*: Ireland under English Eyes', in David Loewenstein and James Grantham Turner (eds), *Politics, Poetics, and Hermeneutics in Milton's Prose* (Cambridge, 1990), pp. 123–34. See further Thomas N. Corns, *John Milton: The Prose Works* (New York, 1998), pp. 75–81.

14 John Kerrigan, *Archipelagic English: Literature, History, and Politics, 1603–1707* (Oxford, 2008), pp. 231–2.

15 See Nicholas McDowell, 'The stigmatizing of Puritans as Jews in early Stuart England: Ben Jonson, Francis Bacon, and the *Books of Sports* Controversy', *Renaissance Studies*, 19 (2005), 348–63.

16 'Milton and the limitations of Englishness', in David Loewenstein and Paul Stevens (eds), *Early Modern Nationalism and Milton's England* (Toronto, 2008), pp. 205–16, at p. 208.

17 Walker, *Anarchia Anglicana, the Second Part* (September 1649), pp. 199–200. I discuss the reaction to Milton's *Tenure*, as well as the various contexts of the *Observations*, in more detail in the introduction to *The Oxford Complete Works of John Milton. Volume VI: Vernacular Regicide and Republican Tracts*, ed. N. H. Keeble and Nicholas McDowell (Oxford, 2012). Walker would die in the Tower.

18 John Milton, *Complete Shorter Poems*, ed. John Carey (2nd edn, Harlow, 1997), pp. 298–300, line 20. All references to Milton's poems are to this edition. The poem, a tailed sonnet, was not published until 1673. For further discussion of the motivating force of anti-Presbyterianism in Milton's writing between 1644 and 1649, see Nicholas McDowell, *Poetry and Allegiance in the English Civil Wars: Marvell and the Cause of Wit* (Oxford, 2008), pp. 53–111.

19 For a helpful summary of the different arguments about when Milton composed the *History of Britain* and the 'Digression', see Martin Dzelzainis, '*Samson Agonistes* and the dating of the "Digression" in Milton's *History of Britain*', *Milton Studies*, 48 (2008), 160–78. For more detailed discussion of the issues, see Nicholas Von Maltzahn, *Milton's 'History of Britain': Republican Historiography in the English Revolution* (Oxford, 1991), pp. 22–48; see chapters 3–5 for February–March 1649 as the context for the composition of the *History*. Lewalski argues the 'Digression' was rather composed in the latter half of 1648, in the aftermath of the second civil war. See Barbara Lewalski, *The Life of John Milton* (Oxford, 2000), pp. 219–20; Blair Worden, however, has ascribed the 'Digression' to the Restoration, even the early 1670s, in his book *Literature and Politics in Cromwellian England: John Milton, Andrew Marvell, Marchamont Nedham* (Oxford, 2007), pp. 410–26.

20 Andrew Hadfield, 'The English and other peoples', in Thomas N. Corns (ed.), *A Companion to Milton* (Oxford, 2001), pp. 171–90, at p. 186.

21 For a cogent summary and explanation of Milton's refusal to extend toleration to Catholics, see Andrew Hadfield, 'Milton and Catholicism', in Achinstein and Sauer (eds), *Milton and Toleration*, pp. 186–99.

22 Thomas Mason, *Christs victorie ouer Sathans tyrannie* (1615), p. 194. This is an abridgement of Foxe's *Actes and Monumentes* (1st English edn, London, 1563).

23 Among the extensive literature detailing the relationship between the *Maske* and *The Tempest*, see e.g. John M. Major, '*Comus* and *The Tempest*', *Shakespeare Quarterly*, 10 (1959), 177–83; Mary Loeffelholz, 'Two Masques of Ceres and Proserpine: *Comus* and *The Tempest*', in Mary Nyquist and Margaret W. Ferguson (eds), *Re-Membering Milton* (London and New York, 1988), pp. 25–42; David Norbrook, '"What Cares These Roarers for the Name of King": language and utopia in *The Tempest*', in Gordon McMullan and Jonathan Hope (eds), *The Politics of Tragicomedy: Shakespeare and After* (London and New York, 1992), pp. 21–54; Jeff Dolven, *Scenes of Instruction in Renaissance Romance* (Chicago, 2007), pp. 242–9.
24 All references to Shakespeare are to *The Arden Shakespeare Complete Works*, ed. Richard Proudfoot, Ann Thompson and David Scott Kastan (rev. edn, 2001).
25 For further discussion, see Nicholas McDowell, 'Milton's regicide tracts and the uses of Shakespeare', in McDowell and Smith (eds), *Oxford Handbook of Milton*, pp. 252–71. For an overview of the connection between idolatry and slavery in Milton, see Barbara K. Lewalski, 'Milton and Idolatry', *Studies in English Literature, 1500–1900*, 43 (2003), 213–42.
26 Dympna Callaghan, 'Irish memories in *The Tempest*', in her *Shakespeare Without Women: Representing Gender and Race on the Renaissance Stage* (London, 2000), pp. 97–138, at p. 100. See also Paul Brown, '"This thing of darkness I acknowledge mine": *The Tempest* and Colonial Criticism', in Jonathan Dollimore and Alan Sinfield (eds), *Political Shakespeare: Essays in Cultural Materialism* (2nd edn, Manchester, 1994), pp. 48–71; David J. Baker, 'Where is Ireland in *The Tempest*?', in Mark Thornton Burnett and Ramona Wray (eds), *Shakespeare and Ireland: History, Politics, Culture* (Basingstoke, 1997), pp. 66–88.
27 See *CPW*, i, 495–6. For discussion, see Willy Maley, 'How Milton and some contemporaries read Spenser's *View*', in Brendan Bradshaw, Andrew Hadfield and Willy Maley (eds), *Representing Ireland: Literature and the Origins of Conflict, 1534–1660* (Cambridge, 1993), pp. 191–208.
28 *The Faerie Queene*, ed. Thomas P. Roche, Jr (Harmondsworth, 1987), V. i. 12, xii. 26.
29 See also the discussion of the reference to Talus in Richard A. McCabe, *Spenser's Monstrous Regiment: Elizabethan Ireland and the Poetics of Difference* (Oxford, 2007), p. 281.
30 See Ó Siochrú, *God's Executioner*, pp. 60–1.
31 For further discussion of the place of *Macbeth* in Milton's 1649 prose, see McDowell, 'Milton's regicide tracts and the uses of Shakespeare', pp. 259–66.
32 For further discussion of this passage, see Dzelzainis, 'Conquest and slavery in Milton's *History of Britain*', pp. 417–18.

Index

1641 Rebellion 1, 5, 7, 12, 98, 112, 113–14, 127, 206–8, 246–7

Abercorn, Earls of *see* Hamilton, Sir George; Hamilton, Sir James
'A bhean fuair faill ar an bhfeart' (Mac an Bhaird) 186–7
Adamson, John 79
'Agallamh na Seanórach' 209–10
agriculture 81, 150, 151, 153, 227, 239
Albanese, Denise 220
Alexander, William 26, 34
Altman, Joel 218
Amends for Ladies (Field) 229
Americas 2, 9, 36, 37, 57, 115, 158, 219–20, 226, 228, 230
Anatomy of Melancholy (Burton) 228–9
Ancient Law-Merchant, The (Malynes) 230–1
Anderson, Patrick 131
Andrews, J.H. 3
Annals of the Four Masters 7, 176, 201–2, 211
'Anois molfam Mág Uidhir' (Ó hEódhusa) 181
'Anothomy of Irelande' (Rich) 227–8
Antrim, County 8, 23, 128, 131, 144, 146–55, 182
Antrim, Earl of *see* MacDonnell, Sir Randal

archaeology 11, 143, 144–8, 152, 153, 154–5
Archer, Ian 10
architecture 143, 144, 147, 150, 152, 153, 154
Ardnamurchan 40, 44
Ards peninsula 22, 61–2, 159, 163, 224, 226
Argyll, Duke of *see* Campbell, John, 2nd Duke of Argyll
Argyll, Earls of *see* Campbell, Archibald, 4th Earl of Argyll; Campbell, Archibald, 5th Earl of Argyll; Campbell, Archibald, 7th Earl of Argyll
Aristotle 164, 165, 166
arithmetic 164, 165
Armagh, County 6, 8, 109, 112–13, 198, 200
Armagh town 9
Armitage, David 219
Artes of Logike and Rhetorike, The (Fenner) 166
Arthur, King 42
Articles of Peace 238, 240
Ashton, Robert 79
assimilation 37, 109–12
Astrophil and Stella (Sidney) 224
Athgeave 85, 87
Audley, Lord *see* Tuchet, George, Lord Audley

Bacon, Sir Francis 81, 219–20, 221, 222–3
  New Atlantis 220
  'Of Plantations' 220
  'A Speech in Parliament touching the Naturalization of the Scottish Nation' 221
Bagenal, Sir Henry 186
Bagenal, Marshall 150
Bagwell, Richard 59
Ballycastle 147, 150, 153–4
barbarity 20, 33, 34, 35, 40, 62–3, 201, 227, 228, 229–30, 239–40, 249
bardic poetry 11, 176–92
bardic schools 11, 205, 210
Barnard, Toby 5–6
Barry, Jonathan 71
Barston, John 56, 64–5, 69
Bartholomew Fair (Jonson) 223
Basilikon Doron (James VI and I) 22, 33, 40, 45, 102
'Batter my Heart' (Donne) 221
Beacon, Richard 56
'Beag mhaireas do mhacraidh Ghaoidheal' 182–3
'Beatha Aodha Rua Uí Dhomhnaill' (Ó Cléirigh) 211
Beaumont, Francis 229
Bedell, William 119, 123, 132, 200
Behan, John 30
Belfast 239, 240, 241–2, 243, 247, 248
Bell, John 100
Bellings, Richard 100
Beresford, Tristram 87–8, 92
Berkeley, George 9
Bingham, Sir Richard 200
Blennerhassett, Thomas 58–9, 63–4, 68–9, 73, 81, 223, 226
  A Direction for the Plantation of Ireland 58, 59, 63–4, 72, 108
  Mirror for Magistrates 59
Blount, Charles, Lord Mountjoy 6, 23, 24, 186, 200
Blundell, Francis 99, 108, 113
Bodin, Jean 165
Bohemia 209

Booke of Falconrie (Turbeville) 230
'Book of O'Conor Don' 176–8, 210
'Book of O'Donnell's daughter' (Mac an Bhaird) 185–6
Book of the Courtier (Castiglione) 165
'Book of the Dean of Lismore' 42
Boyle, Sir Robert 200
Braddick, Michael J. 57
Bradshaw, Brendan 2, 56
Brannon, Nick 145
Breen, Colin 11
Brenner, Robert 79
'Brief Note of Ireland' (Spenser) 160
Britannia (Camden) 165
British history 3–6
Britishness 5–6, 19–20, 41–3, 103–5
Buchanan, George 34
  History of Scotland 165, 169
Burgess, Glen 4
Burke, Richard 114
Burton, Robert 228–9
Bussy D'ambois (Chapman) 229
Butler, James, Marquis of Ormond 238, 246
Bynemann, Henry 163

Caball, Marc 2, 11
Cain, Tom 221
'Cáit ar ghabhadar Gaoidhil?' (Ó Dálaigh) 177–8, 202–3, 223–4
'Caithréim Chlainne Raghnaill' (Mac Muireadhaigh) 207
Calvinism 166, 167
Cambridge University 58, 64, 69, 163, 164–6
Camden, William 165, 211, 219, 226
Campbell, Archibald, 4th Earl of Argyll 41
Campbell, Archibald, 5th Earl of Argyll 42, 43
Campbell, Archibald, 7th Earl of Argyll 22–3, 25–6, 35, 36, 42, 105, 149, 150
Campbell clan 10, 22, 26, 34, 35, 37, 39, 40–5, 48–9
Campbell, Gordon 240

Campbell, John, 2nd Duke of Argyll 41–2
Campion, Edmund 7, 211
Canning, George 85, 86, 87–90, 91
Canny, Nicholas 2, 4, 23, 24, 28, 63, 158, 159
Captain Thomas Stukeley (anon.) 229
Carey, Sir George 218
Carey, Vincent 2
Carney, James 181
Carrickfergus 241
Carroll, Clare 3
Carrough, Edmund 112
Carrough Maguire, Donn 112
Carswell, John 42, 43
Cary, Barony of 147, 148
Castiglione, Baldassare 165
Catholic Confederate Association 5, 135, 238, 247, 248
Catholicism
  Catholic clergy 3, 5, 11, 112, 119–35
  and continental Europe 5, 7, 11, 121, 122, 124, 126, 127, 128, 129, 198, 211
  and English settlers 8, 132, 133–5
  and Gaelic literature 203–4, 205–6, 211
  and identity 122, 127, 199, 204, 211
  and James VI and I 2, 8, 10, 27–8, 29, 104–5
  Milton on 241, 242–3, 244–5, 246–7, 249–50
  and persecution 2, 211
  and poverty 11
  and Protestantism 2–3, 8, 27–8, 110–11, 112, 127, 130, 131, 134–5, 188–9, 246–7
  revival of 11, 112, 113
  in Scotland 119, 121, 129, 130–1, 132, 245
  and Scottish settlers 8, 27–8, 130–31, 132, 133
  see also Counter-Reformation
Cavan, County 8, 101, 113, 129, 132, 133, 198
Cavan town 9
ceannas nan Gàidheal 39, 41, 47, 48

Cecil, Sir Robert, 1st Earl of Salisbury 24, 179, 224
Cecil, William, 2nd Earl of Salisbury 58, 69, 81
Céitinn, Seathrún 7, 210, 211–12
censorship 160, 218
Chapman, George 229
Character of the Long Parliament (Milton) 242–3
Charles I
  and city of London Star Chamber case 78, 79–80, 93
  and the *Eikon Basilike* 244–5, 247
  execution of 210, 238
  financial circumstances 10, 79, 91
  and Gaelic Scotland 44, 47, 132
  Milton criticises 244–5, 246, 247–8
  Personal Rule 79–80, 132
  religious policy 28
  trial of 247–8
Chichester, Sir Arthur
  biographical studies of 2
  and the city of London 91
  and Hugh O'Neill 6, 7, 24–5
  owns Spenser's *View* 162, 169, 170
  proposals for the plantation 81, 101–2, 104, 108, 114, 169
  and Randal MacDonnell 150, 151, 154
  ruthlessness of 200
  views on the plantation 99, 105, 107, 219
church buildings 120, 122–3, 134, 188
church lands 78, 86–8, 101, 107–8, 110, 121, 125, 200–1
Church of Ireland 9, 27, 101, 107–8, 110–11, 112, 122–4, 127, 130, 134
  see also Protestantism
Circe myth 245–6
Cistercian order 124
citizenship 58, 69, 71, 72–3
civility
  and Gaelic Scotland 10, 21, 28, 33–4, 40, 102, 149
  and Ireland 2, 7, 9, 20, 55–6, 62–3, 71, 84, 102, 160–1, 200, 218, 239

civility (cont'd)
　and James VI and I 2, 7, 9, 10, 20, 21,
　　28, 33–4, 40, 102, 149, 200, 218
　Milton on 239
　More on 72–3
　Smith on 72
　Spenser on 62–3, 84, 160–1
　and urbanisation 71, 84
civil war *see* English civil war
Clan Donald South 40, 41, 45
Clann Eoin Mhòir *see* Clan Donald
　South
Clarke, Aidan 1
Clothworkers' Company 82, 83–4
Clotworthy, Sir John 80
coarbs 119, 121
Cockayne, William 82
Coleraine
　archaeology at 145
　compared with Dunluce 153
　construction of 9, 84–5, 93
　planning of 81
　production of *Much Ado About
　　Nothing* 231–2
　revenue from 79, 90
　settlement centres around 109
　stone sourced from 87
　success of 85
Coleraine, County *see* Londonderry,
　County
Collinson, Patrick 55, 57–8
colonialism
　and the aristocracy 222–3
　Bacon on 219–20, 222–3
　classical models of 61, 226
　and corporatism 64
　and historiography 4
　intellectual background to 163–4,
　　168–70
　Irish 36
　and private enterprise 59, 158–9
　Scottish 10, 26–7, 34, 45–7
　sexual metaphors of 223–4
　and the state 2, 57
　and urbanisation 69

common law *see* law
company, concept of 57, 58, 60–1, 63–4,
　65–8, 73
'Comus' (Milton) *see Masque Performed
　at Ludlow Castle* (Milton)
confederate government *see* Catholic
　Confederate Association
Connacht, expulsion to 208–9
Conway, Henry 145
Cooke, John 229
Coote, Sir Charles, Jr. 241
Corns, Thomas N. 240, 241
corporatism 10, 56–73, 91, 105
Counter-Reformation 7, 127, 189, 191,
　198, 204, 211
Coxcomb, The (Beaumont and Fletcher)
　229
Creagh, Richard 7–8
Cromwell, Oliver 29, 208, 209, 211, 239,
　240
cultural heritage 7–8
Curl, James Stevens 2

'Dá ghrádh tréigfead Máol Mórdha'
　(Ó hEódhusa) 183–5
Davies, C.S.L. 72
Davies, Sir John
　and Catholicism 119
　'civilising' of Ireland 200
　describes jury at Limavady 201
　Discoverie of the True Causes 11, 29,
　　162, 168, 221, 222, 228, 230–1, 246–7
　funeral of 220
　and Hugh O'Neill 6, 24
　proposals for plantation 29, 71, 81,
　　102, 106, 108, 224–5, 226
　and Protestantism 28
　Spenser's influence upon 162
Dawson, Jane 43
Daye, Angel 69
De Doctrina Christiana (Milton) 249–50
Dekker, Thomas 229
'Demands of the Irish' 114
De Republica Anglorum (Smith) 58, 60,
　71, 73, 224

Derry, County 8, 9, 113, 121, 125, 130, 198, 200–1, 206, 207
Derry City
  archaeology at 145
  captured by Cahir O'Doherty 144
  compared with Dunluce 153
  construction of 9, 84–5, 93
  planning of 81
  revenue from 79, 90
  settlement centres around 109
  Siege of 241, 248
  success of 85
Derry plantation 1, 2, 9, 71, 78–93, 109, 111, 130, 143
Descrypcyon of Englonde (Higden) 230
'deserving Irish' 3, 8, 9, 88, 102, 107, 112–13, 144
Desmond Rebellion 3, 161, 168
Devereux, Robert, 2nd Earl of Essex 3, 24, 159–60, 218, 227
Devereux, Walter, 1st Earl of Essex 224, 227
Devil is an Ass, The (Jonson) 223
'Dia libh, a uaisle Éireann' (Mac an Bhaird) 191
dialectic 164, 165, 166
Dickson, David 12
'Dibirt go Connachta, An' (Ó Mealláin) 208–9
Direction for the Plantation of Ireland (Blennerhasset) 58, 59, 63–4, 72, 108
Discourse of the Commonweal of the Realm of England (Smith) 61
Discoverie of the True Causes (Davies) 11, 29, 162, 168, 221, 222, 230–1, 246–7
divorce 129, 130, 229, 239, 241
'Docum Glóire Dé agus Onóra na hÉireann' 211
Docwra, Sir Henry 2, 23
Doddington, Sir Edward 145
Dòmhnall Dubh 37, 39, 41
Donegal, County 8, 112, 121, 144, 198, 227
Donegal town 63

Donne, John 220–1
Down, County 8, 23, 102, 107, 128, 130, 132, 152, 182
Drapers' Company 80, 86
dress 111–12
Drogheda 121, 124, 125, 129
'Duanaire Finn' 209–10
Dublin 98, 99
Duffy, Patrick (historical geographer) 3
Duffy, Patrick (vicar-general of Clogher) 124, 126
Dundalk 241
Dundrum 107
Dungannon 6, 9, 105, 201
Dungiven priory 145
Dunineny 147, 152, 153
Dunluce 11, 147, 148, 152–3, 154
Dunluce Castle 130, 131, 144, 149, 150, 152
Duns Scotus, John 206

East India Company 9, 83
education 164–7
  see also Cambridge University
Edwards, David 2
Edwards, Robert Dudley 1
Eikon Basilike 244–5
Eikonoklastes (Milton) 244–8
'Éisd rem égnach, a fhir ghráidh' (Mac an Bhaird) 190–1
Elizabeth I 3, 6, 8, 18, 19, 20, 22, 25, 27, 39, 164, 200, 210
England 69–71, 135, 242–4
English civil war 79, 238, 240, 242
English language 34
English literature *see* Jacobean literature
English Privy Council 10, 78, 83, 91, 100, 101, 102, 160
English settlers
  in Antrim 155
  archaeological evidence of 146
  and Catholicism 8, 132, 133–5
  in Gaelic literature 177–8
  in the Londonderry plantation 10, 89
  in the Munster plantation 159, 161–2

English settlers (cont'd)
  numbers of 158
  in the 'Orders and Conditions' 106, 107
  and proposals for plantation 101, 103–4, 144, 169
  and Protestantism 5–6, 8, 133–5
  solidarity with Scottish settlers promoted 5–6, 103–4
  unsuitability for physical environment 219
  see also London, city of
Enniskillen 9
Epicoene (Jonson) 223
erenaghs 121, 133, 200, 201, 207
Erroll, Earl of see Hay, Francis, 9th Earl of Erroll
Essex, Earl of see Devereux, Robert, 2nd Earl of Essex; Devereux, Walter, 1st Earl of Essex
Europe, continental
  migration to 5, 7, 11, 27, 185, 188–9, 209
  military service in 5, 7
  religious communities in 5, 7, 121, 122, 124, 126, 127, 128, 129, 198, 211
  see also Bohemia; Flanders; France; Louvain; Low Countries; Netherlands; Rome; Spain
Everard, Sir John 28
exclusion 71, 73, 104
exempla 184, 188

'Fada re urchóid Éire' (Ó hEódhusa) 181
Faerie Queene, The (Spenser) 230, 247–8
Fair Quarrel, A (Middleton and Rowley) 229
fairs 105, 107, 151, 154
Faunce, Abraham 167
Feiritéar, Piaras 210
Fenner, Dudley 166
Fenner, George 218
Fermanagh, County 6, 8, 101, 112, 113, 180, 198
Field, Nathan 229
Fife Adventurers 21, 40, 45, 46, 149

fishing 36–7, 46–7, 81, 150, 153, 154, 230
Fishmongers' Company 87, 88
FitzGerald, Thomas 2
Flanders 7, 185–6, 207, 211
  see also Low Countries
Fleming, Patrick 205
Fletcher, Alan 231–2
Fletcher, John 229
Flight of the Earls 6–7, 10, 23, 25–6, 30, 57, 114, 144, 168, 176–8, 182, 211
  see also 'Imeacht na nIarlaí' (Ó Cianáin); O'Donnell, Rory; O'Neill, Hugh
Florio, John 220
Fogarty, Anne 220
'Foras Feasa ar Éirinn' (Céitinn) 7, 211–12
Ford, Alan 2
Ford, John 229, 230
Foster, Roy 211
Four Prentices of London, The (Heywood) 229
Foxe, John 245
France 5, 7, 9, 211, 221
Franciscan order
  in Bohemia 209
  convents 122, 129
  and Hugh O'Neill 128
  at Louvain 120, 122, 124, 127, 128, 135, 205–6
  missions 11, 124, 125, 126, 127, 129, 131, 132, 135
freedom 58, 71–2, 73
freeholders 89, 101–2, 106, 182
Freeman, William 84, 92

Gaelic culture 177, 179, 198–9
Gaelic literature 7, 11, 176–92, 198–212, 223–4
Gainsford, Thomas 230
garrisons 47, 57, 62, 63, 101, 145, 147, 161, 224, 247
'Gearr bhur ccuairt, a chlanna Néill' (Ó Gnímh) 191
geometry 164, 165

Gilbert, Sir Humphrey 163–4, 200
Gillespie, Raymond 2, 10–11, 151
Gillies, William 178
Giraldus Cambrensis 7, 211
Glenarm 147, 148, 152, 154
Glens of Antrim 23, 150, 154
Glenstrae 40
Goodare, Julian 45
Goodlands 146–7
Gordon, George, 6th Earl of Huntly 22–3, 25–6, 105
Gordon, Sir Robert, of Lochinvar 26, 34
grammar 164, 165
'Greater Gaeldom' 10, 34, 38–9
Greenes Tu Quoque (Cooke) 229
Gregory, Donald 33
Grocers' Company 83
Guicciardini, Francesco 165
Gunpowder Plot 10

Haberdashers' Company 83, 84, 92, 111
Hadfield, Andrew 3, 11, 218, 243
Haicéad, Pádraigín 210
Hamilton, Sir Claude 109–10
Hamilton, Sir George 28
Hamilton, Sir James 8, 28, 104, 130
Hamilton family 130–1, 132
Hamner, Meredith 7, 211
Hanratty, Patrick 124
Harington, John 59
Harris, Isle of 40
Harvey, Gabriel 58, 163–4, 165, 167, 169
Hay, Sir Alexander 36
Hay, Francis, 9th Earl of Erroll 22
heathenism 35, 182, 242, 243
Henrietta Maria, Queen 135
Henry VIII 19, 38, 39, 210
Hey for Honesty, Down with Knavery (Randolph) 229
Heylyn, Peter 230
Heywood, Thomas 229
Hiberno-Latin literature 7–8, 11, 209
Higden, Ranulf 230
high-kingship of Ireland 39
Hill, George 99, 103, 201

History of Britain (Milton) 242–3, 250
History of Scotland (Buchanan) 165, 169
Holinshed's *Chronicles* 224
Holme, Thomas 9
Honest Whore, Part Two (Dekker) 229
Honourable the Irish Society *see* Irish Society
Hooker, John 224
Horning, Audrey 146–7
humanism 56, 57–60, 64, 69, 73
Hunt, Lynn 178
Hunter, Robert 1
Huntly, Earl of *see* Gordon, George, 6th Earl of Huntly

identity
    and Catholicism 122, 127, 199, 204, 211
    national 104, 158, 191, 199, 204, 211
    Old English 125, 126
    Protestant 6
idolatry 244–7
If This be not a Good Play, the Divell is in It (Dekker) 229
'Imeacht na nIarlaí' (Ó Cianáin) 211
    see also Flight of the Earls
imperialism *see* colonialism
Iona, Isle of 38
'Ionnmhas ollaimh onóir ríogh' (Mac an Bhaird) 189
Irish language 110, 111, 125, 127, 154, 204–6, 211
    see also Gaelic literature
Irish literature *see* Gaelic literature; Hiberno-Latin literature
Irish Masque at Court (Jonson) 218, 221–2
Irish Privy Council 102
Irish Society 2, 9, 64, 71, 78, 79, 82, 85, 89, 90–1, 92
Ironmongers' Company 82, 85, 86–90, 91, 109
Irving, Sarah 219–20
Islay 36, 40, 44, 45
Islay Rising 34, 35, 44, 45
Iveagh 102

Jacobean literature 11, 218–32
Jacobite Rising (1715) 42
Jacobite Rising (1745) 37
James III 48
James IV 38, 48
James V 38, 41
James VI and I
   Basilikon Doron 22, 33, 40, 102
   and Catholicism 2, 8, 10, 27–8, 29, 104–5
   'civilising' of Ireland 2, 6–7, 40
   and Gaelic Scotland 9–10, 20, 21–3, 33–5, 39–40, 44, 48, 100, 149–50
   and Hugh O'Neill 6, 10, 23–6, 27–8, 144
   Ó hEódhusa's poem celebrating 181
   plantation of Lewis 9–10, 21–2, 100, 149–50
   plantation of Ulster 2, 8–9, 19–20, 27–8, 91, 104–5, 114–15, 144, 158, 169, 178–9, 210, 219
   and Protestantism 2, 8, 27–8
   and Randal MacDonnell 23, 27–8, 147–50, 154
   style of kingship 18–30
Jamestown, Virginia 9, 115
Jaster, Margaret Rose 222
Jefferies, Henry 2
Jesuit order 11, 124, 126, 131
joint-stock companies 9, 60–1, 63, 68
Jolles, John 82
Jones, Sir Baptist 145
Jones, Katherine, Viscountess Ranelagh 239
Jonson, Ben
   Bartholomew Fair 223
   The Devil is an Ass 223
   Epicoene 223
   Irish Masque at Court 218, 221–2
   The New Inn 223
   Prince Henry's Barriers 223
Judaism 241

Kearney, Hugh 5
Keating, Geoffrey *see* Céitinn, Seathrún
Kerrigan, John 240–1
Kilconway 148
Kilkenny 5, 135
Kinsale, battle of 6, 38, 162
Kintyre 21, 40, 44, 149
Knott, Eleanor 187
Knox, Andrew 28, 36, 41, 43–4, 45
Knox, John 42
Knox, Thomas 28
Kohl, Johann Georg 198

Lacey, Brian 145
Lake, Peter 4
Laois, County 158
law 34, 40, 44–5, 57, 71, 104, 106, 110, 130, 161–2, 168
Leslie, John 28
Letter Sent by I.B. (Smith) 58, 59–61, 71, 163
Lewis, Isle of 9–10, 21–2, 33, 36, 37, 40, 45–7, 100, 149
Ley, Sir James 102
Lifford 63
Limavady 78, 92, 201
Lindley, David 222
Lindley, Keith 79
literature *see* Gaelic literature; Hiberno-Latin literature; Jacobean literature
Lithgow, William 226–7
livery companies 9, 79–80, 81–4, 85–91, 145, 146
   see also London, city of
Lochaber 21
logic 164, 165–8
Lombard, Peter 7, 124, 125
London, city of 9, 10, 57, 69, 78–93, 130, 145
Londonderry
   archaeology at 145
   captured by Cahir O'Doherty 144
   compared with Dunluce 153
   construction of 9, 84–5, 93
   planning of 81
   revenue from 79, 90
   settlement centres around 109
   Siege of 241, 248
   success of 85

Londonderry, County 8, 9, 113, 121, 125, 130, 198, 200–1, 206, 207
Londonderry plantation 1, 2, 9, 71, 78–93, 109, 111, 130, 143
Long Parliament 80, 242
Lordship of the Isles 37–8, 39, 41, 47–9, 132
Loughran, Patrick 123
Louvain 120, 122, 124, 127, 128–9, 135, 189–90, 198–9, 205, 211
'Loves Warre' (Donne) 221
Low Countries 63, 190, 209
  see also Flanders; Netherlands; Zeeland
Lownes, Michael 160
Lucas, Scott 59
Luther, Martin 188, 204
Lynch, Michael 48
Lythe, Robert 224

Mac Aingil, Aodh 205, 206
Mac an Bhaird, Eoghan Ruadh 178, 185–9
  'A bhean fuair faill ar an bhfeart' 186–7
  'Book of O'Donnell's daughter' 185
  'Maith an sealad fuair Éire' 188–9
Mac an Bhaird, Fearghal Óg 178, 189–91
  'Éisd rem égnach, a fhir ghráidh' 190–1
  'Ionnmhas ollaimh onóir ríogh' 189
Mac an Bhaird, Uilliam Óg
  'Dia libh, a uaisle Éirionn' 191
Macbeth (Shakespeare) 249
McCabe, Richard 3
McCafferty, John 2
MacCarthy Reagh, Florence 179
Mac Cathmhaoil, Aodh *see* Mac Aingil, Aodh
McCavitt, John 2
MacCoinnich, Aonghas 45
Mac Cuarta, Brían 3, 11
MacDonald, Alasdair Cathanach 41
MacDonald, Sir Seumas 35, 36, 41, 45
MacDonald Clan 37–8, 39, 41, 45, 47–8

MacDonnell, Sir Randal 23, 28, 36, 114, 126, 130, 131, 143, 144, 146–55
MacDonnell, Sorley Boy 149, 209–10
MacDonnell family 131, 144, 149, 150
McDowell, Nicholas 11
Mac Gearailt, Muiris Mac Dháibhí Dubh 210
MacGregor, Martin 10
MacGregor clan 23, 33, 35, 40, 48
McGurk, John 2
Machiavelli, Niccolò 56, 73, 165
MacIain clan 40
MacInnes, John 41
MacKenzie, Cailen, 1st Earl of Seaforth 46, 47
MacKenzie, Coinneach Cam 46
MacKenzie clan 10, 37, 40, 45, 46–8, 49, 149
McKibben, Sarah 223–4
MacLeod, Niall 36
MacLeod, Rory 21
MacLeod clan 40, 45–6, 149
MacMahon, Aodh Óg 182
MacMahon, Sir Brian 182
McMahon, Hugh Roe 6
Mac Muireadhaigh, Niall 207
McNeill, Hugh 153
McNeill, Tom 147
Mac Póilín, Séamus 203–4
McQuillan family 144, 149
Magee, Alexander 146–7
Magee, Donal 146–7
Magennis, Bonaventure 131
Maguire, Cú Chonnacht 6, 179–81
Maguire, Cú Chonnacht Óg 181
Maguire, Hugh 179, 181
'Maith an sealad fuair Éire' (Mac an Bhaird) 188–9
Maitland, John 22, 25
Maley, Willy 3, 11, 240
Malynes, Gerard 230–1
manorial courts 105–6
Manship, Henry 69
maps 3, 92, 146, 147, 224–6
Margey, Annaleigh 3
marginalisation 71, 73, 126, 135

markets 62, 104, 105, 107, 151
marriage 123, 125, 129, 130
Marshall, Tristan 218
Mary, Queen 3, 20, 158
Masque Performed at Ludlow Castle (Milton) 245
Mason, John 36–7, 47
Matthews, Patrick 124
Mellon, Henry 206
Melville, Sir James 21
mercenaries 5, 7, 27, 39–40
Mercers' Company 82, 83, 87, 146
Merchant Tailors' Company 82, 83, 88
Messingham, Thomas 205
Middleton, Thomas 229
Mikrokosmos (Heylyn) 230
military migration 5, 7, 27
military presence 47, 57, 62, 63, 69, 73, 101, 112, 160, 162, 247
  see also garrisons
Miller, Orloff 146
Milton, John 11, 238–50
  Character of the Long Parliament 242–3
  De Doctrina Christiana 249–50
  Eikonoklastes 244–8
  History of Britain 242–3, 250
  Masque Performed at Ludlow Castle 245
  Observations 238–41, 242, 243–4, 246, 247, 248–9
  'On the New Forcers of Conscience' 242
  The Tenure of Kings and Magistrates 241, 249
Mirror for Magistrates (Blennerhassett) 59
'Mochean don loing si tar lear' (anon.) 176–7
Monaghan, County 6, 124, 132, 152, 182
Monaghan plantation 8, 101, 102, 182
monarchical republicanism 10, 55–9, 72, 73
Monck, George 241, 248
Monson, Sir William 47

Montgomery, George 28, 119, 200–1
Montgomery, Hugh 8
Moody, Theodore 1, 90
Moore, Adrian 84, 92
'Mór an t-ainm ollamh flatha' (Ó hEódhusa) 179
More, Sir Thomas 56, 64, 72–3
Morgan, Hiram 59, 61, 72
Morrill, John 3, 5
'Mór theasda dh'obair Óivid' (Ó hEódhusa) 181
Moryson, Fynes 11, 201, 211
'Mo thruaighe mar táid Gaoidhil!' (Ó Gnímh) 178
Mountjoy, Lord see Blount, Sir Charles, Lord Mountjoy
Movanagher 146
Much Ado About Nothing (Shakespeare) 92, 231–2
Mulcaster, Richard 59
Munster plantation 2, 3, 9, 20, 22, 62, 100, 159, 161–3, 199–200, 226
Murphy, Andrew 218
music 164, 165

national identity 104, 158, 191, 199, 204, 211
native Irish 10, 68, 88–90, 102, 105–6, 107–14, 123–4, 127–8, 153
  see also 'deserving' Irish
Navan 126
Necessary Representation 240
Neill, Michael 218
Netherlands 46–7, 122, 127
  see also Low Countries
Netzloff, Mark 219, 225, 232
New Atlantis (Bacon) 220
New British History 3–5
New Description of Ireland, A (Rich) 227
New England 37
Newfoundland 37
New Hampshire 37
New Inn, The (Jonson) 223
New Model Army 242, 248

Newry 9
Nicholls, Kenneth 2
Nine Years War 1, 3, 18–19, 43, 62, 114, 119, 120, 121, 123, 144, 168, 182, 218
Northward Hoe (Dekker and Webster) 229
Nova Scotia 26, 34

O'Boyle, Neil 120–2
O'Boyle, Turlough Roe 112
Observations (Milton) 238–41, 242, 243–4, 246, 247, 248–9
Ó Buachalla, Breandán 2
O'Cahan, Donal 144
Ó Catháin, Niall 209–10
Ó Ceallaigh, Seamus 201
Ochiltree, Lord *see* Stewart, Andrew, Lord Ochiltree
Ó Cianáin, Tadhg Óg 200, 211
Ó Cléirigh, Giolla Riabach 179–80
Ó Cléirigh, Lughaigh 211
Ó Cléirigh, Mícheál 199, 200, 201–2, 204–5, 210
O'Connell, Robert 8
Ó Cuív, Brian 2
O'Cullenan, John 133
Ó Dálaigh, Lochlainn 151
  'Cáit ar ghabhadar Gaoidhil?' 177–8, 202–3, 223–4
O'Devany, Cornelius 8, 121, 123
Ó Dochartaigh, Aodh 209–10
O'Doherty, Sir Cahir 28, 144, 178, 200, 220
Ó Doibhlin, Breandán 2
Ó Doibhlin, Diarmaid 11
O'Donnell, Aodh Ruadh 185, 189, 211
O'Donnell, Cathbharr 186, 187
O'Donnell, Niall Garbh 24, 185
O'Donnell, Nuala 185–7
O'Donnell, Rory
  and Catholicism 28
  death of 186–7, 188
  in Europe 176
  flight of 6–7, 10, 23, 25–6, 30, 57, 114, 144, 168, 176–8, 182, 211

  and James VI and I 23, 24
  and the Nine Years War 186
  Ó hEódhusa as retainer of 185–6
  tomb of 210
O'Dufferne, Patrick Groome 109
'Of Plantations' (Bacon) 220
O'Farrell, Richard 8
Offaly, County 158
Ó Fiaich, Tomás 2
Ó Gadhra, Fearghal 200
Ó Gnímh, Fearflatha
  'Gearr bhur ccuairt, a chlanna Néill' 191
  'Mo thruaighe mar táid Gaoidhil!' 178
Ó hAnnracháin, Tadhg 3
Ó hEodhasa, Bonabhentúra (Giolla Brighde) 205–6
Ó hEódhusa, Eochaidh 178, 179, 180–5, 200
  'Anois molfam Mág Uidhir' 181
  'Beag mhaireas do mhacraidh Ghaoidheal' 182–3
  'Dá ghrádh tréigfead Máol Mórdha' 183–5
  'Fada re urchóid Éire' 181
  'Mór an t-ainm ollamh flatha' 179
  'Mór theasda dh'obair Óivid' 181
  'Suirgheach sin, a Éire ógh' 181
Ohlmeyer, Jane 2, 5, 12, 80, 151
Old English 20, 35, 56, 62, 68, 100, 125, 126, 128, 129, 132, 133, 159, 210, 238, 247
Ó Maoilchonaire, Muirgheas 190
Ó Maolchonaire, Flaithrí 189–91
Ó Mealláin, Fear Dorcha 208–9
Ó Mealláin, Toirealach 206–8
O'Neill, Aodh Óg 186
O'Neill, Conn 8
O'Neill, Cormac McBaron 121
Ó Néill, Eoghan Ruadh 207–8, 248
O'Neill, Hugh
  and Catholicism 2, 27–8
  and Chichester 6, 7, 24–5
  constableship of Duniney fort 147
  defeat at Kinsale 6

O'Neill, Hugh (cont'd)
  in Europe 176–7, 185, 210
  flight of 6–7, 10, 23, 25–6, 30, 57, 114, 144, 168, 176–8, 182, 211
  hoped for return of 128
  and James VI and I 6, 10, 23–6, 27–8, 144
  marriage of daughter to MacMahon 182
  and the Nine Years War 3, 101, 161, 182, 186
  signs Treaty of Mellifont 6, 24, 143–4, 168
  tomb of 210
O'Neill, Sir Phelim 111–12, 131, 207
O'Neill, Shane 224
O'Neill, Toirdhealbhach Luineach 189
'On the New Forcers of Conscience' (Milton) 242
O'Rahilly, Cecile 207, 211
O'Rahilly, T.F. 201
oral tradition 107, 108–9, 180
Ó Rathaille, Aogán 199–200
'Orders and Conditions' 103, 105, 106, 114
O'Reilly, Máol Mórdha 183–4
O'Reilly, Pilib 180
Ormond, Marquis of see Butler, James, Marquis of Ormond

'Pairlement Chloinne Tomáis' (anon.) 211
Pale, the 101, 125–6, 128, 129, 132, 133, 152, 161, 221
Palmer, Patricia 3
Parsons, William 99–100, 107
patents (for land) 105–6, 148, 150
Pearl, Valerie 79
Peltonen, Markku 56, 64
Perceval-Maxwell, Michael 1, 158
Percival, Sir John 9
Perkins, Lieutenant 86–7, 88
Perkin Warbeck (Ford) 229, 230
Phillips, Sir Thomas 78–9, 81, 85, 90, 92, 101
philosophy 164, 165, 206

Pocock, J.G.A. 3–4
poverty 11, 100, 123, 126, 133
praise poetry 11, 178, 179–92
Presbyterianism 11, 27, 41, 238, 240, 241–2, 243–4, 246–7, 248–50
Prince Henry's Barriers (Jonson) 223
Privy Council see English Privy Council; Irish Privy Council; Scottish Privy Council
'Project' 102, 103, 105
Propaganda Fide, Rome 198
Protestantism
  and Catholicism 2–3, 8, 27–8, 110–11, 112, 127, 130, 131, 134–5, 188–9, 246–7
  and civility 34, 71
  and English settlers 5–6, 8, 133–5
  and Gaelic culture 119
  and Gaelic literature 7
  and James VI and I 2, 8, 27–8
  in Scotland 43, 119, 135, 200
  and Scottish settlers 5–6, 8, 27–8, 133, 134–5
  see also Church of Ireland; Reformation
Puritan Widow (Middleton) 229
Pynner, Nicholas 90

Quinn, D.B. 2, 226

Raleigh, Sir Walter 200
Ramism 11, 165–7
Ramus, Peter 165–6
Randolph, Thomas 229
Raphoe 120–1, 133, 188, 189
Rapple, Rory 56, 59
Rathlin Island 154
Raven, Thomas 92, 146, 147
Raymond, Joad 240
recusancy fines 123, 128, 132
Reformation 2, 5, 9, 35, 41, 42–3, 124, 204, 211
religion see Calvinism; Catholicism; Counter-Reformation; Judaism; Presbyterianism; Protestantism; Reformation

Remaines (Camden) 219
Renaissance 55, 58, 204, 211
rhetoric 164, 165–6
Rich, Barnaby 11, 82, 109, 150, 201, 227–8
 'The Anothomy of Irelande' 227–8
 A New Description of Ireland 227
Robinson, Philip 1–2, 12, 178–9
Roche, Lord Maurice 161–2, 168
Rochford, Luke 125
Rome 124, 127, 128, 186, 188, 189, 198, 210
Ross 45, 46, 47–8
Rothe, David 7, 124, 125, 126, 128
Route, the 23, 144, 148–9, 150, 154
Rowley, John 86, 87
Rowley, William 229
Rubin, Miri 184
Russell, Conrad 79

Safeguard of Society (Barston) 64–5
St Anthony's, Louvain *see* Louvain
St Isidore's College, Rome 198
Salisbury, Earls of *see* Cecil, Sir Robert, 1st Earl of Salisbury; Cecil, William, 2nd Earl of Salisbury
Salterstown 146
schools *see* bardic schools; education
Scornful Ladie, The (Beaumont and Fletcher) 229
Scotland
 association with France 221
 and Britishness 5, 19
 and Catholicism 119, 121, 129, 130–1, 132, 245
 'civilising' of 10, 21, 28, 33–4, 40, 102, 149
 Franciscan missions to 129, 131
 James' policies on Gaelic Scotland 10, 21, 22–4, 28, 33–49, 104–5, 149
 plantation of Lewis 9–10, 21–2, 33, 36, 37, 40, 45–7, 100, 149
 and Presbyterianism 242, 243–4
 and Protestantism 43, 119, 135, 200
 settlers from *see* Scottish settlers
 Statutes of Iona 28, 33, 41, 43–5, 103

Scottish Privy Council 21–2, 35, 43–4
Scottish settlers
 in Antrim 143, 146–7, 151, 152–3, 155
 archaeological evidence of 146–7
 and Catholicism 8, 27–8, 130–31, 132, 133
 in Gaelic literature 177–8
 Irish wish to remove 114
 in the Londonderry plantation 89
 Milton on 240–1
 numbers of 158
 in the 'Orders and Conditions' 106, 107
 and Presbyterianism 27–8, 238
 and proposals for plantation 101, 103–4, 144, 169
 and Protestantism 5–6, 8, 27–8, 133, 134–5
 solidarity with English settlers promoted 5–6, 104
 suitability for physical environment 219
Seaforth, Earl of *see* MacKenzie, Cailean, 1st Earl of Seaforth
sectarianism 2–3, 120, 133–5
Selden, John 223
servitors 8, 9, 57, 91–2, 101, 102, 105, 107–8, 144, 152
Seton, Alexander 25
sexualisation 223–4
Shakespeare, William 218
 Macbeth 249
 Much Ado About Nothing 92, 231–2
 The Tempest 231, 245–6
Sharpe, Kevin 79
Shuger, Deborah 71
Sidney, Sir Henry 27, 224
Sidney, Philip 59, 167, 224
Siege of Derry 241, 248
Silken Thomas *see* FitzGerald, Thomas
'Síogaí Rómhánach, An' (anon.) 210
slavery 58, 71–3, 242–3, 246
Skinners' Company 83
Skye, Isle of 40
Smith, George 86

Smith, James 222
Smith, Sir Thomas 22, 58, 59–62, 64, 68–9, 71–3, 163–4, 224, 226
　De Republica Anglorum 58, 60, 71, 73, 224
　A Discourse of the Commonweal of the Realm of England 61
　A Letter Sent by I.B. 58, 59–61, 71, 163
Smith, Thomas (son) 61, 163, 226
Smyth, William J. 3, 57
society, concept of 57, 58, 62, 63–8, 73
Society in Scotland for the Propagation of Christian Knowledge 37
Spain 5, 6, 7, 9, 18–19, 158, 185, 221
'Speech in Parliament touching the Naturalization of the Scottish Nation' (Bacon) 221
Speed, John 3, 225–6
Spenser, Edmund
　anti-Irishness 7, 29, 201, 211
　'Brief Note of Ireland' 160
　and corporatism 62, 64, 68–9, 84
　educational background 11, 58, 59, 163–7
　The Faerie Queene 230, 247–8
　influence upon Davies 162
　influence upon Milton 247
　and the Munster plantation 62, 159, 161–3
　proposals for plantation 29, 58, 62, 64, 68–9, 71, 161
　View of the Present State of Ireland 11, 58, 62–3, 84, 159–63, 165, 167–70, 227, 231, 247
Spert, Richard 100
Springham, Matthew 85, 86
Stanihurst, Richard 7, 211, 231
Star Chamber, Court of 10, 78–80, 90, 92
state-formation 2, 57
Statutes of Iona 28, 33, 41, 43–5, 103
stereotyping 222, 229–30, 239–40
　see also barbarity
Stewart, Andrew, Lord Ochiltree 36

stigmatisation 71, 73
Stornoway 21, 37, 45, 46, 149
Strabane 28, 85, 104, 130–1
'Suirgheach sin, a Éire ógh' (Ó hEódhusa) 181
Suranyi, Anna 223
surveys 90, 91, 92, 99, 104, 147, 224–5
syllogisms 166–7, 168

Tanner, John 87
Tempest, The (Shakespeare) 231, 245–6
Tenure of Kings and Magistrates, The (Milton) 241, 249
Thirty Years War 27
Titles of Honor (Selden) 223
tithes 123
Treaty of Mellifont 6, 24, 143–4, 168
Trinity College, Dublin 9
Triumphs of Honour and Industry, The (Middleton) 229
Trumbull, Sir William 199
Tuchet, George, Lord Audley 103, 104
Tullyhogue 24, 207
Turbeville, George 230
Two Wise Men and all the rest Fools (Chapman) 229
Tyrconnell, Earl of *see* O'Donnell, Rory
Tyrone, County 8, 101, 104, 109–10, 112, 113, 144, 198, 200, 207, 230
Tyrone, Earl of *see* O'Neill, Hugh

Uí Mheallain family 200
Uist, Isle of 40
undertakers 8–10, 57, 68, 73, 85–6, 93, 101–2, 105–8, 144, 159
　see also English settlers; Scottish settlers
urban communities 62–3, 69–71, 73, 80, 84–5, 105
Utopia (More) 72–3

Vagrancy Act 72, 73
View of the Present State of Ireland (Spenser) 11, 58, 62–3, 84, 159–63, 165, 167–70, 227, 231, 247

Vintners' Company 145
Virgil 165, 226
Virginia 9, 115, 219–20, 226, 228, 230

Wadding, Luke 7
Wales 5, 18, 29
Walker, Clement 241
Walsh, Paul 199
Ward, Hugh 205
Ware, Sir James 162
Webster, John 229
Welsh Embassador, The (Middleton and Rowley) 229

Wentworth, Thomas 78, 79, 80, 114
White Devil, The (Webster) 229
Wilson, Thomas 229
Wilson-Okamura, David Scott 226
Withington, Philip 10, 91
Wood, Robert 239
woodkerne 72, 108–9, 110, 112, 226
Wormald, Jenny 9–10

Yeats, W.B. 158
Young, Rose Maud 208

Zeeland 46–7

EU authorised representative for GPSR:
Easy Access System Europe, Mustamäe tee 50,
10621 Tallinn, Estonia
gpsr.requests@easproject.com

www.ingramcontent.com/pod-product-compliance
Lightning Source LLC
Chambersburg PA
CBHW021341230426
43666CB00006B/365